Field Guide to the Wildflowers of Georgia and Surrounding States

FIELD GUIDE TO THE WILDFLOWERS OF GEORGIA AND SURROUNDING STATES

Linda G. Chafin

HUGH AND CAROL NOURSE,
CHIEF PHOTOGRAPHERS

Published in cooperation with the State Botanical Garden of Georgia

THE UNIVERSITY OF GEORGIA PRESS | ATHENS

Athens, Georgia 30602
www.ugapress.org

Set in Garamond Premier Pro & LeBeaune by Graphic Composition, Inc.
Printed and bound by Versa Press, Inc.

The paper in this book meets the guidelines for permanence and durability
of the Committee on Production Guidelines for Book Longevity
of the Council on Library Resources.

Most University of Georgia Press titles are available from popular e-book vendors.

Printed in the United States of America
21 P 5 4 3

Library of Congress Cataloging-in-Publication Data

Names: Chafin, Linda G., author.
Title: Field guide to the wildflowers of Georgia and surrounding states /
Linda G. Chafin ; Hugh and Carol Nourse, chief photographers.
Description: Athens : Published in cooperation with the State Botanical Garden
of Georgia [by] The University of Georgia Press, [2016]
| Includes bibliographical references and index.
Identifiers: LCCN 2015037520 | ISBN 9780820348681 (pbk. : alk. paper)
Subjects: LCSH: Wild flowers—Georgia—Identification.
| Flowering trees—Georgia—Identification.
Classification: LCC QK155 .C43 2016 | DDC 582.1309758—dc23
LC record available at http://lccn.loc.gov/2015037520

THIS WILDFLOWER GUIDE IS DEDICATED TO:

Dr. Samuel B. Jones Jr., 1933–2016, friend, teacher, and mentor

AND TO

the members of the Georgia Botanical Society

whose passion for the wildflowers of Georgia inspired this guide

AND TO

the memory of Wilda Gaye Walters Riley,

who loved all the flowers.

CONTENTS

ACKNOWLEDGMENTS

This field guide owes its existence to the Friends of the State Botanical Garden of Georgia and the Board of Advisors of the State Botanical Garden of Georgia, whose support was both essential and inspirational. Many thanks to all of you for your hard work on behalf of Georgia's flora and for your support of the Garden and this book.

I would also like to thank Wilf Nicholls, Director of the State Botanical Garden, for his support and enthusiasm throughout this project. Special thanks to my colleagues in the Garden's Plant Conservation Program—Jim Affolter, Heather Alley, Jennifer Ceska—for their support, patience, inspiration, and kindness. It is a privilege and a joy to work with you all. Thanks to other Garden staff, past and present, who were enthusiastic supporters of this project and contributed in ways large and small: Andrea Parris, Anne Shenk, Cora Keber, Connie Cottingham, James Gilstrap, Jason Burdette, Lisa Kennedy, Mary McCoy, and Shirley Berry.

The State Botanical Garden is privileged to be a part of the University of Georgia's Public Service and Outreach unit. I am grateful for this opportunity to contribute to PSO's mission of spreading university knowledge to all corners of the state of Georgia.

The heart of this guide is its photographs—the wonderful images that were donated by more than 75 photographers. The quality of these images is testimony to the many hours each of the photographers has spent in the field, learning about and photographing plants and their habitats. Chief among these are Hugh and Carol Nourse, my coauthors and master photographers. I thank them for their artistic vision, their powers of observation, their passion for plants and conservation, and their friendship. Also many thanks to these highly talented photographers who gladly donated a large number of photographs: Alan Cressler, Richard and Teresa Ware, Dan Tenaglia, and Eleanor Dietrich. I also thank the following photographers who donated images or made them freely available in the public domain: Allen Norcross (nhgardensolutions.wordpress.com), Anna Armitage (Texas A&M at Galveston), Arthur Haines (New England Wildflower Society), Bernd Haynold, BJO, Bob Peterson, Bob Smith, Bobby Hattaway, BotGardBLN, Bruce Sorrie, Carmen Champagne, Cary Paynter, Chris Evans, Christa Hayes, Corey Raimond, Dalgial, Dan Mullen, David Smith (delawarewildflowers.org), Dendrofil, Dendroica Cerulea, Elise Smith, Ellen Honeycutt, Eric Hunt (Arkansas Native Plant Society), Felinus Noir, Forest and Kim Starr, Fornax, Frank Mayfield, Fritz Flohr Reynolds, Gary Fleming, Gary Knight, GDFL, George F. Mayfield, H. Zell, Hal Massie, Hans Hillwaert, Ilona L. Woolcarderbee, Irvine Wilson (Virginia Department of Conservation and Recreation), Jacqueline Donnelly, James

Holland, James van Kley, Jan Coyne, Janie Marlowe, Jason Hollinger, Jason Sharp, Jerry Oldenettel, Jim Drake, Jim Fowler, John Gwaltney, John Oyston, Jomegat, Joseph A. Marcus (Lady Bird Johnson Wildflower Center), Júlio Reis, K. Draper, Karan Rawlins, Keith Bradley, Linda Lee (University of South Carolina Herbarium), Loret, Mac Alford, Marco Iocchi, Margie Hunter, Mary Keim, Michael Apel, Pieter Pelser, Rebekah D. Wallace (UGA-Bugwood.org), Richard Canton, Rob Routledge, Roy Cohutta, S. Honeytart, Sam Fraser-Smith, Steve Baskauf, Suzanne Cadwell, Thayne Tuason, Tigerente, and Tom Potterfield.

Bobby Hattaway meticulously reviewed all the species accounts and introductory essays, and provided invaluable intellectual support throughout the writing process. Dr. Hattaway spent countless hours ensuring that measurements are correct, nomenclature is current, and descriptions are accurate. For all this and for his enthusiastic support of this project I am eternally grateful. Thanks too to Kevin Tarner for his review of the carnivorous species accounts and to Wilf Nicholls for reviewing and improving the family descriptions and short essays. Gary Knight and George Baughman provided invaluable advice and suggestions. Jim Drake reviewed the Gentian Family, and I am most appreciative of his expertise. Despite the best efforts of reviewers, there are bound to be errors; they are all mine.

Many thanks to Wendy Zomlefer and David Giannasi, of the Plant Biology Department at the University of Georgia, whose updated version of the "Distribution of the Vascular Flora of Georgia," first published in 1988 by Sam Jones and Nancy Coile, provided accurate information about species distributions in Georgia. Thanks also for their friendship and interest in this project.

My thanks to Alan Weakley and the many other botanists who have contributed to his ongoing publication, *Flora of the Southern and Mid-Atlantic States*. This work, which I consulted on a daily if not hourly basis, was an invaluable aid in the creation of this guide.

Thanks also to the great folks at the University of Georgia Press whose support and professionalism has been invaluable to this project from the outset.

Any work related to the flora of Georgia, and especially this guide, owes an enormous debt to the Georgia Botanical Society. This organization has been advocating on behalf of Georgia's native plants since 1926, leading field trips and wildflower pilgrimages, producing high-quality publications, monitoring plant populations and communities, discovering new plant populations, funding research scholarships, and offering educational workshops. Thanks to Bot Soc, I continue to learn something new about Georgia's native plants and natural communities every month of the year. Anyone interested in learning more about Georgia's wildflowers is invited to join the Botanical Society—it's a lot of fun, a continuing education, and a great way to learn the back roads and byways of our great state. Check us out at http://www.gabotsoc.org.

Last but not least, I acknowledge my dear friends and hiking buddies, especially Gary Knight, Jan Coyne, and Molly Mitchell, who get me out in the woods and share with me their enthusiasm for the wonders of the natural world.

Field Guide to the Wildflowers of Georgia and Surrounding States

INTRODUCTION

More than 4,000 species of plants are native to Georgia, a level of species diversity surpassed only by larger states such as Texas and California and equaled in the east only by Florida, North Carolina, and Alabama. Georgia owes its botanical richness to a geological history that stretches from 400 million years ago, when the Appalachian Mountains were formed, to the recent creation (in geologic terms) of barrier islands that harbor subtropical species. Several large rivers and their tributaries have carved out Georgia's landscape, including the third-largest watershed on the Atlantic Seaboard, the Altamaha River. With temperatures ranging from subzero in the mountains to more than 100° F in the lowlands, and rainfall as much as 68 inches per year in some corners, Georgia provides fertile ground for many types of plants to flourish. Ecologists have captured this mix of geology, climate, and physiography by recognizing 5 physiographic regions and 60 natural communities in the state. Though the 5 regions have many plants in common, each is characterized by unique assemblages of plants described in detail in *The Natural Communities of Georgia* (Edwards, Ambrose, and Kirkman 2013), an encyclopedic and indispensable resource that inspired the creation of this wildflower guide.

The Field Guide to the Wildflowers of Georgia and Surrounding States is the first field guide devoted to Georgia's wildflowers, although numerous guides and technical manuals include a portion of Georgia's plants. I am especially indebted to *Vascular Flora of the Southeastern United States: Asteraceae* (Cronquist 1980), *Wildflowers of the Eastern United States* (Duncan and Duncan 1999), *Aquatic and Wetland Plants of the Southeastern United States* (Godfrey and Wooten 1979, 1981), *Wildflowers of Tennessee, the Ohio Valley, and the Southern Appalachians* (Horn et al. 2005), *Vascular Flora of the Southeastern United States: Leguminosae* (Isely 1990), *Atlantic Coastal Plain Wildflowers* (Nelson 2006), *Manual of the Vascular Flora of the Carolinas* (Radford, Ahles, and Bell 1968), *Field Guide to Wildflowers of the Sandhills Region: North Carolina, South Carolina, Georgia* (Sorrie 2011), and *Flora of the Southern and Mid-Atlantic States* (Weakley 2012, 2014). A complete list of sources is found in the list of references.

This field guide draws upon centuries of fieldwork by dedicated botanists and plant lovers, from Anna Kleist Gambold, who cataloged plants in northwest Georgia in the early 19th century, to members of the Georgia Botanical Society and the Georgia Plant Conservation Alliance, who make new discoveries and safeguard new plant populations every year. It is a privilege to dedicate this guide to all those who have devoted their time, energy, and passion to identifying and conserving Georgia's plants.

Who Can Use This Guide?

This book is designed to be used by anyone with an interest in learning to identify wildflowers in the southeastern United States, whether a novice or an experienced field botanist, including native plant gardeners, amateur botanists, naturalists, hikers, and professionals such as environmental consultants and land managers. The guide uses English measurements rather than the metric system and, where possible, common English words rather than technical botanical terms. The Glossary (pp. 471–474) defines the few unavoidable technical terms. The species descriptions are grouped alphabetically by the scientific name of plant families. The color thumbnail photos on pages 431–470 will facilitate identification for those unfamiliar with the scientific classification of plants. The most common species found in natural areas in Georgia and surrounding states are included as well as a number of species found in readily accessible areas such as roadsides and woodland edges.

This guide will be useful to anyone living in the southeastern United States. More than 90% of the featured species also occur in Alabama, North Carolina, and South Carolina; more than 80% also occur in Florida, Louisiana, Mississippi, Tennessee, and Virginia; and more than 70% occur in Kentucky and eastern Texas.

What Is a Wildflower?

Wildflowers are flowering plants that grow in wild or seminatural habitats without benefit of human cultivation. That simple definition includes a broad, complex, and diverse range of plants. Although wildflowers are sometimes defined as plants with soft stems, called herbs, the term "wildflower" can be stretched to include some woody plants, such as shrubs and vines, with showy or interesting flowers. This broad definition includes grasses and grasslike flowering plants, but not ferns, which do not have flowers. The term "wildflower" is sometimes reserved for native plants—species that were present in a given region in North America before the arrival of Europeans. However, a small number of our most beloved wildflowers are European plants that have become naturalized in disturbed areas, are not a threat to native species, and belong in a guide such as this; Queen Anne's Lace, Ox-eye Daisy, and Cornflower come to mind. Another group of nonnative plants that are not yet considered invasive but are increasingly obvious on our roadsides and rights-of-way, such as Cat's Ear, Beefsteak Plant, and Common Camphorweed, should be identified and monitored for their potential to become invasive.

About 80% of Georgia's plant species meet this broad definition of wildflower. This guide offers detailed information and photographs for 770 wildflower species and in-

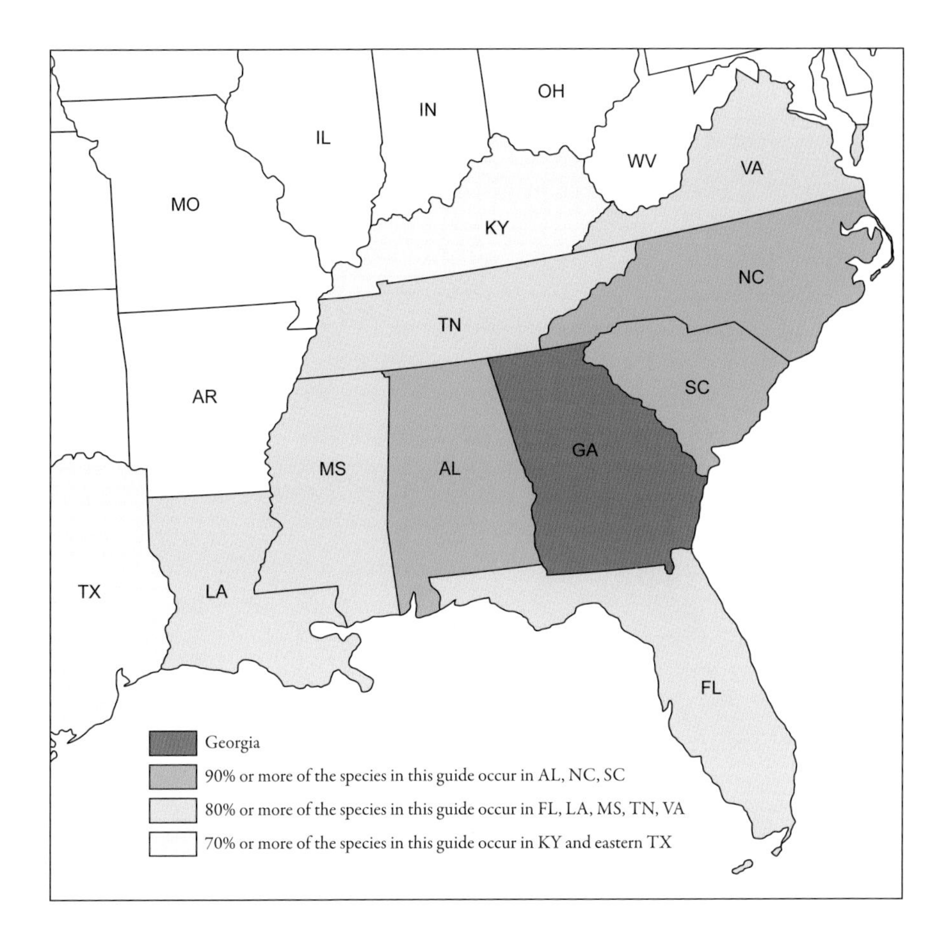

cludes brief descriptions for another 530 species that resemble or are closely related to the featured species. Many of Georgia's rare plants are wildflowers; these are not featured in this guide, though some are included as "similar to" species. For detailed information on many of Georgia's rare plants, see the *Field Guide to the Rare Plants of Georgia* (Chafin 2007).

Wildflower Conservation

A shocking percentage—more than 20%—of the world's plants are threatened with extinction. Sadly, Georgia is more than keeping pace; 27% of our flora is in trouble to one degree or another—155 species are formally listed by the state of Georgia as Endangered, Threatened, or Rare; almost 600 additional species are monitored as Special Concern; and 330 have been placed on the state's watch list. What can we do to keep our wildflowers off these frighteningly long lists?

- Support land conservation and ecological land management efforts with your time, energy, and resources. Volunteer!
- Join the Georgia Botanical Society, the Georgia Native Plant Society, Coastal Wildscapes, The Nature Conservancy, Riverkeepers, and other organizations dedicated to preserving Georgia's native plants, animals, and natural communities.
- Eradicate invasive pest plants on your own property and volunteer to control them on public lands.
- Plant native Georgia species in your yard, garden, and back forty. If you are a farmer, plant native hedgerows. You will not only be helping plants, you'll be providing food, shelter, and nesting habitat for wildlife.
- Urge wildlife management agencies to reduce the size of Georgia's deer herd. Browsing by deer is a major threat to many native plants.
- Support efforts to eradicate feral hogs, which do enormous damage to plant communities throughout the state.
- Encourage land management agencies to conduct prescribed burning in fire-maintained natural communities.
- Lobby landowning companies and agencies to plant native species instead of exotics for erosion control, wildlife food, landscaping, and roadside beautification; and to limit the use of herbicides.
- Don't poach (dig up) plants, and don't buy poached plants from vendors. Plants dug from the wild rarely survive long after transplanting into gardens; without the cross-pollination supplied by their wild neighbors, they are doomed to wink out.
- If you like to eat wild-harvested plants or otherwise collect specimens of native plants, never take more than 10% of a population, and harvest only the leaves and aboveground stems, leaving the roots and rhizomes in the ground. Before taking native plants from the wild, consider how many natural communities and native plant populations have already been destroyed. Consider taking photos instead!
- Enjoy native plants wherever and whenever you can: botanical gardens, state parks, national forests, and your own front yard or back forty. In the immortal words of Edward Abbey:

> Be as I am—a reluctant enthusiast . . . a part-time crusader, a half-hearted fanatic. Save the other half of yourselves and your lives for pleasure and adventure. It is not enough to fight for the land; it is even more important to enjoy it. While you can. While it's still here. So get out there . . . explore the forests, climb the mountains . . . sit quietly for a while and contemplate the precious stillness, the lovely, mysterious, and awesome space.

How This Guide Is Organized

Detailed descriptions for 770 of the most common wildflowers found in Georgia and the surrounding states are found on the following pages. Additional information for

530 "similar to" species are included within them. The species descriptions are divided into the two widely recognized groups of flowering plants: the dicots (Dicotyledons, pp. 19–354) and the monocots (Monocotyledons, pp. 355–428). Within each group, plant families are arranged alphabetically. Within each plant family the plants are arranged alphabetically by genus and then by species name. Although this organization scheme is artificial (it does not reflect the evolutionary relationships among the species), it follows the practice of a number of other field guides and technical manuals currently in use in the Southeast. With use, this arrangement will become familiar and convenient for readers of this field guide, especially when used in conjunction with the flower color thumbnail photos, pages 431–470.

What Is a Dicot?

Dicots are plants with two seed leaves, or cotyledons. Seed leaves provide the newly germinated plant with sugars and other nutrients until the true leaves (which usually look very different) develop. Dicot flowers typically have 4 or 5 petals or multiples of these, such as 8 or 10. Of course, there are exceptions; for example, Pawpaw and May-apple have 6 petals. Most dicots have netted leaf venation; that is, the veins in their leaves form a branching network rather than running parallel from the base of the leaf to the tip. Recent molecular research has established that most plants would be better classified as eudicots ("true dicots"), with a smaller evolutionary branch called the dicots. However, for the purposes of this field guide, the familiar and inclusive term "dicots" will be used.

What Is a Monocot?

Monocots are plants with one seed leaf, or cotyledon. Typically, their flowers have three or six petals and three or six sepals. Most monocots have parallel leaf venation; that is, the veins in their leaves run parallel for the length of the leaf. There are, of course, exceptions—for example, trilliums and Jack-in-the-pulpit have netted veins. Monocots do not develop true woody tissue, although some, such as yuccas and palms, are able to grow quite tall.

What Is a Plant Family and, for That Matter, a Species?

Botanists have organized the plant kingdom into a hierarchy, with the largest group at the top of the hierarchy, descending down through division, class, order, family, genus, and species, with each level reflecting the evolutionary relationships among the different groups. A plant family brings together all the plants that share an evolutionary past and similar traits, such as shape of the flowers, number of seeds in each fruit, chemical makeup, and other genetically determined features. Within families, plants are grouped into genera, with each genus containing plants that are closely related and share a more tightly defined set of traits. Each genus contains one to many species, and each species contains plants that are essentially alike in their appearance and genetic makeup. Plants within a species may be affected by environmental conditions, isolation from other plants of the same species, hybridization, and other factors, sometimes leading to genetic changes and the recognition of subspecies and varieties; these are plants that may be on their way to becoming a separate species in the future.

Natural Communities in Georgia

Georgia's diverse landscape supports 60 different types of natural communities, ranging from salt marshes along the coast to high-elevation rocky summits in the Blue Ridge to prairies and flatwoods in the Coosa River valley. This diversity reflects a long and complex geological history dating back hundreds of millions of years. Georgia consists of five physiographic regions that reflect this geologic history: from northwest to southeast, they are the Cumberland Plateau, the Ridge & Valley, the Blue Ridge, the Piedmont, and the Coastal Plain (including the barrier islands). Georgia shares these regions with surrounding states but is the only state south of Virginia to have all five regions. Within these regions there is much variation in soil type and topography. Elevations range from nearly 5,000 feet to sea level, with corresponding variations in temperature and rainfall patterns. Georgia has an extensive system of rivers and streams that have cut gorges and ravines and deposited vast floodplains and bottomlands. The interaction among these physical factors has produced a varied landscape with many natural community types, each with its complement of animals and the native plants that support them.

Wildflowers can be found in almost all of Georgia's natural communities at one season or another. The following descriptions briefly characterize a few natural communities that are notable for their wildflower displays. Public lands (and private conservation lands open to the public) with outstanding displays of wildflowers are listed afterward, with the best seasons for discovering wildflowers there. For in-depth descriptions of Georgia's natural communities, *The Natural Communities of Georgia* (Edwards, Ambrose, and Kirkman 2013) is an outstanding resource.

A word about conservation: it is impossible to protect and preserve Georgia's wildflowers without protecting the habitats that support them. Commercial, industrial, and residential development; conversion to silviculture and agriculture; fire suppression; and pollution pose increasing threats to natural communities in all parts of the state. The Georgia Department of Natural Resources, the U.S. Fish and Wildlife Service, The Nature Conservancy of Georgia, land trusts, and other private and public conservation agencies are making laudable efforts to preserve areas with diverse plants and animals. These agencies and organizations deserve your support. Continuing efforts supported and financed by both public and private dollars are essential if the plants presented in this guide are to survive the 21st century.

Ridge & Valley and Cumberland Plateau Regions

CEDAR GLADES

Cedar glades occur where limestone or dolomite is at or near the surface, forming pavements of bare rock that are open, sunny "islands" in hardwood-pine forests. They are named for their most conspicuous tree, Eastern Red Cedar, which forms a patchy canopy over a ground layer of wildflowers and grasses. Cedar glades are noted for their plant species diversity, with more than 400 species in such habitats in Georgia, Tennessee, Alabama, Kentucky, and Virginia. Cedar glade wildflowers include Buffalo Clover, Hoary Puccoon, Lime-barren Sandwort, Nettle-leaf Sage, Purple Coneflower, Rattlesnake Master, Southern Sundrops, and Thimbleweed. Cedar glades are also known for their high number of rare and endemic plant species. SYNONYMS: Calcareous glades and barrens, limestone glades and barrens.

COOSA VALLEY PRAIRIES

Coosa Valley prairies occur in Floyd County, Georgia, and adjacent areas of Alabama in the Coosa River watershed. They are open grasslands that have developed over high pH soils derived from limestone or calcareous shale. Although much smaller, they resemble prairies found farther west and support a number of midwestern prairie plants such as Big Bluestem, Switch Grass, Indian Grass, and Little Bluestem. They also support many species of sun-loving wildflowers such as goldenrods, coneflowers, sunflowers, asters, and milkweeds. SYNONYM: Calcareous prairies and barrens.

LIMESTONE SLOPE AND RAVINE FORESTS

The northwest corner of Georgia intersects the Cumberland Plateau in the Sand-Lookout-Pigeon Mountains area. These long, flat-topped mountains consist of strata of limestone, shale, conglomerate, and sandstone. Rich, moist, deciduous forests develop where limestone is near the surface on lower slopes, coves, ravines, valley floors, and sinkholes. The forests are dominated by American Beech, Basswood, Bigleaf Magnolia, Black Cherry, Buckeye, Northern Red Oak, Shagbark Hickory, Sugar Maple, Tulip Tree, White Ash, White Oak, and Yellow Buckeye. Striking displays of spring-flowering herbs such as Black Cohosh, Blue Phlox, Celandine Poppy, Eastern Columbine, Fern-leaf Phacelia, May-apple, Twinleaf, Virginia Bluebell, Wild Geranium, Yellow Mandarin, and several trillium species occur here. SYNONYMS: Mesic forest, mixed mesophytic forest, mesic hardwood forest.

Blue Ridge Region

MOUNTAIN COVE FORESTS

The ravines, hollows, and cool, moist slopes of the southern Appalachians support some of the richest plant communities in the world. Typical trees include Basswood, Black Birch, Black Cherry, Carolina Silverbell, Northern Red Oak, Tulip Tree, White Ash, and Yellow Buckeye. The soils are rich in nutrients and organic matter and support a wide diversity of spring-blooming wildflowers such as Blue Cohosh, Clinton's Bead Lily, Doll's Eyes, Solomon's Plume, Solomon's Seal, Trout Lily, Wild Ginger, Yellow Mandarin, and numerous bellworts, orchids, trilliums, and violets. Bellflower, Black Cohosh, Appalachian Bunchflower, Carolina Lily, Rattlesnake Plantain Orchid, Starry Campion, Turk's-cap Lily, and Yellow Jewelweed bloom in mid- to late summer. SYNONYMS: Cove forest–fertile variant, hardwood cove forest, broadleaf deciduous cove forest, rich cove forest, mixed mesophytic forest.

NORTHERN HARDWOOD FORESTS AND BOULDERFIELD FORESTS

Northern hardwood and boulderfield forests are found in Georgia above 3,500 feet on some cool, moist, north-facing mountain slopes. The tree canopy usually includes a large number of deciduous hardwood species such as American Beech, Basswood, Black Cherry, Northern Red Oak, Sugar Maple, White Ash, Yellow Birch, and Yellow Buckeye. Boulderfields are relics of the extremely cold conditions that prevailed in Georgia during the last ice age; they occur on the upper slopes of many of the north-facing coves and have a similar tree canopy. Northern hardwood and boulderfield forests are especially interesting in the spring when the ground is carpeted with spring ephemerals such as bellworts, Blue Monks-hood, Clinton's Bead Lily, Dutchman's Britches, Spring-beauty, Solomon's Seal, trilliums, Trout Lily, and Umbrella-leaf. Coves and boulderfields support more than 70 species of Special Concern, including the southernmost populations of several Appalachian species. SYNONYM: High-elevation cove forest.

Piedmont Region

PIEDMONT MOIST FORESTS

Piedmont moist forests occur on north-facing and lower-lying slopes and in well-drained stream bottoms where soils are moist and richer in nutrients than the surrounding uplands. They support American Beech, Basswood, Hop Hornbeam, Northern Red Oak, Red Mulberry, Silverbell, Tulip Poplar, White Ash, and many other species. Before the canopy leafs out in early spring, many showy flowering herbs such as Bloodroot, Green-and-gold, Heartleaf, May-apple, Pipsissewa, Round-lobed Hepatica, Rue-anemone, trilliums, Trout Lily, and violets bloom. SYNONYMS: Mesic forest, mixed hardwood forest, broadleaf deciduous forest, southern piedmont mesic forest.

PIEDMONT GRANITE OUTCROPS

Piedmont granite outcrops occur from Virginia to Alabama, with 90% occurring in Georgia. Usually composed of granite or granitic gneiss, outcrops occur as flatrocks, low domes, and monadnocks (isolated "mountains" such as Stone Mountain). All support a diverse mosaic of microhabitats, including lichen-encrusted rocks, moss mats, ephemeral pools, and tree "islands" of Eastern Red Cedar supported by soils formed by the weathering of rock over thousands of years. Each microhabitat has its own suite of wildflowers and rare and endemic plants that are adapted to the extreme temperature and moisture conditions that occur on granite outcrops throughout the year—cool and seepy in winter and spring, hot and dry in summer and fall. Outcrops are most spectacular in April and late August but are interesting year-round.

Coastal Plain Region

COASTAL PLAIN SANDHILLS AND DRY LONGLEAF PINE–WIRE GRASS WOODLANDS

Longleaf Pine–Wire Grass woodlands occur on sandy and sandy-loamy soils on gently rolling hills in the Coastal Plain, on the ancient dunes of the Fall Line sandhills, and on sand ridges that occur on the eastern and northeastern banks of the Altamaha, Ohoopee, Flint, and Canoochee Rivers. The canopy consists of widely scattered Longleaf Pines with an understory of Turkey Oak and other scrub oaks. The most ecologically important plant in this community is the fire-carrying Wire Grass, but the ground cover also includes showy wildflower species such as Blue Sage, Clasping Milkweed, Dog-tongue, Fringed Blue Star, Honeycomb-head, Rabbit Bells, Spurred Butterfly Pea, Tread-softly, Wild Buckwheat, and Wild Indigo, as well as blazing stars, goldenasters, goldenrods, lupines, and sunflowers. SYNONYMS: Dry and/or mesic upland Longleaf Pine forests and woodlands, Turkey Oak scrub, upland pine forest, high pine, Longleaf Pine–Wire Grass savanna, Longleaf Pine–scrub oak woodland.

COASTAL PLAIN MOIST SLOPE FORESTS

These relatively cool, shady forests occur on the well-drained but moist soils of steep, north-facing slopes, in ravines and steepheads, on river bluffs, and around the edges of deep limesinks. The nutrient-rich soils there support a diversity of trees, including American Beech, Basswood, Florida Maple, Shumard Oak, Southern Magnolia, Spruce Pine, and White Oak; and many wildflowers, including Atamasco Lily, Blue Phlox, Green Dragon, Indian Pink, Sarsaparilla Vine, and several species of orchids, toothworts, and trilliums. The combination of deep shade, cooler temperatures, and rich soils creates conditions that support plants that are more typical of the Blue Ridge than the Coastal Plain, such as Bloodroot, Doll's Eyes, May-apple, Round-lobed Hepatica, Solomon's Seal, Spring-beauty, Trout Lily, and Wild Ginger. SYNONYMS: Mesic slope forest, beech-magnolia forest, bluff and slope forest, Torreya ravines, Chattahoochee ravines, beech-magnolia hammock, basic mesic forest, mesic hammock.

WET SAVANNAS AND WET FLATWOODS

These wetland communities occur on flat, poorly drained terrain with sandy, acidic soils overlying a hardpan that traps water during the wettest parts of the year. They have a sparse canopy of Pond Cypress and Longleaf Pine, Pond Pine, or Slash Pine and a dense sward of grasses, sedges, and herbs, including many showy wildflowers such as Bog Buttons, Carolina Milkweed, Deer Tongue, Georgia Tickseed, Orange Milkwort, Pine Lily, Pipewort, Savanna Eryngo, Savannah Sneezeweed, Skyflower, Toothache Grass, and White-top Sedge, as well as several species of blazing-stars, meadow-beauties, orchids, sunflowers, and violets. Carnivorous plants that occur in these communities include pitcherplants, butterworts, dew threads, and sundews. Fires every two to three years during the growing season are necessary to prevent shrub and tree invasion and to promote flowering of the grasses and herbs. At least 60 species of rare plants are known from wet savannas and flatwoods. SYNONYMS: Pine flatwoods, wet pine savanna, cypress savannah, moist pine barren, wet pine flatwoods, wet flatwoods, wet prairies.

SEEPAGE SLOPES AND PITCHERPLANT BOGS

Seepage slopes and pitcherplant bogs are open, sunny wetlands that develop on lower slopes or in swales within rolling, Longleaf Pine–dominated communities. Soils are acidic, nutrient-poor, and saturated most of the year by water seeping constantly from the sandy soils of surrounding communities; a subsurface hardpan confines seepage to the upper layers of soil. Trees are sparse, with an occasional stunted Pond Pine, Slash Pine, or Pond Cypress. Frequent fires prevent shrub and tree invasion and support an amazing diversity of grasses, sedges, and wildflowers, including Bog Buttons, butterworts, Carolina Milkweed, colic roots, hatpins, milkworts, Pipewort, rose-pinks, sunbonnets, Toothache Grass, and yellow-eyed grasses; and dozens of showy composites, lilies, meadow-beauties, mints, and orchids. Five pitcherplant species and at least 20 other species of Special Concern occur in bogs and hillside seepages. SYNONYMS: Pitcherplant bogs, hillside seeps, herb bog, herb and shrub bog, sandhill seeps.

OKEFENOKEE SWAMP

The Okefenokee Swamp, encompassing more than 700 square miles of floating peat mats, pine uplands, cypress and bay swamps, scrub-shrub wetlands, and vast freshwater prairies, is among the most important wetlands in the United States. Lightning-set fires play a major role in creating and maintaining the shifting mosaic of plant communities that make up the "swamp." The acidic, tea-colored waters of the prairies provide spectacular views of aquatic plants such as bladderworts, Blue Flag, Golden Club, Spatterdock, and White Water Lily. The *okeefenokeensis* variety of Hooded Pitcherplant that occurs in the swamp on peat mats is many times larger than the typical variety. The swamp also harbors numerous rare and endangered plants and animals.

Best Places and Months to See Wildflowers in Georgia

Places marked with an asterisk (*) are described in detail in Hugh and Carol Nourse's *Favorite Wildflower Walks in Georgia* (2007). For directions, accessibility information, and opening and closing hours for other sites, check the internet.

Cumberland Plateau and Ridge & Valley

Black's Bluff Preserve, trail to spring at south end of the preserve, Floyd County, Apr.
*Carter's Lake Rereg Dam Public Use Area, Hidden Pond Trail, Murray County, Apr–May.
*Chickamauga and Chattanooga National Military Park, cedar glade walks, Catoosa County, Apr.
Cloudland Canyon State Park, Sittons Gulch Trail, Dade County, late Mar–late Apr.
*Keown Falls Trail, Chattahoochee National Forest, Walker County, Apr–May.
*Shirley Miller Wildflower Trail, Crockford–Pigeon Mountain Wildlife Management Area, Walker County, late Mar–mid-Apr.

Blue Ridge

Amicalola Falls State Park, various trails, Dawson County, mid-Apr–mid-May.
Appalachian Trail (north and south of Woody Gap), Lumpkin County, mid-Apr–late May.

Blood Mountain, Chattahoochee National Forest, Lumpkin and Union Counties, May and Aug.
*Brasstown Bald, Arkaquah Trail and Wagon Train Trail, Union and Towns Counties, mid-Apr–Jun.
Cohutta Wildlife Management Area, Forest Service Road 68, roadside botanizing, Murray County, mid-Apr–late May.
Cooper Creek Scenic Area, Old Growth Forest Trail, Chattahoochee National Forest, Union County, mid-Apr–late May.
*Fort Mountain State Park, Gahuti Trail, Murray County, late Apr–May.
Grassy Ridge Trail to Till Ridge Cove, north of Patterson Gap, Chattahoochee National Forest, Rabun County, May.
Lake Winfield Scott Recreation Area, Jarrard Gap Trail, Union County, mid-Apr–late May; Lake Trail, late Jul–Aug.
*Sosebee Cove Wildflower Trail, Chattahoochee National Forest, Union County, mid-Apr–mid-May.
*Tennessee Rock Trail, Black Rock Mountain State Park, Rabun County, mid-Apr–mid-May.

Piedmont

Callaway Gardens, Lady Bird Johnson Wildflower Trail, Harris County, Mar–Apr.
Chicopee Woods, Elachee Nature Preserve, Ed Dodd Trail, Hall County, Apr–May.
*Davidson–Arabia Mountain Park, Arabia Mountain Trail, DeKalb County, mid-Apr and early Sep.
Kennesaw Mountain National Battlefield Park, Kennesaw Mountain Trail and Burnt Hickory Loop, Cobb County, Apr–May.
*Panola Mountain State Park, Outcrop Trail, DeKalb County, mid-Apr.
*Panther Creek Trail, Chattahoochee National Forest, Habersham County, Mar–May.
State Botanical Garden of Georgia, White and Orange Trails, Clarke County, Mar–Apr.
Sweetwater Creek State Park, White Trail, Douglas County, Apr.
*West Palisades Trail, Chattahoochee River National Recreation Area, Fulton County, Mar–Apr.

Coastal Plain

Big Hammock Natural Area, Big Hammock Trail, Tattnall County, Apr–Jun.
Black Creek Wildlife Management Area, Taylor County, mid-Sep–mid-Oct.
*Doe Run Pitcher Plant Bog Natural Area, Ann Barber Wildflower Trail, Colquitt County, mid-Mar–Apr, Aug–Oct.
Fall Line Sandhills Wildlife Management Area, Taylor County, mid-Sep–mid-Oct.
GA Highway 177, roadsides in Waycross State Forest, mid–late Apr, early–mid-Jul.

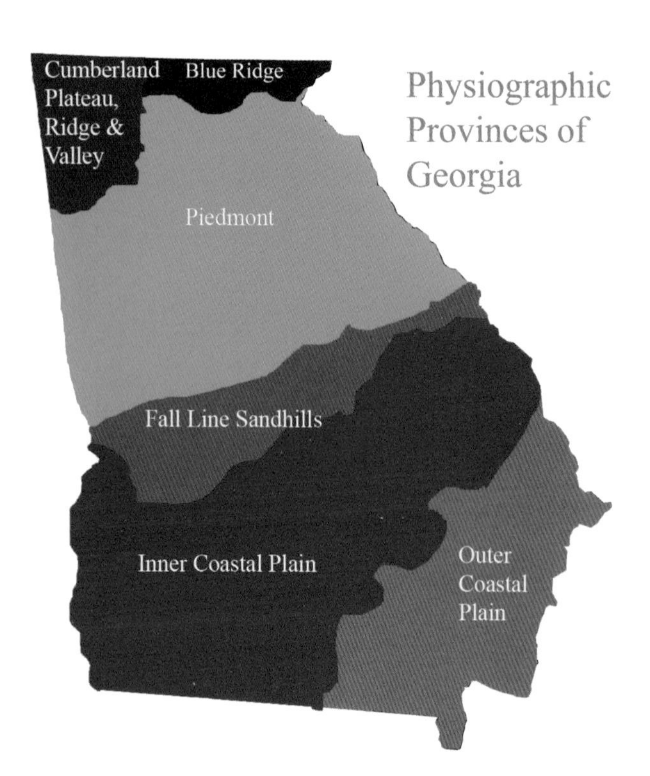
Physiographic
Provinces of
Georgia
Cumberland
Plateau,
Ridge &
Valley
Blue Ridge
Piedmont
Fall Line Sandhills
Inner Coastal Plain
Outer
Coastal
Plain

DADE
CATOOSA
FANNIN
TOWNS
UNION
RABUN
WALKER
WHITFIELD
MURRAY
GILMER
WHITE
CHATTOOGA
GORDON
LUMPKIN
HABERSHAM
STEPHENS
PICKENS
DAWSON
HALL
BANKS
FRANKLIN
HART
FLOYD
BARTOW
CHEROKEE
FORSYTH
JACKSON
MADISON
ELBERT
POLK
PAULDING
COBB
GWINNETT
BARROW
CLARKE
OGLETHORPE
HARALSON
DOUGLAS
DEKALB
WALTON
OCONEE
WILKES
LINCOLN
CARROLL
FULTON
ROCKDALE
NEWTON
MORGAN
GREENE
CLAYTON
HENRY
TALIAFERRO
MCDUFFIE
COLUMBIA
HEARD
COWETA
FAYETTE
JASPER
PUTNAM
WARREN
RICHMOND
SPALDING
BUTTS
HANCOCK
GLASCOCK
TROUP
MERIWETHER
PIKE
LAMAR
MONROE
JONES
BALDWIN
WASHINGTON
JEFFERSON
BURKE
UPSON
BIBB
WILKINSON
JENKINS
HARRIS
TALBOT
CRAWFORD
TWIGGS
JOHNSON
SCREVEN
TAYLOR
PEACH
EMANUEL
MUSCOGEE
HOUSTON
LAURENS
CHATTA-
HOOCHEE
MARION
MACON
BLECKLEY
TREUTLEN
CANDLER
BULLOCH
EFFINGHAM
SCHLEY
DOOLY
PULASKI
DODGE
WHEELER
MONTGOMERY
TOOMBS
EVANS
BRYAN
STEWART
WEBSTER
SUMTER
WILCOX
CRISP
TELFAIR
TATTNALL
CHATHAM
QUITMAN
RANDOLPH
TERRELL
LEE
BEN HILL
JEFF DAVIS
APPLING
LIBERTY
TURNER
LONG
CLAY
CALHOUN
DOUGHERTY
WORTH
IRWIN
COFFEE
BACON
WAYNE
TIFT
MCINTOSH
EARLY
BAKER
MITCHELL
BERRIEN
ATKINSON
PIERCE
MILLER
COLQUITT
COOK
BRANTLEY
GLYNN
WARE
SEMINOLE
DECATUR
GRADY
THOMAS
BROOKS
LOWNDES
LANIER
CLINCH
CHARLTON
CAMDEN
ECHOLS

Kolomoki Mounds State Park, Early County, Mar–Apr.
Montezuma Bluffs Natural Area, River Bluff Trail, Macon County, mid-Mar–mid-Apr.
Moody Forest Natural Area and Wildlife Management Area, Tavia's Trail, Appling County, Sep–Oct.
Oaky Woods Wildlife Management Area, Houston County, Apr–Jun.
*Ohoopee Dunes Wildlife Management Area, McLeod Bridge/L. F. Gambrell Trail, Emanuel County, Sep–Oct.
*Okefenokee National Wildlife Refuge, Swamp Island Drive, boardwalk and canoe trails, Ware and Charlton Counties, mid-Apr–mid-May.
Silver Lake Wildlife Management Area, Decatur County, early Jun and late Oct.

A Few Words about Pronouncing Plant Names

The biggest obstacle to the use of scientific names by newcomers to botany is pronunciation and the fear of sounding like an idiot. But, really, you already use a lot of Latin names without a second thought: *Magnolia*, *Camellia*, *Wisteria*, *Nasturtium*, *Petunia*, *Gardenia*, *Poinsettia*, and so on. There's the Latin you learned in high school, church Latin, British Latin, and the Latin you make up on the spur of the moment. No one knows how the Romans pronounced Latin, so you can actually pronounce scientific names any way you want. But following a few rules will give you confidence and promote communication, which is one of the main goals of any naming system. Just remember there are exceptions to every rule, and try your best not to argue with folks who pronounce things differently than you do!

In most scientific names, you stress the third syllable from the end of the word; for example, mag-NO-lee-ah.

If there are two identical consonants toward the end of the name, stress the second syllable from the end. For example, *heterophylla* is pronounced het-uh-ro-FIL-la, *perennis* is pronounced puh-REN-nis, and *capsella* is pronounced cap-SEL-lah.

Generally, every syllable of a Latin word is pronounced and every vowel represents a syllable; for example, *Symphyotrichum* is pronounced sim-fee-AH-tri-kum. There are plenty of exceptions to this rule, such as when the scientific name is based on a person's name. In that case, you may choose to follow this rule or pronounce the name as you normally would. Plants named for Stephen Hales are pronounced either HALES-ee-ah or ha-LEE-see-ah.

A single consonant following a vowel goes with the following syllable; for example, *Tridens* is pronounced TRY-dens.

Double consonants are usually split between the preceding and following syllables. The exceptions, where the two consonants are kept together, are really just common-sense pronunciation: bl, cl, gl, kl, pl, tl, br, cr, dr, gr, kr, pr, tr, ch, ph, th. *Oblongifolia* is pronounced ah-blon-guh-FOE-lee-ah; *Conoclinium* is pronounced con-o-CLIN-ee-um.

If there are two consonants at the beginning of the word, the first is silent. For example, *Cnidosculus* is pronounced nie-DOS-cue-lus; *Ptilimnium* is pronounced tuh-LIM-nee-um.

"Ch" is pronounced as a hard "k." *Chrysopsis* is pronounced CRY-sop-sis.

"G" and "c" are usually pronounced hard; "guh," not "gee"; and "kuh" not "cee." Both letters, however, are soft when followed by e, i, y, ae, oe. So, *Cakile* is pronounced kah-KYE-lee, and *Coccinea* is pronounced kahk-SIN-ee-ah. This is a rule I often ignore in favor of "what sounds right."

The word "species" is both singular and plural—there is no such word as "specie" in the biological world.

That's enough for starters. For more on this topic, there are several newer books such as Sara Mauritz's *Fearless Latin*. I also recommend *Stearn's Dictionary of Plant Names for Gardeners* and *Botanical Latin*, both by William Stearn, who advised, "Botanical Latin is essentially a written language.... How [names] are pronounced really matters little provided they sound pleasant and are understandable by all concerned." The website *Name That Plant* (www.namethatplant.net) includes links to audio versions of nearly 1,600 plant names in the Southeast.

How to Read the Species Descriptions

Each species description begins with the plant's COMMON NAME. Common names for plants are not assigned or regulated by any official botanical organization and often differ from region to region, book to book, and botanist to botanist. The common names used in this guide usually echo names used in other field guides, widely used websites, and technical manuals. In a few cases where there are no commonly used names, a name was assigned based on appearance, habitat, or names of closely related plants.

SCIENTIFIC NAMES are, with some exceptions, those used in the *Flora of the Southern and Mid-Atlantic States* (Weakley 2012, 2014). Ideally, scientific names reflect evolutionary relationships among plants, and they are subject to change as plant taxonomists learn more about those relationships. The scientific name consists of a genus name—for example, *Rosa*; a species name, correctly known as the "specific epithet"—for example, *carolina*; and the last name of the person who first described the species—for example, Linnaeus. If the original name given the plant was subsequently changed, the last name of the person who made the change is added and the first author's name is placed inside parentheses—for example, (Linnaeus) Smith. Genus and species names are always italicized because they are based on Latin; the genus is always capitalized, the species is not. In this guide, species are arranged alphabetically by genus name within their families.

SYNONYMS are scientific names that have been used in the past or are in current use in another botanical source. Research in molecular plant taxonomy has led to a fair amount of reclassification and renaming in recent years, and some familiar names are now treated as synonyms. Most of the synonyms found in this guide are the scientific names given in the *Manual of the Vascular Flora of the Carolinas* (Radford, Ahles, and Bell 1968).

The FAMILY designation is the basis for the organization of this guide. A plant family comprises plants that share a close evolutionary past and important traits such as shape of the flowers, the number of seeds in each fruit, the chemical makeup, and other genetically determined features. The families are arranged alphabetically, and both the common family name and the technical botanical family name (ending with "aceae") are given. The historical name still used for some families is given in the introduction sections to those families (these names end in "ae," as in Cruciferae and Leguminosae). Learning to recognize the characteristics held in common by members of a plant family makes identification easier and widens our understanding of a plant's biology.

The first sentence following the family name provides information on the plant's LIFE HISTORY (annual, perennial, or biennial) and LIFE FORM (herb, shrub, or vine).

HABITAT—where on the landscape the species lives in Georgia—is also described in the first sentence. To learn more about Georgia's plant habitats, see the *Natural Communities of Georgia* (Edwards, Ambrose, and Kirkman 2013).

GEORGIA DISTRIBUTION, also provided in the first sentence, refers to the physiographic regions and counties where the species is known to occur in Georgia, based on the *Vascular Plant Atlas of Georgia* (Giannasi and Zomlefer 2010). The general term "north Georgia" includes all of the state north of the Fall Line. The term "mountains" refers to the Blue Ridge mountains of northeast Georgia *and* mountains in the Ridge & Valley and Cumberland Plateau of northwest Georgia. "Throughout" means that a species has been found in at least a few counties in each of Georgia's five physiographic regions. Keep in mind that some sections of Georgia are not well known botanically and that some groups of plants, such as grasses and sedges, are often overlooked. For a map of Georgia's physiographic regions, see page 12.

STEMS are the part of the plant that supports its leaves and branches. Unless otherwise noted, stems are assumed to be erect and aboveground. In this guide, the word "stem" includes the technical term "scape."

LEAVES are the main photosynthetic organ of most plants; they come in many sizes, shapes, and colors. Unless described as "evergreen," all leaves are assumed to be deciduous.

FLOWER CLUSTERS consist of several to many flowers held more or less closely together. The technical term is "inflorescence."

FLOWERS are the reproductive organs of plants. They usually consist of a whorl of green sepals, a whorl of colorful petals, a whorl of pollen-producing stamens, and a fruit-producing pistil, although not all flowers have all these parts. For more details, see these terms in the Glossary.

FLOWERING times are the months when the plant is usually in bloom. This may vary depending on latitude, elevation, and the effects of climate change.

RANGE describes the species' regional, national, or global distribution. Species are native to the southeastern United States unless otherwise noted. US states and Canadian provinces are indicated by their postal codes. Canadian province postal codes used in this guide are: AB—Alberta, BC—British Columbia, MB—Manitoba, NB—New Brunswick, NL—Newfoundland and Labrador, NS—Nova Scotia, ON—Ontario, PEI—Prince Edward Island, QC—Quebec, SK—Saskatchewan, YT—Yukon.

SIMILAR TO species are those that most resemble or are closely related to the featured species.

A Word about Using a Hand Lens

Throughout this guide, you'll see frequent references to a 10× hand lens, a small magnifier you can carry in your pocket or around your neck on a long cord. I can't recommend this instrument too highly! Using a hand lens opens up a new vista of botanical wonders. Scales on the lower surfaces of leaves, hairs on the stalks of stamens, the size and number of lots of botanical details—all spring into view under a 10× lens. In the spot-on words of one of my students, "It's a whole 'nuther world down there!"

A 10× hand lens magnifies 10 times, plenty for most botanical uses. Its focal length is 1 inch. If you decide to specialize in grasses or mosses, which have minute features, you may benefit from a 20× lens, but these are hard to use because their focal length is only ½ inch and the field of view is smaller. A 10× lens is the best for looking at most plants, insects, rocks, and other small elements of the natural world.

To use a hand lens, hold it steady in front of your eye, using either side of the lens. Rest your thumb against your cheekbone to place the object about 1 inch from your eye, then move the object back and forth until it comes into focus. Don't move the lens! This may take a bit of practice, but it is so worth it! Look for hand lenses at biological and geological websites and at nature stores.

SPECIES DESCRIPTIONS

DICOTS

Blue Twinflower

Dyschoriste oblongifolia (Michaux) Kuntze
Acanthus Family (Acanthaceae)

PERENNIAL HERB found in sandhills, pine flatwoods, and dry, upland forests in Georgia's Coastal Plain and a few eastern Fall Line counties. **STEMS:** Up to 20 inches tall, very hairy. **LEAVES:** Up to 1¾ inches long and ½ inch wide, opposite, oblong to elliptic, rounded at the base, hairy, with a very short or no stalk. **FLOWERS:** About 1 inch long, funnel-shaped with a short tube and 5 spreading lobes; blue or lavender with purple markings on the inner surface; the 5 green calyx lobes are very narrow, pointed, and ½–¾ inch long; the anthers have sharp points at the base. **FLOWERING:** Apr–May. **RANGE:** FL west to AL, north to NC. **SIMILAR TO** Swamp Twinflower (*Dyschoriste humistrata*), which occurs in floodplains and bottomlands in southeast Georgia; its stems and leaves are mostly smooth, and its flowers are less than ½ inch long.

Carolina Scaly-Stem

Elytraria caroliniensis (J. F. Gmelin) Persoon
Acanthus Family (Acanthaceae)

PERENNIAL HERB found in swamp forests, usually over marl or limestone bedrock in the Coastal Plain. **STEMS:** 4–20 inches tall, covered with many small, hard, pointed, overlapping bracts. **BASAL LEAVES:** Up to 8 inches long and 3 inches wide, broadly oval, usually wider above the middle, smooth or hairy; a raised midvein and deeply inset side veins give the leaves a textured look; there are no stem leaves except the scalelike bracts. **FLOWER CLUSTERS:** Held erect at the top of the stem, ¾–2 inches high, covered with bracts like those on the stem. **FLOWERS:** Up to ½ inch long, blue or white, with a slender tube and 5 spreading lobes; only 1 flower per spike blooms at a time, emerging from under a bract. **FLOWERING:** Jun–Aug. **RANGE:** GA, FL, SC. **SIMILAR TO** no other wildflower in Georgia.

American Water-Willow

Justicia americana (Linnaeus) Vahl
Acanthus Family (Acanthaceae)

PERENNIAL HERB found in rocky streambeds and on river banks, sandbars, and pond margins in north Georgia. **STEMS:** Up to 3 feet tall, 4-angled, smooth; forms large colonies by the spread of underground stems and by rooting at nodes. **LEAVES:** 2–8 inches long and up to 1 inch wide, opposite, lance-shaped or elliptic, smooth, with very short or no stalk. **FLOWERS:** In short spikes that arise from the leaf axils, each flower about ½ inch wide, with a curved upper lip and a 3-lobed lower lip, white (or pale pink) marked with purple; the stamens are tipped with purple anthers. **FLOWERING:** Jun–Oct. **RANGE:** FL west to TX, north to NE and QC. **SIMILAR TO** Coastal Plain Water-willow (*Justicia ovata*), which occurs in swamps and marshes; it has an elongated flower spike with widely spaced flowers.

Carolina Wild Petunia

Ruellia caroliniensis (J. F. Gmelin) Steudel
Acanthus Family (Acanthaceae)

PERENNIAL HERB found in moist to dry woodlands and forests throughout Georgia. **STEMS:** 4–24 inches tall, hairy. **LEAVES:** Up to 4 inches long and 1¾ inches wide, opposite, oval to lance-shaped, with a short stalk. **FLOWERS:** 1–2 inches long, blue-violet, funnel-shaped with 5 spreading lobes; sepals ½–1 inch long, very narrow and pointed; the anthers are blunt-tipped. **FLOWERING:** May–Sep. **RANGE:** FL west to TX, north to IL and NJ. **NOTES:** Carolina Wild Petunia is not related to the garden petunia, which is in the Nightshade Family. **SIMILAR TO** Sandhills Wild-petunia (*Ruellia ciliosa*), which occurs in sandhills in Georgia's Coastal Plain; its flowers are similar but it has a basal rosette of purplish leaves and stem leaves that are somewhat spoon-shaped.

Elderberry

Sambucus canadensis Linnaeus
Moschatel Family (Adoxaceae)

SHRUB found in open or disturbed wetlands, floodplains, and swamps throughout Georgia. **STEMS:** Up to 13 feet tall, branched from the base, somewhat woody with soft pith and grayish-brown bark with many lenticels. **LEAVES:** About 1 foot long, opposite, with 5–11 toothed, lance-shaped leaflets up to 6 inches long and 2½ inches wide. **FLOWER CLUSTERS:** 8–16 inches across, flat-topped with many small, white, 5-lobed flowers. **FRUIT:** A small, round, purplish-black berry. **FLOWERING:** Apr–Jul. **RANGE:** Most of North America. **NOTES:** The leaves and stems are poisonous and foul-smelling when crushed. Stalked glands at the bases of some of the leaflets secrete an ant-attracting nectar that protects the leaves from other insects. **SIMILAR TO** Smooth Sumac (*Rhus glabra*) and Winged Sumac (*R. copallina*) in general appearance, but both have alternate leaves and yellowish flowers in elongated clusters.

Maple-Leaf Viburnum

Viburnum acerifolium Linnaeus
Moschatel Family (Adoxaceae)

SHRUB found in upland hardwood forests in north Georgia and a few Coastal Plain counties. **STEMS:** Up to 6 feet tall, with opposite branches; new growth very hairy, older twigs smooth. **LEAVES:** Up to 3 inches long and wide, opposite, with 3 pointed, toothed lobes and 3 main veins radiating from the base of the leaf, resembling a maple leaf. **FLOWER CLUSTERS:** 1–2½ inches across, flat-topped, with many small, white, 5-lobed flowers. **FRUIT:** About ⅓ inch long, fleshy, round or oval, blue-black. **FLOWERING:** Apr–May. **RANGE:** FL west to TX, north to ON and NB. **NOTES:** The leaves turn a rich maroon in the fall. **SIMILAR TO** no other shrub in Georgia; it is the only *Viburnum* with maple leaf–shaped leaves.

Walter's Viburnum, Small-Leaved Viburnum

Viburnum obovatum Walter
Moschatel Family (Adoxaceae)

SHRUB found in floodplains and on stream banks in the Coastal Plain. **STEMS:** 6–30 feet tall, with opposite twigs and branches. **LEAVES:** 1–2 inches long and ½–1¼ inches wide, opposite; elliptic, spoon-shaped, or nearly round; entire or toothed near the tip, smooth, deciduous or evergreen. **FLOWER CLUSTERS:** 1½–3 inches wide, flat-topped with many small, white, 5-lobed flowers; the flowers are stalked, but the cluster sits directly on the tip of the twig. **FRUIT:** About ⅓ inch long, oval to round, red turning to black. **FLOWERING:** Mar–Apr. **RANGE:** FL west to AL, north to SC. **SIMILAR TO** Southern Wild Raisin (*Viburnum nudum*), which occurs in similar habitats; its leaves are 2–6 inches long, and the flower cluster is held on a stalk 1–2 inches long.

Southern Black Haw

Viburnum rufidulum Rafinesque
Moschatel Family (Adoxaceae)

SMALL TREE found in dry upland forests and edges of disturbed areas throughout Georgia. **STEMS:** Up to 18 feet tall, with opposite twigs; young twigs and buds with curly, rust-colored hairs. **LEAVES:** Up to 4 inches long and 2 inches wide, opposite, oval to nearly round, somewhat leathery, shiny dark green on the upper surface, pale green on the lower, finely toothed, with rusty-colored hairs on the lower midvein and stalk. **FLOWER CLUSTERS:** 2–4 inches across, flat-topped, with many small, white, 5-lobed flowers. **FRUIT:** About ½ inch long, oval, waxy, bluish-black on red stalks. **FLOWERING:** Mar–Apr. **RANGE:** FL west to TX, north to KS, OH, and VA. **SIMILAR TO** Black Haw (*Viburnum prunifolium*), which occurs in floodplains in north and central Georgia; its leaves are not leathery or shiny, the stalks and veins are mostly smooth, and the fruit is black.

Carpetweed

Mollugo verticillata Linnaeus
Fig-Marigold (Aizoaceae) or
Carpetweed (Molluginaceae) Family

ANNUAL HERB found in disturbed areas throughout Georgia. **STEMS:** 1–18 inches long, round or angled, smooth, sprawling and spreading from a central taproot, often forming a circular mat but not rooting at the nodes. **LEAVES:** Up to 1½ inches long and ¼ inch wide, in whorls of 3–8, linear to lance- or spoon-shaped, entire, smooth, with a single prominent midvein and no stalk. **FLOWERS:** On stalks arising from the leaf nodes, up to ¼ inch wide with 5 white, spreading, green-striped sepals (there are no petals) and a large, oval, green ovary in the center. **FLOWERING:** May–Nov. **RANGE:** Probably native to the tropical Americas, now spread throughout most of North America. The growth form is **SIMILAR TO** the bedstraws (*Galium* spp., pp. 317–318), which have 4-petaled flowers.

Large Sea-Purslane

Sesuvium portulacastrum (Walter) Britton
Fig-Marigold Family (Aizoaceae)

PERENNIAL HERB found on beach dunes, sandbars, and salt flats on Georgia's barrier islands. **STEMS:** Rooting at the leaf nodes and forming mats up to 6 feet wide and 1 foot high; branches succulent, smooth, green or reddish. **LEAVES:** Up to 3 inches long and 1 inch wide, opposite, clasping the stem, smooth, succulent, blue-green, with blunt tips. **FLOWERS:** Up to ¾ inch wide, solitary in the leaf axils, with 5 pointed, spreading sepals and no petals; sepals are pink on the inside, green on the outside, and last only a few hours; flower stalks are ¼–1 inch long. **FLOWERING:** May–Dec. **RANGE:** FL west to TX, north to NC; coastal zones worldwide. **SIMILAR TO** Small Sea-purslane (*Sesuvium maritimum*), found in similar habitats; it has very short or no flower stalks, and the stems do not root at the nodes.

Spiny Amaranth

Amaranthus spinosus Linnaeus
Amaranth Family (Amaranthaceae)

ANNUAL HERB found on roadsides and in other disturbed areas throughout Georgia. **STEMS:** 3 or more feet tall, much-branched, mostly smooth, with 2 long spines at each leaf node. **LEAVES:** Up to 4 inches long and 2½ inches wide, alternate, broadly oval, entire, smooth, with long stalks. **FLOWER CLUSTERS:** Female flowers in clusters in the leaf axils; male flowers in elongated, nodding spikes at the top of the stem. **FLOWERS:** Tiny, with 5 green-and-white tepals and 3 green bracts that are shorter than the tepals. **FLOWERING:** Jul–Oct. **RANGE:** FL west to TX, north to MB and ME; native to tropical America. **SIMILAR TO** Slender Amaranth (*Amaranthus viridis*), which occurs in disturbed areas mostly in the Coastal Plain; its flowers are in slender, interrupted spikes and have only 2 or 3 tepals; the stems do not have spines.

Lamb's-Quarters, Pigweed

Chenopodium album Linnaeus
Amaranth (Amaranthaceae) or Goosefoot (Chenopodiaceae) Family

ANNUAL HERB found in disturbed areas throughout Georgia. **STEMS:** 1–10 feet tall, branched, angled, blue-green, sometimes with red or purple stripes, covered with fine, white, mealy powder. **LEAVES:** Up to 5 inches long and 3 inches wide, alternate, oval to lance-shaped, with irregularly toothed or lobed margins and long leaf stalks, variable in shape and size even on the same plant; younger leaves with a white, mealy coating. **FLOWER CLUSTERS:** Small, round balls held in spikes arising from the axils of upper leaves. **FLOWERS:** Tiny, green, with 5 mealy, white, pointed sepals. **FLOWERING:** Jun–Nov. **RANGE:** Throughout North America and worldwide. **SIMILAR TO** Epazote (*Dysphania ambrosioides*, synonym *Chenopodium ambrosioides*), which occurs in disturbed habitats throughout Georgia; its leaves are gland-dotted on the lower surface and have a pungent smell.

Tall Cottonweed, Snake Cotton

Froelichia floridana (Nuttall) Moquin-Tandon
Amaranth Family (Amaranthaceae)

ANNUAL HERB found in sandhills and dry, disturbed areas. **STEMS:** Up to 6½ feet tall, single-stemmed or bushy-branched, wiry, covered with tangled, white or brown hairs. **LEAVES:** Up to 8 inches long and 1¾ inches wide, opposite, elliptic to lance-shaped, silky-hairy. **FLOWER CLUSTERS:** A spike 1–4 inches long, densely packed with cottony, white flowers. **FLOWERS:** Less than ¼ inch long, covered with tangled, white, cottony hairs and tipped with red, pink, or yellowish teeth. **FLOWERING:** Jun–Oct. **RANGE:** FL west to NM, north to ND and NJ. **SIMILAR TO** Slender Cottonweed (*Froelichia gracilis*), a western species known from disturbed areas in a few east Georgia counties; it is less than 2 feet tall with low branches sprawling across the ground; its flower spikes are less than 1½ inches long.

Russian-Thistle, Saltwort

Salsola kali Linnaeus
SYNONYMS: *Kali tragus* (Linnaeus) Scopoli, *Salsola caroliniana* Walter
Amaranth (Amaranthaceae) or Goosefoot (Chenopodiaceae) Family

ANNUAL HERB found on beaches and in salt marshes in coastal Georgia. **STEMS:** 1–4 feet tall, erect or sprawling, much-branched and bushy, green or red and succulent when young, hard and spiny with age. **LEAVES:** Up to 2½ inches long, very narrow, alternate, linear, fleshy, with a stiff, spiny tip. **FLOWERS:** Held in the leaf axils, tiny, green, with 5 papery, winged sepals and a pair of spine-tipped bracts. **FRUITS:** Up to ⅓ inch wide, enclosed by the enlarged, winged sepals, remaining on the plant after it dies. **FLOWERING:** Jun–frost. **RANGE:** Native to Eurasia, now common on the coast in the United States, Mexico, and North Africa. **NOTES:** The entire plant may turn pinkish-purple in the fall. Dead plants break off and become seed-spreading tumbleweeds. **SIMILAR TO** Southern Sea-blite (*Suaeda linearis*), page 27.

Woody Glasswort

Sarcocornia pacifica (Standley) A. J. Scott
SYNONYM: *Sarcocornia perennis* (P. Miller) A. J. Scott
Amaranth (Amaranthaceae) or Goosefoot (Chenopodiaceae) Family

PERENNIAL HERB found in salt and brackish marshes in coastal Georgia. **STEMS:** Horizontal, woody, usually buried in the sand, with many aboveground branches; 4–20 inches tall, succulent, jointed, waxy, dull green tinged with red. **LEAVES:** 2 tiny, opposite, triangular blades at the top of each joint. **FLOWER CLUSTERS:** Fleshy, reddish, jointed spikes at the tips of some branches. **FLOWERS:** Each joint of the flowering spike has 3 tiny, scalelike flowers with minute yellow anthers. **FLOWERING:** Jul–Oct. **RANGE:** FL north to NH; CA, Mexico. **SIMILAR TO** two annual species of Glasswort (*Salicornia bigelovii* and *S. virginica*), which also occur in salt and brackish marshes; they have a single, erect, much-branched stem that looks like a tiny, succulent tree; each of the branches is tipped with a flowering spike.

Southern Sea-Blite

Suaeda linearis (Elliott) Moquin
Amaranth (Amaranthaceae) or Goosefoot (Chenopodiaceae) Family

ANNUAL OR PERENNIAL HERB found on beaches, dunes, and in salt marshes. **STEMS:** Up to 3 feet long, woody at the base, much-branched, green or red. **LEAVES:** Up to 2 inches long and ¾ inch wide, flat with parallel sides or nearly round in cross-section, succulent, green or red, entire, smooth, without stalks. **FLOWER CLUSTERS:** Leafy spikes up to 5 inches long, the flowers in small clusters in the leaf axils. **FLOWERS:** Less than ⅛ inch across with 5 red or green sepals enclosing the fruit; there are no petals. **FLOWERING:** Aug–frost. **RANGE:** FL west to TX, north to ME; West Indies. **SIMILAR TO** Russian-thistle (*Salsola caroliniana*), page 26; Dune Marsh-elder (*Iva imbricata*), page 93; and Saltwort (*Batis maritima*), page 127.

Small-Flowered Pawpaw

Asimina parviflora (Michaux) Dunal
Custard-Apple Family (Annonaceae)

DECIDUOUS SHRUB found in upland hardwood forests in the Piedmont and Coastal Plain. **STEMS:** 3–6 feet tall, young twigs with rust-colored hairs. **LEAVES:** Up to 8 inches long and 4 inches wide, alternate, widest above the middle, tapering at both ends, with rust-colored hairs along the veins on the lower surface; smells like green bell pepper when crushed. **FLOWERS:** About ½ inch across, with 6 fleshy, veiny, maroon petals, the outer 3 petals longer and curved strongly backward; flowers are produced on last year's twigs. **FRUIT:** Up to 2½ inches long and ½ inch wide, smooth, yellow-green, oblong, tasty if you can beat the squirrels to them. **FLOWERING:** Apr–May. **RANGE:** FL west to TX, north to AR and VA. **SIMILAR TO** Pawpaw Tree (*Asimina triloba*), which has similar though larger leaves, flowers, and fruits; it is a small tree up to 25 feet tall.

Spoon-Leaf Pawpaw

Asimina spatulata (Kral) D. B. Ward
SYNONYM: *Asimina longifolia* Kral var. *spatulata* Kral
Custard-Apple Family (Annonaceae)

DECIDUOUS SHRUB found in sandhills, dry pine flatwoods, and sandy disturbed areas in the Coastal Plain. **STEMS:** 3–5 feet tall, new growth mostly smooth. **LEAVES:** Up to 8 inches long and 1¼ inches wide, alternate, widest near the tip, leathery, more or less smooth, foul-smelling when crushed, with raised veins on the lower surface. **FLOWERS:** Nodding from leaf axils on this year's growth, opening as leaves are nearly or fully expanded; the outer 3 petals are white or pink and up to 3 inches long, the inner 3 are white, pink, or maroon and about half that length. **FLOWERING:** Apr–Jun. **RANGE:** GA, AL, and FL. **SIMILAR TO** Slim-leaf Pawpaw (*Asimina angustifolia*, synonym *A. longifolia* var. *longifolia*), which occurs in similar habitats; its outer petals are white, the leaves are widest at or just above the middle, and new twig growth is hairy.

Carrot Family (Apiaceae or Umbelliferae)

The Carrot Family, also known as the Parsley Family or the Umbel Family, is one of the easiest *families* to recognize in the field, but its *species* are among the most difficult to identify. The Carrot Family is known for its divided, compound leaves; hollow, ridged stems; expanded leaf bases that wrap the stem; and ribbed or winged fruits—all with strong-smelling and -tasting compounds. Humans have used these powerful compounds for thousands of years, incorporating carrots, parsnips, celery, parsley, cilantro, fennel, anise, cumin, dill, and many others into our meals in one form or another. At the same time, Water Hemlock (p. 31), and Poison Hemlock are among the most toxic of all plants.

The defining feature of the Carrot Family is its flower cluster, the umbel. In an umbel, the flowers are held at the tips of slender stalks (sometimes called rays) that originate at a single point and spread laterally like the ribs of an umbrella. If the stalks are different lengths, with the shortest stalks in the center and the longest at the edge, the umbel is flat-topped; if the stalks are more or less the same length, the umbel is a rounded or dome-shaped cluster (a head). Only a few of our Carrot Family species have flowers held in compact, round heads (see Coin-leaf, p. 30; Rattlesnake Master, p. 34; and Canada Sanicle, p. 37). Umbels are usually compound, with the first round of stalks holding smaller umbelets whose stalks hold the flowers. The flowers are nondescript, with 5 white or yellow (rarely blue or green) petals.

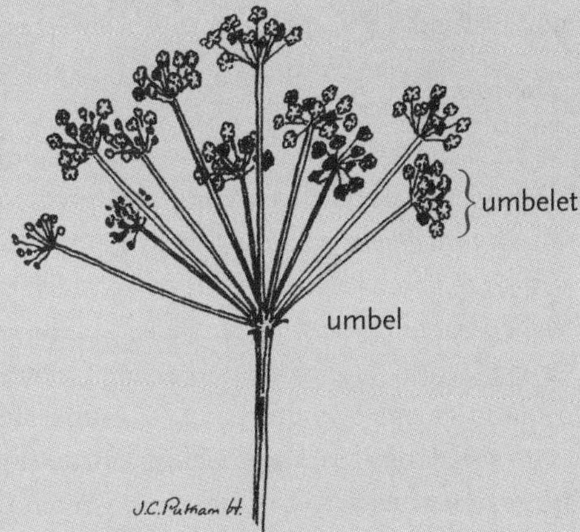

Hairy Angelica

Angelica venenosa (Greenway) Fernald
Carrot Family (Apiaceae)

PERENNIAL HERB found in moist to dry woodlands in north and southwest Georgia. **STEMS:** 1½–5 feet tall, unbranched, finely hairy. **LEAVES:** Alternate, divided into many oval, finely toothed leaflets, up to 2¼ inches long and 1 inch wide, dark green, leathery, finely hairy; the base of the leaf stalk expands into a ½-inch-wide sheath that clasps the stem. **FLOWER CLUSTERS:** Many small, rounded clusters at the tips of slender, hairy stalks forming an umbrella-shaped cluster up to 5 inches across. **FLOWERS:** Small, with 5 white petals and 5 long, white stamens. **FLOWERING:** Jun–Aug. **RANGE:** FL west to LA and OK, north to MN and ON. **NOTES:** The species name *venenosa* means "very poisonous," referring to the toxic roots. **SIMILAR TO** other white-flowered species in the Carrot Family; look for the leathery, oval leaflets and overall hairiness.

Coin-Leaf

Centella asiatica Lamarck
SYNONYM: *Centella erecta* (Linnaeus f.) Fernald
Carrot Family (Apiaceae)

PERENNIAL HERB found in bogs, wet savannas, Carolina bays, ditches, and pond margins in the Coastal Plain. **STEMS:** Slender, spreading across the ground, with clusters of several leaves arising from widely spaced nodes. **LEAVES:** ½–2 inches long, heart-shaped with rounded tips, toothed, dull green, somewhat succulent, hairy or smooth; the leaf stalk is attached at the edge of the leaf. **FLOWERS:** Held in rounded clusters on short stalks, each flower with 5 pointed, white or greenish petals tinged with rose. **FLOWERING:** Jun–Aug. **RANGE:** FL west to TX, north to NJ; globally throughout the tropics. **NOTES:** Known as Gotu Kola, Coin-leaf has been used for thousands of years in Asia to treat a variety of ailments. **SIMILAR TO** Dune Pennywort (*Hydrocotyle bonariensis*), page 42, and other species in that genus.

Southern Chervil

Chaerophyllum tainturieri Hooker
Carrot Family (Apiaceae)

ANNUAL HERB found in moist, disturbed areas and floodplains throughout Georgia. **STEMS:** Up to 3 feet tall, densely hairy, branching in the upper half. **LEAVES:** Up to 4¾ inches long and wide, divided into many very hairy leaflets that are divided into many narrow, pointed segments; sweet-smelling when crushed. **FLOWER CLUSTERS:** Up to 2½ inches wide, composed of several smaller flat-topped clusters, each with 3–10 flowers; held in the leaf axils on short stalks. **FLOWERS:** Tiny, with 5 white petals. **FLOWERING:** Mar–Apr. **RANGE:** FL west to AZ, north to NE and MD. **SIMILAR TO** Spreading Chervil (*Chaerophyllum procumbens*), which occurs in floodplains in northwest Georgia; it branches from the base of the stem and has nearly hairless stems and leaves.

Water Hemlock

Cicuta maculata Linnaeus
Carrot Family (Apiaceae)

PERENNIAL OR BIENNIAL HERB found in wetlands throughout Georgia. **STEMS:** Up to 7 feet tall, stout, hollow, smooth, green mottled with purple. **LEAVES:** Up to 1 foot long and wide, alternate, divided into many toothed, narrowly lance-shaped leaflets up to 4 inches long and 1¼ inches wide. **FLOWER CLUSTERS:** Up to 6 inches wide, dome-shaped, with hundreds of tiny, white, 5-petaled flowers. **FLOWERING:** May–Aug. **RANGE:** Most of North America. **NOTES:** Water Hemlock has been called the most poisonous plant in North America; the roots are especially toxic to both humans and livestock. **SIMILAR TO** Southern Water Hemlock (*Cicuta mexicana*), which occurs in wetlands throughout Georgia; it has wider, oval leaflets and a stouter stem, 1½–2 inches in diameter, that is striped green-and-purple but appears solid purple from a distance. Also see Lovage (*Ligusticum canadense*), page 34.

Honewort

Cryptotaenia canadensis (Linnaeus) A. P. de Candolle
Carrot Family (Apiaceae)

PERENNIAL HERB found in moist, rich forests and along streams in north Georgia and a few Coastal Plain counties. **STEMS:** Up to 3 feet tall, smooth, branched. **LEAVES:** Alternate, divided into 3 oval, toothed, usually lobed leaflets up to 6 inches long and 3 inches wide; the base of the leaf stalk expands into a sheath that clasps the stem. **FLOWERS:** In loose, open clusters at the top of the plant, each with 5 tiny, white petals. **FLOWERING:** May–Jun. **RANGE:** FL west to TX, north to MB and NB. **SIMILAR TO** Smooth Sweet Cicely (*Osmorhiza longistylis*), page 35, which smells strongly of anise or licorice. Honewort has no distinctive smell.

Queen Anne's Lace

Daucus carota Linnaeus
Carrot Family (Apiaceae)

BIENNIAL HERB found in disturbed areas throughout Georgia. **STEMS:** 1–6 feet tall, much-branched, ribbed, covered with long, white hairs. **LEAVES:** 2–8 inches long, including the stalks, alternate, divided into many finely cut leaflets. **FLOWER CLUSTERS:** 2–5 inches wide, flat-topped, with many tiny, white, 5-petaled flowers; the central flower is usually maroon; twice-divided bracts surround the base of the cluster, which folds up into a bird's nest shape after pollination. **FLOWERING:** May–Sep. **RANGE:** Native to Europe, spread throughout North America. **NOTES:** The edible carrot is a cultivar of *Daucus carota* ssp. *sativa*. **SIMILAR TO** American Queen Anne's Lace (*Daucus pusillus*), a smaller, unbranched annual plant found in similar habitats; it does not have the central maroon flower, and the bracts surrounding the base of the flower cluster are only once divided.

Savanna Eryngo

Eryngium integrifolium Walter
Carrot Family (Apiaceae)

PERENNIAL HERB found in wetlands and flatwoods throughout Georgia. **STEMS:** 12–32 inches tall, branching near the top. **STEM LEAVES:** Up to 2¾ inches long and 1½ inches wide, alternate, oval to lance-shaped, toothed, with short or no leaf stalks; larger, lower leaves are usually gone by flowering time. **FLOWER CLUSTERS:** Round heads less than ½ inch wide, blue or lavender, surrounded by a whorl of spiny bracts. **FLOWERS:** Tiny, with 5 blue or greenish petals, long, blue stamens and style branches, and a small, 3-toothed bract below each flower. **FLOWERING:** Aug–Oct. **RANGE:** FL west to TX, north to OK and VA. **SIMILAR TO** Fragrant Eryngo (*Eryngium aromaticum*), which occurs in dry pinelands in southeast Georgia; its flower clusters are greenish, the stem leaves are deeply divided into 3 or more segments, and the basal leaves have 5 or more pointed lobes.

Creeping Eryngo

Eryngium prostratum Nuttall ex A. P. de Candolle
Carrot Family (Apiaceae)

PERENNIAL HERB found in floodplains, pond margins, and other wetlands throughout Georgia. **STEMS:** Up to 16 inches long, sprawling and forming mats. **LEAVES:** Up to 2¾ inches long and 1 inch wide, oval or lance-shaped, sometimes 3-lobed, entire or toothed, smooth. **FLOWER CLUSTERS:** Held at the tips of 1-inch-long stalks that arise from the leaf axils, each cluster about ⅓ inch long, cylinder-shaped, surrounded by 5–10 green, drooping, entire bracts that are about ⅓ inch long. **FLOWERS:** Tiny, with 5 blue petals and long, bluish-purple stamens. **FLOWERING:** May–Oct. **RANGE:** FL west to TX, north to KS and DE. **SIMILAR TO** Baldwin's Eryngo (*Eryngium baldwinii*), which occurs in similar habitats in the lower Coastal Plain; its flower heads are smaller and round with very short bracts.

Rattlesnake Master

Eryngium yuccifolium Michaux
Carrot Family (Apiaceae)

PERENNIAL HERB found in prairies, glades, flatwoods, and savannas throughout Georgia. **STEMS:** Up to 3 feet tall, stout, pale green. **LEAVES:** Up to 40 inches long and 2½ inches wide near the base of the stem, much smaller up the stem, alternate, thick, straplike, with parallel veins and bristly margins. **FLOWER CLUSTERS:** Round heads held at the tips of branches at the top of the stem, about 1 inch wide, bristly, greenish-white, surrounded by sharp-tipped bracts. **FLOWERS:** Tiny, with 5 white petals and 5 green sepals. **FLOWERING:** Jun–Sep. **RANGE:** FL west to TX, north to MN and CT. **SIMILAR TO** Marsh Eryngo (*Eryngium aquaticum*), which occurs in marshes, swamps, wet flatwoods, and shallow waters around ponds and lakes in the Coastal Plain; its leaves have netted veins, and its flowers are blue or lavender.

Lovage

Ligusticum canadense (Linnaeus) Britton
Carrot Family (Apiaceae)

PERENNIAL HERB found in moist forests and creek bottoms in north Georgia. **STEMS:** 2–6 feet tall, green with a whitish, waxy coating. **LEAVES:** Up to 18 inches long and wide, alternate, usually with a sheathing base; divided into many sharply toothed leaflets 2–4 inches long and 1–2 inches wide; the lower ⅓ of each leaflet is straight and toothless. **FLOWER CLUSTERS:** Up to 5 inches wide, flat-topped, composed of many smaller flat-topped clusters, each with about 15 flowers. **FLOWERS:** About ¼ inch wide, with 5 white petals. **FLOWERING:** May–Jul. **RANGE:** GA west to MS, north to MO and PA. **NOTES:** All parts of the plant smell like celery or licorice when crushed. **SIMILAR TO** Mountain Angelica (*Angelica triquinata*), which has red or purple stems and occurs in moist, high-elevation forests in Georgia's Blue Ridge. Also see Water Hemlock (*Cicuta maculata*), page 31.

Smooth Sweet Cicely, Anise-Root

Osmorhiza longistylis (Torrey) A. P. de Candolle
Carrot Family (Apiaceae)

PERENNIAL HERB found in moist, rich forests in north Georgia. **STEMS:** Up to 3 feet tall, usually smooth. **LEAVES:** Up to 1 foot long, divided into many oval, coarsely toothed leaflets ½–3½ inches long. **FLOWER CLUSTERS:** Up to 5 inches wide, flat-topped, composed of many smaller flat-topped clusters, each with 9–18 flowers. **FLOWERS:** Less than ¼ inch wide, with 5 white petals; the style (an extension of the green ovary at the center of the flower) is longer than the petals and easily seen. **FLOWERING:** Apr–May. **RANGE:** GA west to NM, north to AB and NS. **NOTES:** All parts of the plant smell like anise. **SIMILAR TO** Sweet Cicely (*Osmorhiza claytonii*), which occurs in similar habitats; its flowers have inconspicuous styles shorter than the petals; it has a fainter anise smell and usually hairy stems.

Water Dropwort

Oxypolis filiformis (Walter) Britton
SYNONYM: *Tiedemannia filiformis* (Walter) Feist & S. R. Downie
Carrot Family (Apiaceae)

PERENNIAL HERB found in cypress ponds and savannas, wet pine flatwoods, seepage slopes, and wet ditches in the Coastal Plain. **STEMS:** Up to 6 feet tall, slender, finely ribbed, with a whitish, waxy coating. **LEAVES:** Lowest leaves up to 2 feet long, most leaves much shorter; tubelike and hollow except for regularly spaced cross-walls; the base is expanded, flattened, and clasping the stem; leaves at the lower nodes persist through flowering time. **FLOWER CLUSTERS:** About 3 inches wide, flat-topped, composed of several wide-spreading, smaller, flat-topped clusters. **FLOWERS:** Tiny, with 5 white petals. **FLOWERING:** Jul–Sep. **RANGE:** FL west to TX, north to NC. **SIMILAR TO** Canby's Dropwort (*Oxypolis canbyi*), a federally listed species, which drops its lower leaves by flowering time. Also see Eastern Bishopweed (*Ptilimnium capillaceum*), page 36.

Cowbane

Oxypolis rigidior (Linnaeus) Rafinesque
Carrot Family (Apiaceae)

PERENNIAL HERB found in floodplains, seepages, wet meadows, and stream banks in north Georgia and a few Coastal Plain counties. **STEMS:** Up to 5 feet tall. **LEAVES:** Up to 12 inches long, divided into 5–11 oblong or lance-shaped leaflets up to 5 inches long and 1½ inches wide, with widely spaced teeth above the middle. **FLOWER CLUSTERS:** Up to 6 inches wide, flat-topped, composed of many smaller flat-topped clusters. **FLOWERS:** About ⅛ inch wide, with 5 white, upwardly curved petals. **FLOWERING:** Aug–Oct. **RANGE:** FL west to TX, north to MN and NY. **NOTES:** As the common name indicates, Cowbane is poisonous to livestock. **SIMILAR TO** many other species in the Carrot Family; the relatively wide leaflets that are toothed only toward the tips are distinctive.

Eastern Bishopweed

Ptilimnium capillaceum (Michaux) Rafinesque
Carrot Family (Apiaceae)

ANNUAL HERB found in marshes, beaver ponds, bottomlands, and wet ditches throughout Georgia. **STEMS:** ½–5 feet tall, usually much-branched. **LEAVES:** Up to 4 inches long, alternate, divided into 3 leaflets that are whorled around the leaf axis; the leaflets are divided into very narrow, thread-like segments. **FLOWER CLUSTERS:** Held at the tips of branches, up to 2 inches wide, flat-topped, composed of many widely spreading smaller clusters, each with 4–20 flowers. **FLOWERS:** Tiny, with 5 white, upwardly curved petals and pinkish-purple anthers. **FLOWERING:** Jun–Aug. **RANGE:** FL west to TX, north to MO and MA. **SIMILAR TO** Marsh-parsley (*Cyclospermum leptophyllum*), which occurs in marshes and disturbed wetlands throughout Georgia (except the mountains); its flower clusters sit directly on the stem across from a leaf base.

Canada Sanicle, Black Snakeroot

Sanicula canadensis Linnaeus
Carrot Family (Apiaceae)

PERENNIAL HERB found in dry to moist hardwood forests, stream banks, and disturbed areas throughout Georgia. **STEMS:** ½–2 feet tall. **LEAVES:** Up to 8 inches long and wide, alternate, usually divided into 3 or 5 toothed, hairless segments; leaf margins are whitish or have a transparent border (use 10× lens). **FLOWER CLUSTERS:** Held at the tips of long, grooved stalks; clusters look bristly because of the long stamens. **FLOWERS:** Tiny, with 5 greenish-white petals. **FRUIT:** In round, burlike clusters covered with hooked bristles. **FLOWERING:** Apr–May. **RANGE:** FL west to TX, north to SD and QC. **SIMILAR TO** Yellow-flowered Snakeroot (*Sanicula odorata*, synonym *Sanicula gregaria*), which has yellow petals. Several other green-flowered species of *Sanicula* occur in Georgia; the differences are technical and hard to discern in the field.

Hairy-Jointed Meadow Parsnip

Thaspium barbinode (Michaux) Nuttall
Carrot Family (Apiaceae)

PERENNIAL HERB found in moist forests in north Georgia and a few Coastal Plain counties. **STEMS:** Up to 3½ feet tall, smooth except for hairs at the base of the leaf stalks. **LEAVES:** Mostly basal, up to 12 inches long and 6 inches wide, divided, parsley-like, into 6 or 9 coarsely toothed leaflets with hairy margins; leaf bases expand to a hairy, purple-tinged sheath that clasps the stem. **FLOWER CLUSTERS:** Up to 3 inches wide, with a rounded top and 12–20 smaller clusters. **FLOWERS:** Tiny, with 5 golden yellow petals. **FLOWERING:** Apr–May. **RANGE:** FL west to TX, north to MN, ON, and NY. **SIMILAR TO** Meadow Parsnip (*Thaspium trifoliatum*) and Mountain Golden Alexander (*Zizia trifoliata*), page 38.

Meadow Parsnip

Thaspium trifoliatum (Linnaeus) A. Gray
Carrot Family (Apiaceae)

PERENNIAL HERB found in upland hardwood forests in north and southwest Georgia. **STEMS:** Up to 3 feet tall, smooth, ribbed. **BASAL LEAVES:** Up to 4 inches long and wide, heart- or kidney-shaped, *or* divided into 3 toothed leaflets with long stalks. **STEM LEAVES:** Alternate, divided into 3 toothed, lance-shaped or oval leaflets that are toothed and have a narrow, translucent border (use 10× lens). **FLOWER CLUSTERS:** 1–3 inches wide, composed of several smaller, widely spreading, flat-topped clusters, with the central flower of each small cluster on a tiny stalk. **FLOWERS:** Tiny, with 5 maroon or golden yellow petals. **FLOWERING:** Apr–Jun. **RANGE:** FL west to TX, north to MN and ON. **NOTES:** Yellow-flowered plants are variety *aureum*; maroon-flowered plants are variety *trifoliatum*. **SIMILAR TO** Mountain Golden Alexander (*Zizia trifoliata*), below.

Mountain Golden Alexander

Zizia trifoliata (Michaux) Fernald
Carrot Family (Apiaceae)

PERENNIAL HERB found in moist, hardwood forests and creek bottoms in north and southwest Georgia. **STEMS:** Up to 3 feet tall. **LEAVES:** Alternate, divided into 3–7 leaflets, each leaflet up to 2 inches long and 1 inch wide, somewhat leathery; margins of the leaflets are coarsely toothed (5–10 teeth per inch) and have a narrow, translucent border (use 10× lens). **FLOWER CLUSTERS:** Up to 3 inches wide, composed of 6–12 smaller, flat-topped clusters; the central flower of the small clusters has no stalk while the surrounding flowers have stalks about 1/16 inch long. **FLOWERS:** About 1/8 inch wide, with 5 golden yellow petals. **FLOWERING:** Apr–May. **RANGE:** FL north to KY and MD. **SIMILAR TO** Common Golden Alexander (*Zizia aurea*), which has finely toothed leaflet margins (more than 10 teeth per inch) and thin-textured leaflets. Heartleaf Golden Alexander (*Z. aptera*) has undivided, heart-shaped basal leaves, finely toothed leaflet margins, and thin-textured leaflets.

Fringed Blue Star, Sandhills Blue Star

Amsonia ciliata Walter
Dogbane Family (Apocynaceae)

PERENNIAL HERB found in sandhills, sandy woodlands, and clearings mostly in the Coastal Plain. **STEMS:** 1–2 feet tall, branched, leafy, hairy. **LEAVES:** Up to 3 inches long and ¼ inch wide, usually much narrower, alternate, very hairy. **FLOWERS:** About ¾ inch wide, with a narrow, ¼-inch tube and 5 spreading lobes, pale blue with a white or pale yellow center. **FRUIT:** Up to 7 inches long, green, slender, erect, in pairs. **FLOWERING:** Apr. **RANGE:** FL west to TX (but not MS and LA), north to MO and NC. **NOTES:** Leaves and stems of all Apocynaceae species contain a milky latex that is toxic to deer and other herbivores. **SIMILAR TO** Eastern Blue Star (*Amsonia tabernaemontana*), below.

Eastern Blue Star

Amsonia tabernaemontana Walter
Dogbane Family (Apocynaceae)

PERENNIAL HERB found in moist, deciduous forests and bottomlands in north and southwest Georgia. **STEMS:** 1–3 feet tall, smooth. **LEAVES:** Up to 6 inches long and 2½ inches wide, alternate, oval to lance-shaped, tapering to a very short stalk; surfaces are smooth, margins have long hairs; upper surface dull green with a white midvein, lower surface pale green. **FLOWERS:** About ¾ inch wide, blue with a white or pale yellow center, with a short tube and 5 spreading lobes; a dense patch of white hairs blocks the top of the tube. **FRUIT:** Up to 4 inches long, slender, erect, green, in pairs. **FLOWERING:** Apr–May. **RANGE:** FL west to TX, north to KS, IL, and NY. **SIMILAR TO** Stiff Blue Star (*Amsonia rigida*), which occurs in wetlands in the Coastal Plain; its stem is hairy, and the flowers are smooth.

Indian Hemp, Common Dogbane

Apocynum cannabinum Linnaeus
Dogbane Family (Apocynaceae)

PERENNIAL HERB found in dry woodlands, roadsides, and other disturbed areas throughout Georgia. **STEMS:** 1–4½ feet tall, smooth or hairy, reddish-green, with a few widely spreading branches. **LEAVES:** Up to 5½ inches long and 2 inches wide, opposite, oval to lance-shaped, smooth or hairy, with a pale midvein. **FLOWER CLUSTERS:** Held at the tips of branches or in leaf axils, with many flowers. **FLOWERS:** Less than ¼ inch wide, greenish-white, with 5 pointed, erect lobes. **FRUIT:** 4–8 inches long, slender, drooping, green, in pairs; each seed is tipped with a tuft of silky hairs. **FLOWERING:** May–Aug. **RANGE:** Most of North America. **NOTES:** Leaves and stems contain a toxic, milky latex. **SIMILAR TO** Spreading Dogbane (*Apocynum androsaemifolium*), which occurs in similar habitats in north Georgia; it has pinkish-white flowers with spreading or curved lobes.

Climbing Dogbane

Thyrsanthella difformis (Walter) Pichon
SYNONYM: *Trachelospermum difforme* (Walter) A. Gray
Dogbane Family (Apocynaceae)

PERENNIAL WOODY VINE found in bottomlands, marshes, and wet thickets in the Coastal Plain and lower Piedmont. **STEMS:** Long, slender, twining over other plants, smooth, reddish. **LEAVES:** Up to 5½ inches long and 3 inches wide, opposite, oval to linear, variable in size and shape, smooth or hairy, deciduous. **FLOWERS:** In stalked clusters held in the leaf axils, each about ½ inch wide, with a short tube and 5 spreading lobes, pale yellow with orange stripes inside. **FRUIT:** 4–8 inches long, slender, green, drooping, in pairs. **FLOWERING:** May–Jul. **RANGE:** FL west to TX, north to MO and DE. **NOTES:** Leaves and stems contain a toxic, milky latex. The leaves are **SIMILAR TO** those of Yellow Jessamine (*Gelsemium sempervirens*), page 251, which has evergreen leaves with clear sap.

Wild Sarsaparilla

Aralia nudicaulis Linnaeus
Aralia Family (Araliaceae)

PERENNIAL HERB found in upland forests in Georgia's Blue Ridge. **STEMS:** Up to 5 inches long, underground with only the tip visible. **LEAVES:** Usually 1 per plant (rarely 2), rising from the tip of the buried stem, divided into 9–21 oval, finely toothed leaflets, each up to 7 inches long and 4 inches wide. **FLOWER CLUSTERS:** Usually 3 per plant, held at the top of a stalk 3–8 inches long that arises from the tip of the buried stem, round, and about 2 inches wide. **FLOWERS:** About ⅛ inch wide, with 5 white petals and 5 long stamens. **FRUIT:** A round, dark blue berry about ¼ inch wide. **FLOWERING:** May–Jul. **RANGE:** GA west to WA, north to YT and NL. **SIMILAR TO** Spikenard (*Aralia racemosa*), found in moist forests in north Georgia; it has an aboveground stem several feet tall, heart-shaped leaflets, and elongated flower clusters.

Devil's Walking Stick

Aralia spinosa Linnaeus
Ginseng Family (Araliaceae)

SHRUB OR SMALL TREE found in moist forests, thickets, and on roadsides throughout Georgia. **STEMS:** 5–25 feet tall, with pale tan bark, stout prickles, and leaf scars that nearly encircle the stem. **LEAVES:** 1–4 feet long with a stout stalk up to 1 foot long, clustered at the top of the stem, composed of many toothed leaflets up to 4 inches long and 3 inches wide; leaf stalks and midveins usually with prickles. **FLOWER CLUSTERS:** 1–4 feet wide, much-branched, at the top of the stem, with many small, white, 5-petaled flowers. **FRUIT:** About ¼ inch wide, purple-black, in large, red-branched clusters. **FLOWERING:** Jun–Sep. **RANGE:** FL west to TX, north to IL and ME. **NOTES:** The large, prickly compound leaves are the largest in North America. **SIMILAR TO** no other plant in Georgia.

Dune Pennywort

Hydrocotyle bonariensis Lamarck
Ginseng (Araliaceae) or Carrot (Apiaceae) Family

PERENNIAL HERB found in dunes and interdune swales on Georgia's barrier islands. STEMS: Spreading across the ground, rooting at leaf nodes and forming mats. LEAVES: 1–4 inches wide, alternate, more or less round, the stalk attached to the center of the leaf, toothed, glossy, somewhat succulent. FLOWER CLUSTERS: Compound umbels 1–4 inches wide, round or pyramid-shaped. FLOWERS: Small, with 5 oval, white or greenish petals. FLOWERING: Apr–Sep. RANGE: FL west to TX, north to VA; South and Central America. SIMILAR TO two pennyworts found in freshwater ponds and wetlands in the Coastal Plain and a few Piedmont counties: Water Pennywort (*Hydrocotyle umbellata*) has flat-topped umbels about 1 inch wide; Whorled Pennywort (*H. verticillata*) flower clusters are widely spaced, forming an interrupted spike, and the leaves are less than 2½ inches wide.

Ginseng

Panax quinquefolius Linnaeus
Aralia Family (Araliaceae)

PERENNIAL HERB found in rich forests in north and southwest Georgia. STEMS: Up to 2 feet tall. LEAVES: In a whorl at the top of the stem, each with 3–5 oval, toothed, pointed leaflets 2⅓–6 inches long and 1½–2¾ inches wide, the 2 lower leaflets much smaller. FLOWER CLUSTERS: 1 per plant, round, held at the tip of a stalk that rises from the top of the stem. FLOWERS: Tiny, with 5 greenish-white petals. FRUIT: A round, red berry about ⅓ inch wide. FLOWERING: May–Jun. RANGE: GA west to LA, north to SD and QC. NOTES: Ginseng is threatened by overcollection for the Asian medicinal herb trade. SIMILAR TO Virginia Creeper (*Parthenocissus quinquefolia*), whose leaflets are all the same size; and Yellow Buckeye (*Aesculus flava*) seedlings, whose leaves arise from a large, brown seed or a short, woody stem.

Wild Ginger

Asarum canadense Linnaeus
Birthwort Family (Aristolochiaceae)

PERENNIAL HERB found in rich, upland, deciduous hardwood forests in north Georgia. **LEAVES:** 3–6 inches long and wide, arising in pairs from an underground stem, heart-shaped, hairy, uniformly green, deciduous; leaf stalk 2–8 inches long, hairy. **FLOWERS:** 1 per plant, held on hairy stalks at the base of the leaves, hidden in the leaf litter and pollinated by crawling insects; each flower about 1 inch long, with a fleshy, 3-lobed calyx (there are no petals); outer surface smooth and maroon, inner surface hairy, white or green. **FLOWERING:** Apr–May. **RANGE:** GA west to LA, north to MB and QC. **NOTES:** Native Americans and pioneers used the rhizomes for seasoning and medicine. **SIMILAR TO** Heartleaf (*Hexastylis arifolia*), below; and Large-flowered Heartleaf (*Hexastylis shuttleworthii*), page 45.

Heartleaf, Little Brown Jugs, Wild Ginger

Hexastylis arifolia (Michaux) Small
Birthwort Family (Aristolochiaceae)

PERENNIAL HERB found in dry to moist forests throughout Georgia, mostly in north Georgia. **STEMS:** Underground, leaves and flowers emerging from the tip, with a spicy fragrance when broken. **LEAVES:** Up to 8 inches long and 6 inches wide, with a long stalk; arrowhead-shaped, leathery, smooth, evergreen, dark green or mottled with silvery patches (overwintering leaves turn bronzy purple), with a spicy fragrance when crushed. **FLOWERS:** About 1 inch long, cup-shaped with 3 short, pointed lobes; thick-walled and brittle, purple or brown, usually hidden in the leaf litter near the base of a leaf stalk. **FLOWERING:** Mar–May. **RANGE:** FL west to LA, north to KY and VA. **SIMILAR TO** Large-flowered Heartleaf (*Hexastylis shuttleworthii*), page 45.

The Skinny on Fat Bodies—Elaiosomes

Plants have evolved ingenious ways to disperse their seeds. Winged seeds are borne on the wind. Rough seeds such as Beggar-ticks and Spanish Needles stick to fur, feathers, and clothing. Other seeds pass unharmed (or even enhanced) through the guts of fruit-eating birds and mammals, and are deposited along with nutrient-rich fertilizer. And some are distributed by ants. Famed for their hardworking ways but not particularly for moving long distances, ants seem unlikely vehicles for long-distance seed dispersal. But a large number of early-blooming wildflowers have seeds that bear elaiosomes—tiny, fat-rich "handles" that are prized by ants. With many mouths to feed, foraging ants carry the elaiosomes back to their nests. After their larvae eat the fat bodies, the worker ants discard the seeds in their waste dumps, which make nutrient-rich seed beds. Ant-dispersed species have fruits that ripen in May through July, the same time that ant populations are peaking. This mutually beneficial relationship between spring wildflowers and ants is called "myrmecochory" and is another example of the fascinating way plant-insect relationships have evolved. Below is a partial list of Georgia wildflowers whose seeds bear elaiosomes:

Bloodroot
Cow-wheat
Dutchman's Britches
Green-and-gold
Heartleaf
Hepatica
milkworts
Purple Dead-nettle
Rue-anemone
Spring-beauty
Squirrel Corn
Trailing Arbutus
trilliums
Trout Lily
violets
Virginia Bluebell
Wild Ginger
Wood Poppy

Bloodroot seeds with elaiosomes

Trillium seeds with elaiosomes

Large-Flowered Heartleaf

Hexastylis shuttleworthii (Britten & Baker f.) Small
Birthwort Family (Aristolochiaceae)

PERENNIAL HERB found in rhododendron thickets, creek banks, and bog edges. **STEMS:** Underground. **LEAVES:** Up to 4 inches long and 3½ inches wide, heart-shaped, entire, leathery, evergreen, glossy, dark green, often with pale patches outlining the veins. **FLOWERS:** Up to 1½ inches long and 1 inch wide, cup-shaped with 3 flaring, triangular lobes at the top, thick-walled, purple or brown on the outside, mottled purple on the inside, partially hidden in the leaf litter. **FLOWERING:** Apr–Jul. **RANGE:** GA west to MS, north to TN and NC. **NOTES:** Variety *shuttleworthii* occurs in the mountains; variety *harperi* occurs in the Piedmont and Coastal Plain. **SIMILAR TO** Variable-leaf Heartleaf (*Hexastylis heterophylla*), which occurs in northeast Georgia, often under Mountain Laurel. Its flower tube is cylindrical and about as long as wide.

Dutchman's Pipe, Pipe-Vine

Isotrema macrophyllum (Lamarck) C. F. Reed
SYNONYM: *Aristolochia macrophylla* Lamarck
Birthwort Family (Aristolochiaceae)

WOODY VINE found in moist, hardwood forests in the Blue Ridge. **STEMS:** Up to 65 feet long, woody, climbing high into trees. **LEAVES:** 3–10 inches long and wide, alternate except on spur shoots near the flowers, kidney- or heart-shaped, nearly hairless. **FLOWERS:** Produced on new growth; up to 1½ inches long, tubular, smooth, pale reddish-brown or yellowish-green, U-shaped with a spreading, 3-lobed, golden brown rim at the tip. **FRUIT:** 4 inches long, 1 inch wide, green, cylindrical, ribbed, splitting from the base when ripe. **FLOWERING:** May–Jun. **RANGE:** GA north to PA in the Appalachian Mountains; ON. **NOTES:** Pipe-vine Swallowtail butterflies lay their eggs on the leaves. **SIMILAR TO** Woolly Dutchman's Pipe (*Isotrema tomentosum*), which occurs infrequently in floodplains in the Coastal Plain; its leaves, stems, and flowers are softly hairy, and the flower is pale yellow.

Got Milkweed?

Milkweeds, including the much-loved Butterfly Weed and the closely related milkvines, are among the most intricately beautiful of wildflowers. They are so named because their stems and leaves exude a milky white latex if damaged. Only Butterfly Weed lacks the milky latex, and it is also the only milkweed in our area with alternate leaves.

Milkweeds are in the genus *Asclepias*, named in honor of Asklepios, the Greek god of healing. They have been used to treat everything from blindness to pleurisy to snakebite, but they contain highly toxic steroids called cardenolides and should never be ingested by humans. Monarch butterflies use the toxins for self-defense. Adult Monarchs lay their eggs on milkweed leaves, and the brightly colored caterpillars that emerge from the eggs eat the leaves. The toxic cardenolides are incorporated into the caterpillars' bodies and passed on to the developing adult butterflies during metamorphosis. Predators quickly learn how toxic and distasteful the Monarchs are and leave them (and their look-alikes, the Queen and Viceroy butterflies) alone. But the Monarchs pay a price for their specialized defense systems: a good many larvae die while eating their milkweed host, gummed up by the sticky latex or sickened by the cardenolides.

Milkweed flowers are unique. The 5 petals are usually reflexed (curved downward and away from the base of the flower), hiding the 5 sepals. A corona sits above the petals: it is usually composed of 5 erect hoods and 5 pointed horns that curve over the center of the flower. Nectar produced within the hoods attracts pollinators—butterflies, moths, bees, ants, and wasps—to the flowers. Fertilized flowers produce a tapering, okra-like fruit, usually 3–7 inches long, held erect on a downcurving stalk. Most milkweed seeds are tipped with a tuft of hairs that catches the wind (and also provides nest-lining materials to goldfinches). Twenty-two species of milkweed and nine species of milkvine occur in Georgia.

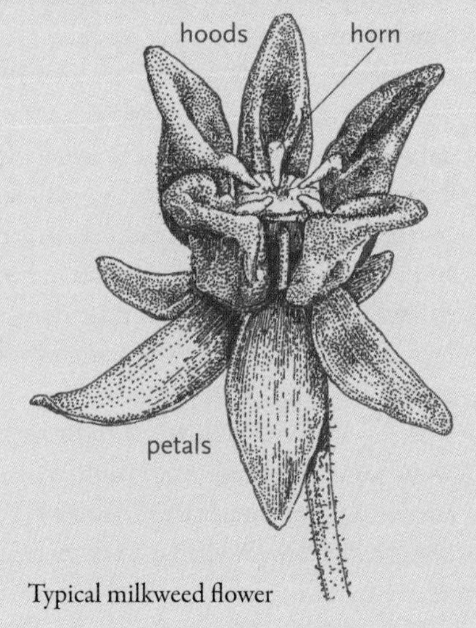

Typical milkweed flower

Clasping Milkweed

Asclepias amplexicaulis J. E. Smith
Milkweed (Asclepiadaceae) or
Dogbane (Apocynaceae) Family

PERENNIAL HERB found in sandhills, open woodlands, and sandy roadsides throughout Georgia. **STEMS:** 16–40 inches tall, stout, smooth. **LEAVES:** 3–6 inches long and 1½–3 inches wide, opposite, broadly oval or oblong, clasping the stem at the base; the margins are wavy, and both surfaces are waxy and hairless; the midvein is white, other veins are often pink. **FLOWERS:** About ½ inch long, with 5 pinkish-green, reflexed petals, 5 dark pink or purple horns, and 5 dark pink hoods enclosing the bases of the horns; the flowers smell like cloves and roses. **FLOWERING:** May–Jul. **RANGE:** FL west to TX, north to MN and NH. **SIMILAR TO** Pineywoods Milkweed (*Asclepias humistrata*), page 48.

Carolina Milkweed

Asclepias cinerea Walter
Milkweed (Asclepiadaceae) or
Dogbane (Apocynaceae) Family

PERENNIAL HERB found in pine flatwoods and savannas and dry, sandy uplands in the Coastal Plain. **STEMS:** Up to 28 inches tall, mostly smooth. **LEAVES:** Up to 4 inches long and very narrow, opposite, linear, smooth, with very short or no stalk. **FLOWER CLUSTERS:** At the top of the stem and sometimes in upper leaf axils, each flower on a slender, nodding stalk up to 1 inch long. **FLOWERS:** About ⅜ inch long, with 5 lavender, downcurved petals and 5 erect, pinkish-lavender hoods, each hood surrounding a pointed horn. **FLOWERING:** Jun–Jul. **RANGE:** GA, FL, and SC. **SIMILAR TO** Michaux's Milkweed (*Asclepias michauxii*), page 49; and Longleaf Milkweed (*Asclepias longifolia*).

Large-Flowered Green Milkweed

Asclepias connivens Baldwin
Milkweed (Asclepiadaceae) or Dogbane (Apocynaceae) Family

PERENNIAL HERB found in wet pine flatwoods, bogs, and edges of cypress ponds in the Coastal Plain. **STEMS:** 1–3 feet tall, stout, unbranched, mostly hairy. **LEAVES:** Up to 2¾ inches long and 2½ inches wide, opposite, somewhat waxy and slightly hairy, oval to oblong, with a short or no stalk. **FLOWERS:** About ¾ inch long, with 5 green, downcurved petals and 5 greenish-white or greenish-yellow hoods that arch over the center of the flower (there are no horns). **FLOWERING:** Jun–Aug. **RANGE:** GA, FL, AL, MS, and SC. **SIMILAR TO** Green Milkweed (*Asclepias viridis*), which has erect, green petals and small, purple hoods; it is found in a few Coastal Plain and Ridge & Valley counties in prairies and dry woodlands.

Pineywoods Milkweed

Asclepias humistrata Walter
Milkweed (Asclepiadaceae) or Dogbane (Apocynaceae) Family

PERENNIAL HERB found in sandhills and other dry, sandy uplands in the Coastal Plain and Fall Line. **STEMS:** Up to 2½ feet tall, leaning or sprawling on the ground, smooth, waxy, pinkish. **LEAVES:** Up to 4 inches long and 3 inches wide, opposite, clasping the stem, oval, waxy, blue-green with pink veins, held perpendicular to the ground. **FLOWERS:** About ¾ inch long, with 5 dull pink, downcurved petals and 5 erect, pinkish-white hoods, each enclosing a tiny horn. **FLOWERING:** May–Jul. **RANGE:** FL west to LA, north to NC. **SIMILAR TO** Clasping Milkweed (*Asclepias amplexicaulis*), page 47.

Few-Flowered Milkweed

Asclepias lanceolata Walter
Milkweed (Asclepiadaceae) or Dogbane (Apocynaceae) Family

PERENNIAL HERB found in wet pine flatwoods and savannas, cypress swamps, and fresh and brackish marshes in the Coastal Plain. **STEMS:** 2–4 feet tall, unbranched, smooth. **LEAVES:** 4–10 inches long and less than ½ inch wide (lower leaves much shorter), opposite, linear to narrowly lance-shaped. **FLOWERS:** About ¾ inch long, with 5 red, downcurved petals and 5 erect, red to yellow-orange hoods, each enclosing a pointed horn that is slightly shorter than the hood. **FLOWERING:** Jun–Aug. **RANGE:** FL north to NJ, west to TX. **SIMILAR TO** Red Milkweed (*Asclepias rubra*), which has reddish-purple petals and orange-tinged hoods; it has been found in one Piedmont county in Georgia, although it occurs in Coastal Plain flatwoods in other states.

Michaux's Milkweed

Asclepias michauxii Decaisne
Milkweed (Asclepiadaceae) or Dogbane (Apocynaceae) Family

PERENNIAL HERB found in seepage slopes, moist to wet pine flatwoods and savannas, and drier sandy uplands in the Coastal Plain. **STEMS:** 4–16 inches tall, unbranched, purplish with hairs in vertical lines. **LEAVES:** 2–4½ inches long, less than ¼ inch wide, mostly opposite, linear, with no stalk, smooth except sometimes with hairy veins on the lower surface. **FLOWERS:** About ⅜ inch long, with 5 green or white, pink-tinged, downcurved petals and 5 erect, rose or purple hoods, each surrounding a tiny, pointed horn. **FLOWERING:** May–Aug. **RANGE:** FL west to LA, north to SC. **SIMILAR TO** Longleaf Milkweed (*Asclepias longifolia*), which occurs in wet savannas and flatwoods in the Coastal Plain; its flowers have no horns inside the hoods, and the central green, flat-topped column rises well above the hoods.

Pineland Milkweed

Asclepias obovata Elliott
Milkweed (Asclepiadaceae) or
Dogbane (Apocynaceae) Family

PERENNIAL HERB found in sandhills, pine woodlands, and dry, sandy disturbed areas in the Coastal Plain. **STEMS:** Up to 2½ feet tall, unbranched, very hairy. **LEAVES:** Up to 3½ inches long and 1½ inches wide, opposite, oval to elliptic, very hairy on the lower surface, with a sharp point at the tip. **FLOWER CLUSTERS:** 1–1½ inch wide, held in the upper leaf axils on very short stalks. **FLOWERS:** About ½ inch long, with 5 pale green, strongly downcurved petals, 5 erect, yellowish-green, pink-tinted hoods surrounding the central column; the tiny horns are shorter than the hoods and curve over the column. **FLOWERING:** Jun–Sep. **RANGE:** FL west to AZ, north to SC. Similar to Green-flowered Milkweed (*Asclepias viridiflora*), page 53.

Swamp Milkweed

Asclepias perennis Walter
Milkweed (Asclepiadaceae) or
Dogbane (Apocynaceae) Family

PERENNIAL HERB found in swamp forests and other wetlands in the Coastal Plain. **STEMS:** 1–2 feet tall, mostly smooth, dark purplish-green. **LEAVES:** 2½–5½ inches long and ½–1½ inches wide, opposite, oval to lance-shaped, smooth, waxy white on the lower surface. **FLOWERS:** About ¼ inch long, white, sometimes tinged pale lavender or pink, with 5 downcurved petals and 5 erect hoods, each hood enclosing a pointed horn. **FLOWERING:** Jun–Aug. **RANGE:** FL west to TX, north to IL and SC. **NOTES:** In bud, the tips of the petals are pink, creating a pink spot in the center of the white flowers. **SIMILAR TO** Whorled Milkweed (*Asclepias verticillata*), page 52.

Four-Leaf Milkweed

Asclepias quadrifolia Jacquin
Milkweed (Asclepiadaceae) or
Dogbane (Apocynaceae) Family

PERENNIAL HERB found in low-elevation oak forests and cove forests in the north Georgia mountains. **STEMS:** Up to 20 inches tall, slender, solitary, unbranched. **LEAVES:** Up to 5 inches long and 2 inches wide, oval or lance-shaped, thin-textured; upper and lower leaves opposite, midstem leaves in whorls of 4. **FLOWERS:** About ⅜ inch long, pale pink to greenish (rarely white), with 5 spreading or downcurved petals and 5 erect hoods, each surrounding a short, pointed horn. **FLOWERING:** May–Jun. **RANGE:** GA west to KS, north to MN, ON, and NH. **SIMILAR TO** Poke or Mountain Milkweed (*Asclepias exaltata*), which occurs in similar habitats; it has pink or white flowers on long, drooping flower stalks, and all of its leaves are opposite.

Butterfly Weed

Asclepias tuberosa Linnaeus
Milkweed (Asclepiadaceae) or
Dogbane (Apocynaceae) Family

PERENNIAL HERB found on roadsides and in dry clearings, prairies, and woodland edges throughout Georgia. **STEMS:** Usually up to 2½ feet tall, often in a clump of stems arising from a large, woody taproot. **LEAVES:** Up to 4 inches long and 1 inch wide, alternate, slightly hairy, oblong to lance-shaped, with no or very short stalks. **FLOWERS:** About ½ inch high, orange, with 5 downcurved petals and 5 erect hoods that curve inward over 5 sharply pointed horns. **FLOWERING:** May–Aug. **RANGE:** Most of eastern and central North America. **SIMILAR TO** other milkweeds, but Butterfly Weed is the only milkweed in our area with alternate leaves and clear, not milky, sap.

White Milkweed

Asclepias variegata Linnaeus
Milkweed (Asclepiadaceae) or Dogbane (Apocynaceae) Family

PERENNIAL HERB found in open upland forests and forest edges in north Georgia and a few Coastal Plain counties. **STEMS:** Up to 3 feet tall, stout, slightly hairy. **LEAVES:** Up to 5½ inches long and 3 inches wide, opposite, oval or oblong, usually stalked; the upper surfaces dark green and smooth, the lower surfaces paler green and hairy. **FLOWERS:** In "snowball" clusters of overlapping flowers, each flower about ⅜ inch long, with 5 white, spreading petals and 5 erect, white hoods with purple rings around the base, each hood surrounding a tiny, pointed horn. **FLOWERING:** May–Jun. **RANGE:** FL west to TX, north to IL, ON, and NY. With its "snowball" flower clusters, White Milkweed is **SIMILAR TO** none of the other milkweeds.

Whorled Milkweed

Asclepias verticillata Linnaeus
Milkweed (Asclepiadaceae) or Dogbane (Apocynaceae) Family

PERENNIAL HERB found in dry woodlands, sandhills, prairies, and openings over mafic and ultramafic rocks throughout Georgia. **STEMS:** Up to 3 feet tall, slender, often branched in the upper half, with narrow, vertical lines of hairs. **LEAVES:** Up to 3 inches long and very narrow, needle-like, with 3–6 whorled leaves per node. **FLOWERS:** About ⅜ inch long, with 5 downcurved petals, white or green tinged with pink or purple, and 5 erect, white hoods, each surrounding the base of a pointed horn that curves over the center of the flower. **FLOWERING:** Jun–Sep. **RANGE:** FL west to AZ, north to SK and VT. **SIMILAR TO** other white-flowered milkweeds, but the narrow, whorled leaves are distinctive.

Green-Flowered Milkweed

Asclepias viridiflora Rafinesque
Milkweed (Asclepiadaceae) or Dogbane (Apocynaceae) Family

PERENNIAL HERB found in dry woodlands, glades, and disturbed areas, especially over mafic and limestone bedrock, in north Georgia and a few Fall Line counties. **STEMS:** Up to 3 feet tall, unbranched, smooth or slightly hairy. **LEAVES:** Up to 4 inches long and 1¾ inches wide, opposite, nearly round to lance-shaped, usually hairy, with wavy margins and a white midvein. **FLOWER CLUSTERS:** About 1 inch wide, nodding from the upper leaf axils, with 15–30 crowded flowers. **FLOWERS:** About ½ inch long, with 5 green, strongly downcurved petals, 5 erect, green hoods that are shorter than the central column, and no horns. **FLOWERING:** Jun–Aug. **RANGE:** FL west to AZ, north to BC, ON, and CT. **SIMILAR TO** Pineland Milkweed (*Asclepias obovata*), page 50.

Climbing Milkweed, Eastern Anglepod

Gonolobus suberosus (Linnaeus) R. Brown
SYNONYM: *Matelea gonocarpa* (Walter) Shinners
Milkweed (Asclepiadaceae) or Dogbane (Apocynaceae) Family

PERENNIAL HERBACEOUS VINE found in moist woods, bottomlands, and thickets throughout Georgia. **STEMS:** Several feet long, slender, twining over other plants, covered with short, soft hairs. **LEAVES:** Up to 8 inches long and 4 inches wide, heart-shaped, hairy on both surfaces, stalks up to 4 inches long. **FLOWERS:** About 1 inch wide, with 5 spreading, pointed petals, yellowish-green at the tips and edges and maroon toward the center, sometimes entirely green, maroon, or brown. **FRUIT:** Up to 3 inches long and 1 inch wide, a smooth (not warty or spiny), tapering pod with 5 angles. **FLOWERING:** Apr–Aug. **RANGE:** FL west to TX, north to KS and MD. **SIMILAR TO** Carolina Spinypod (*Matelea caroliniensis*), which has maroon flowers with rounded petals, round flower buds, and a spiny pod.

Coastal Swallow-Wort, Sand-Vine

Seutera angustifolia (Persoon) Fishbein & W. D. Stevens

SYNONYMS: *Cynanchum angustifolium* Persoon, *Cynanchum palustre* (Pursh) Heller

Milkweed (Asclepiadaceae) or Dogbane (Apocynaceae) Family

PERENNIAL HERBACEOUS VINE found in coastal salt and brackish marshes and maritime forests (hammocks). **STEMS:** 10 feet long or more, slender, smooth, twining over other plants. **LEAVES:** 1–3 inches long, very narrow, opposite, downcurved, evergreen. **FLOWERS:** Held in long-stalked clusters arising from the leaf axils, each flower about ¼ inch wide, greenish-white, often rose-tinged, with 5 pointed petals and 5 shorter, rounded hoods. **FRUIT:** A slender, dangling pod about 2½ inches long. **FLOWERING:** Jun–Jul. **RANGE:** FL west to TX, north to NC; Bahamas, West Indies, Central America. **NOTES:** Queen butterfly larvae feed on the leaves. **SIMILAR TO** Leafless Swallow-wort (*Orthosia scoparia*, synonym *Cynanchum scoparium*), which occurs in maritime forests in Camden County; its leaves fall quickly and the vines usually appear leafless.

Aster Family (Asteraceae or Compositae)

Take a close look at a member of the Aster Family such as a sunflower or a daisy. What appears to be a single flower with several petals and a busy center is in fact a composite of many flowers gathered into a single head. The "petals" are actually individual flowers called ray flowers—each with a single petal. The center disk or cone is a collection of tiny flowers called disk flowers, each with 4 or 5 tiny, pointed lobes and the usual set of stamens and pistils. All these flowers are collected into a head that is surrounded at its base by whorls of small, green bracts called phyllaries or involucral bracts. The bracts come in a variety of shapes and sizes that are often key to identifying species in this family. Some Aster Family species lack the ray flowers and have only disk flowers. Others lack the disk flowers and have only ray flowers. But the basic idea is the same: many flowers collected into a single head. During a single landing an insect can pollinate hundreds of separate flowers. This efficient pollination strategy is one reason why the Aster Family is the largest and most diverse of all the plant families. Worldwide, there are about 23,000 Aster Family species; Georgia has about 400 species, more than 350 of which are native.

Typical Aster Family Flower Heads

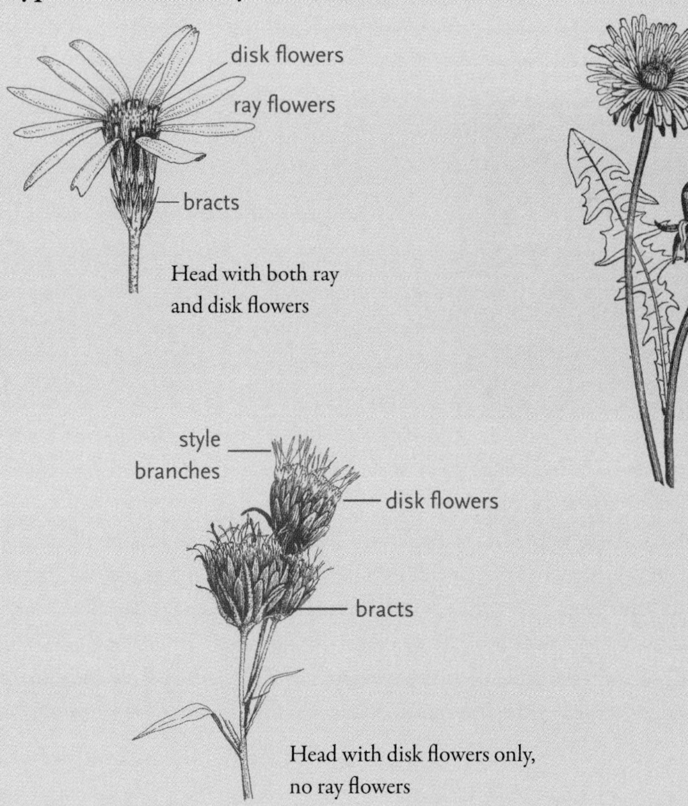

Head with both ray and disk flowers

Head with disk flowers only, no ray flowers

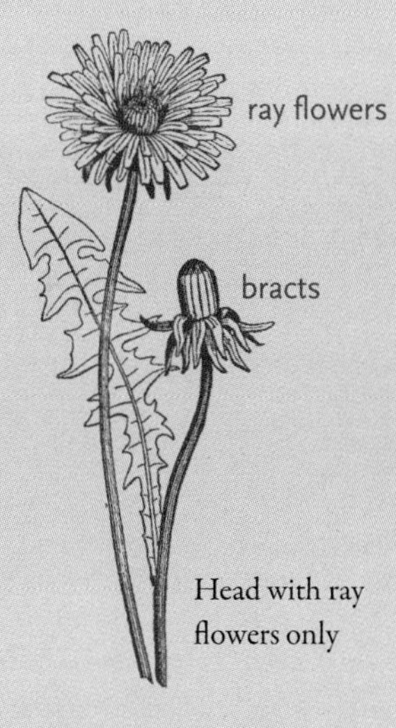

Head with ray flowers only

Yarrow, Milfoil

Achillea millefolium Linnaeus
Aster Family (Asteraceae)

PERENNIAL HERB found along roadsides and in other disturbed areas throughout Georgia. **STEMS:** 1–3 feet tall, usually woolly. **LEAVES:** Up to 6 inches long and 1½ inches wide, alternate, divided into many narrow segments, usually very hairy. **FLOWER CLUSTERS:** Flat-topped or rounded, at the tips of branches, with many white (rarely pink) heads. **FLOWER HEADS:** With 3–8 ray flowers and a central disk of 10–20 tiny, white flowers. **FLOWERING:** Apr–Nov. **RANGE:** Widespread in North America and Eurasia. **NOTES:** Some botanists consider North American plants a separate species, American Yarrow (*Achillea borealis*). **SIMILAR TO** members of the Carrot Family that have white, flat-topped flower clusters and much divided leaves, but these are never woolly-hairy.

White Snakeroot, Milk-Sick

Ageratina altissima King & H. E. Robinson
SYNONYM: *Eupatorium rugosum* Houttuyn
Aster Family (Asteraceae)

PERENNIAL HERB found in moist, deciduous forests and bottomlands in north Georgia. **STEMS:** 1–4 feet tall, hairy. **LEAVES:** 2–7 inches long and 1–4½ inches wide, opposite, oval to lance-shaped, with toothed margins, 3 prominent veins arising from the leaf base, and a stalk up to 2½ inches long. **FLOWER HEADS:** About ⅓ inch high, with many tiny, bright white flowers, each with 2 long style branches, giving the heads a fuzzy look. **FLOWERING:** Jul–Oct. **RANGE:** FL west to TX, north to ND and QC. **NOTES:** White Snakeroot contains a potentially lethal toxin that is passed to humans through the milk of cows that have eaten the plants. **SIMILAR TO** Small-leaved White Snakeroot (*Ageratina aromatica*), which occurs in dry woodlands and disturbed areas throughout Georgia; it has somewhat leathery leaves that are less than 2¾ inches long.

Common Ragweed

Ambrosia artemisiifolia Linnaeus
Aster Family (Asteraceae)

ANNUAL HERB found in fields, pastures, roadsides, and other disturbed areas throughout Georgia. **STEMS:** 1–6 feet tall, branched, green or reddish, hairy. **LEAVES:** 1–4 inches long and 1–2 inches wide, opposite on lower stem, alternate on upper stem; deeply divided into many narrow, blunt-tipped segments; smooth, gland-dotted, fragrant when crushed. **FLOWERS:** Male flowers in small, green, nodding, bowl-shaped heads in spikelike clusters at the tips of branches; female flowers in round, green clusters in the axils of upper leaves with 2 or 3 flowers surrounded by tiny, leaflike bracts. **FLOWERING:** Aug–frost. **RANGE:** Most of North America. **NOTES:** The male flowers release large amounts of allergy-producing pollen. **SIMILAR TO** Giant Ragweed (*Ambrosia trifida*), which occurs in moist disturbed areas in north Georgia; it has large, 3-lobed leaves and can reach 15 feet in height.

Plantain Pussy-Toes

Antennaria plantaginifolia (Linnaeus) Richardson
Aster Family (Asteraceae)

PERENNIAL HERB found in dry, open woodlands and on road banks in north Georgia and a few Coastal Plain counties. **STEMS:** 2½–8 inches tall, covered with tufts of white hairs, in colonies formed by runners. **BASAL LEAVES:** Up to 3 inches long (including the stalk) and 1½ inches wide, spoon-shaped with 3 main veins, the upper surface dark green, the lower surface densely covered with white hairs; stem leaves much smaller, pointed, and alternate. **FLOWER CLUSTERS:** Several heads in a compact cluster at the top of the stem, each head with white-tipped bracts and tiny, white flowers. **FLOWERING:** Apr–May. **RANGE:** FL west to OK, north to SK and NS. **NOTES:** Female and male flowers are on separate plants. **SIMILAR TO** Solitary Pussy-toes (*Antennaria solitaria*), which occurs in moist north Georgia forests; it has only a single flower head at the top of each stem.

Pale Indian-Plantain

Arnoglossum atriplicifolium (Linnaeus) H. E. Robinson
SYNONYM: *Cacalia atriplicifolia* Linnaeus
Aster Family (Asteraceae)

PERENNIAL HERB found in woodland borders and clearings in north Georgia and a few Coastal Plain counties. **STEMS:** 3–10 feet tall, smooth, faintly ribbed, pale green or purplish, with a waxy, white coating. **LEAVES:** Up to 7 inches long and wide, with stalks up to 6 inches long, alternate, broadly oval, triangular, kidney- or heart-shaped, with pointed lobes or coarse teeth, lower surface pale green to white. **FLOWER CLUSTERS:** Large, branched, flat-topped, with many small, cylindrical flower heads. **FLOWER HEADS:** About ¾ inch high, with 5 white or pale green, tubular flowers; 5 pale, erect bracts surround the base of the head. **FLOWERING:** Jun–Nov. **RANGE:** FL west to LA, north to NE, MN, and NY. **SIMILAR TO** Great Indian Plantain (*Arnoglossum reniforme*, synonym *A. muhlenbergii*), which occurs in rich mountain forests; its leaves are green on both surfaces.

Groundsel Tree

Baccharis halimifolia Linnaeus
Aster Family (Asteraceae)

DECIDUOUS SHRUB found near coastal marshes and beaches, recently spread to roadsides and disturbed areas throughout Georgia. **STEMS:** Up to 14 feet tall, much-branched. **LEAVES:** Up to 3 inches long and 2½ inches wide, alternate, oval, coarsely toothed, grayish-green with tiny amber glands, thick and firm. **FLOWER CLUSTERS:** Many small, stalked heads held at the tips of twigs, each head with 20–30 tiny, white or yellow flowers and whorls of tiny, green bracts around the base; female and male flowers are on separate plants. **FRUIT:** Tiny, seedlike, topped with long, cottony bristles; in fruit, female shrubs seem covered with small, white plumes. **FLOWERING:** Aug–Oct, fruiting Sep–Nov. **RANGE:** FL west to TX and OK, north to MA. **SIMILAR TO** False-willow (*Baccharis angustifolia*), which has narrow, needle-like leaves, and Silverling (*B. glomeruliflora*), which has stalkless flower heads scattered along the twigs. Both are strictly coastal.

Yellow Buttons, Coastal Honeycomb Head

Balduina angustifolia (Pursh) B. L. Robinson
Aster Family (Asteraceae)

ANNUAL OR BIENNIAL HERB found in sandhills, dunes, and other dry, sandy sites in Georgia's southeastern Coastal Plain. **STEMS:** Up to 3 feet tall, branched, reddish, gland-dotted. **LEAVES:** Up to 2½ inches long, very narrow and needle-like, alternate, inrolled, lower stem leaves withered by flowering time. **FLOWER HEADS:** 1–2 inches wide, with 5–13 (usually 8) toothed, yellow ray flowers surrounding a dome-shaped center of many tiny, golden yellow disk flowers; several series of small, yellow and green bracts surround the base of each head. **FLOWERING:** Summer–fall. **RANGE:** GA, FL, AL, and MS. **NOTES:** Removing the disk flowers exposes the honeycomb pattern on the receptacle. **SIMILAR TO** Bitter-weed (*Helenium amarum*), page 84.

Yellow Honeycomb Head

Balduina uniflora Nuttall
Aster Family (Asteraceae)

PERENNIAL HERB found in dry to wet pine flatwoods and sandhills in the Coastal Plain. **STEMS:** 1–3 feet tall, usually unbranched, ribbed, hairy, topped with a single flower head. **LEAVES:** Up to 4 inches long and ½ inch wide, smaller and alternate upward, narrowly spoon-shaped. **FLOWER HEAD:** Up to 2 inches wide, with 5–22 (usually 13) bright yellow, toothed ray flowers around a center of tiny, yellow to reddish-orange disk flowers; several series of small, green, pointed bracts surround the base of the head. **FLOWERING:** Jul–Sep. **RANGE:** FL west to LA, north to NC. **NOTES:** Removing the disk flowers exposes the honeycomb pattern of the receptacle. **SIMILAR TO** Purple Honeycomb Head (*Balduina atropurpurea*), a Georgia Special Concern species found in savannas and flatwoods; it has purple disk flowers. Also see Autumn Sneezeweed (*Helenium autumnale*), page 85.

Green-Eyes

Berlandiera pumila (Michaux) Nuttall
Aster Family (Asteraceae)

PERENNIAL HERB found in sandhills and dry, disturbed areas in the Coastal Plain. **STEMS:** Up to 3 feet tall, usually branched, reddish, densely covered with fine white hairs. **LEAVES:** 2–5 inches long and ¾–2¾ inches wide, alternate, more or less triangle-shaped with scalloped margins, covered with velvety gray hairs, midvein and stalk often reddish. **FLOWER HEADS:** Up to 2 inches wide, with 8 yellow ray flowers with notched tips surrounding a disk of tiny, green flowers (aging to maroon); 3 series of thick, green bracts surround the base of the head. **FLOWERING:** Apr–Nov. **RANGE:** FL west to TX, north to SC. With its green disk, triangular leaves, and densely hairy leaves and stems, Green-eyes is **SIMILAR TO** no other wildflower in Georgia.

Spanish Needles

Bidens bipinnata Linnaeus
Aster Family (Asteraceae)

ANNUAL HERB found in roadsides, pastures, and other disturbed areas throughout Georgia. **STEMS:** 1–3 feet tall, slightly 4-angled, smooth. **LEAVES:** 1½–8 inches long including the stalk, opposite, smooth, divided into many lobed and blunt-toothed segments. **FLOWER HEADS:** With 3–5 yellow, oval ray flowers (often missing) around a center of 10–20 dark yellow disk flowers; the base of the head is surrounded by several green, spreading bracts that are often longer than the rays. **FRUIT:** About ½ inch long, very narrow, brown, with 2–4 barbed teeth at the tip. **FLOWERING:** Jul–Oct. **RANGE:** FL west to AZ, north to ON and NB; Mexico, East Asia. **SIMILAR TO** Devil's Beggar-ticks (*Bidens frondosa*) and Smooth Beggar-ticks (*Bidens laevis*), page 61.

Devil's Beggar-Ticks

Bidens frondosa Linnaeus
Aster Family (Asteraceae)

ANNUAL HERB found in low woods, thickets, fields, and other disturbed areas in north and southwest Georgia. **STEMS:** 1–4 feet tall, often purple. **LEAVES:** Opposite, divided into 3 or 5 leaflets, each leaflet up to 4 inches long and 1¼ inches wide, lance-shaped, toothed, hairy. **FLOWER HEADS:** About 1 inch wide, with tiny, yellow disk flowers and no ray flowers (rarely 1–3 yellow rays); the head is surrounded by an outer whorl of 5–10 (usually 8) green, spreading, hairy-edged bracts and an inner whorl of erect, brownish-green bracts. **FRUIT:** About ¼ inch long, seedlike, flat or 4-sided, dark brown, with 2 barbed bristles at the tip. **FLOWERING:** Jun–Oct. **RANGE:** Most of North America. **SIMILAR TO** Beggar-ticks (*Bidens discoidea*), which occurs in wetlands throughout Georgia; it has 3–5 (usually 4) smooth-edged bracts in the outer whorl.

Smooth Beggar-Ticks

Bidens laevis (Linnaeus) B.S.P.
Aster Family (Asteraceae)

ANNUAL HERB found in marshes, pond edges, and wet ditches in the Coastal Plain and Ridge & Valley. **STEMS:** 1–4 feet tall or long, smooth, often forming large mats by falling over and rooting at the nodes. **LEAVES:** 2–4 inches long and ½–1 inch wide, opposite, elliptic or lance-shaped, toothed, fleshy, smooth, without stalks. **FLOWER HEADS:** About 2½ inches across, nodding with age, with 8 golden yellow ray flowers around a center of tiny, dark yellow disk flowers; a whorl of oval, yellowish-brown bracts and a whorl of 5–8 narrower, green bracts surround the base of the head. **FLOWERING:** Aug–Nov. **RANGE:** FL west to CA, north to NE, MO, and ME. **SIMILAR TO** Coastal Plain Beggar-ticks (*Bidens mitis*), which is found in similar habitats; its leaves have stalks and are divided into many narrow, toothed segments.

Pineland Rayless Goldenrod

Bigelowia nudata (Michaux) A. P. de Candolle
Aster Family (Asteraceae)

PERENNIAL HERB found in wet pine savannas and flatwoods, seepage bogs, and cypress pond borders in the Coastal Plain. **STEMS:** Up to 3 feet tall, usually with several stems forming a clump. **BASAL LEAVES:** Up to 5½ inches long and ½ inch wide, lance- or spoon-shaped, entire, gland-dotted; stem leaves smaller, alternate, linear. **FLOWER CLUSTERS:** Flat-topped, held at the tip of each stem, with many slender, elongated, yellow heads, each with 2–6 yellow disk flowers; there are no ray flowers. **FLOWERING:** Aug–Nov. **RANGE:** FL west to TX, north to NC. **SIMILAR TO** Nuttall's Rayless Goldenrod (*Bigelowia nuttallii*), which forms colonies on Altamaha Grit outcrops in southwest Georgia and granite outcrops in the Piedmont; its basal leaves are about 1⁄16 inch wide. Also see Tall Pine Barren Milkwort (*Polygala cymosa*), page 286.

Southern Doll's-Daisy

Boltonia diffusa Elliott
SYNONYM: *Boltonia caroliniana* (Walter) Fernald
Aster Family (Asteraceae)

PERENNIAL HERB found in woodland borders, sunny wetlands, and clearings in the Coastal Plain and one northwest Georgia county. **STEMS:** 1–6½ feet tall, smooth, widely branched. **LEAVES:** Up to 4½ inches long and ¾ inch wide, alternate, linear to narrowly lance-shaped, entire, smooth. **FLOWER HEADS:** Less than 1 inch wide, with 20–40 white to lavender ray flowers around a central disk of tiny, yellow flowers. **FLOWERING:** Jul–Nov. **RANGE:** FL west to TX, north to OK, IL, and NC. **SIMILAR TO** Eastern Doll's-Daisy (*Boltonia asteroides* variety *glastifolia*), which occurs in marshes and ditches near the coast; its flower heads are larger, 1–1½ inches wide. Also see fleabanes, pages 76–77, and "Confusing White-flowered Fall Asters," page 120.

Sea Ox-Eye Daisy

Borrichia frutescens (Linnaeus) A. P. de Candolle
Aster Family (Asteraceae)

PERENNIAL HERB found on slightly elevated areas in salt and brackish marshes around barrier islands. **STEMS:** ½–5 feet tall, succulent, usually densely covered with gray hairs. **LEAVES:** Up to 3 inches long and 1 inch wide, opposite, widest above the middle, succulent, gray-hairy on both surfaces, sometimes toothed for about half the length of the leaf, with 1–3 main veins. **FLOWER HEADS:** Held on erect stalks up to 2½ inches long, each head with 12–30 yellow ray flowers, a central disk of tiny, yellow flowers, and a series of small, green bracts surrounding the base of the head. **FLOWERING:** Spring–fall. **RANGE:** FL west to TX and north to VA. **NOTES:** The disk flowers are interspersed with tiny, green, pointed bracts that become spiny as the seed head matures. **SIMILAR TO** no other plant in the salt marshes.

False Boneset

Brickellia eupatorioides (Linnaeus) Shinners
SYNONYM: *Kuhnia eupatorioides* Linnaeus
Aster Family (Asteraceae)

PERENNIAL HERB found in dry woodlands, Longleaf Pine forests, prairies, and glades in north Georgia and a few Coastal Plain counties. **STEMS:** 1–5 feet tall, smooth or softly hairy. **LEAVES:** Up to 4 inches long and 1½ inches wide, mostly opposite, lance-shaped, softly hairy, gland-dotted on the lower surface, with 3 prominent veins and toothed margins. **FLOWER HEADS:** In a much-branched cluster, each head about ⅓ inch high, cylindrical, with 6–15 creamy white disk flowers and several series of narrow, green, white, or purplish bracts around the base; there are no ray flowers. **FLOWERING:** Aug–Oct. **RANGE:** FL west to TX, north to IN and NJ. **SIMILAR TO** Pink Thoroughwort (*Fleischmannia incarnata*), which occurs in rich, moist forests in north Georgia; its flower heads are pink or white, and its stems are lax and spreading.

Musk Thistle, Nodding Thistle

Carduus nutans Linnaeus
Aster Family (Asteraceae)

BIENNIAL HERB found on roadsides and in pastures and other disturbed areas throughout Georgia. **STEMS:** Up to 6½ feet tall, with spiny wings. **LEAVES:** Up to 10 inches long and 4 inches wide, deeply lobed and spiny, tapering to winged stalks, hairy or smooth, the lower surface green. **FLOWER HEADS:** 1½–3½ inches tall, with many dark pink disk flowers surrounded by narrow, purplish, spine-tipped bracts; the head nods at a 90-degree angle from the top of the stem when mature. **FLOWERING:** May–Nov. **RANGE:** Native to Eurasia, this aggressive pest plant has spread throughout North America and should be eradicated wherever it appears. **SIMILAR TO** Field Thistle (*Cirsium vulgare*), another European weed found in disturbed areas; its lower leaf surfaces are densely covered with white hairs. Also see Bull Thistle (*Cirsium horridulum*) and Tall Thistle (*Cirsium altissimum*), page 68.

Vanilla-Leaf, Deer's Tongue

Carphephorus odoratissimus (J. F. Gmelin) Hebert
SYNONYM: *Trilisa odoratissima* (J. F. Gmelin) Cassini
Aster Family (Asteraceae)

PERENNIAL HERB found in pinelands and savannas in the Coastal Plain. **STEMS:** 1½–6½ feet tall, often reddish-purple, smooth, with a whitish, waxy coating. **BASAL LEAVES:** 2–14 inches long and 1–5 inches wide, spatula-shaped, entire, fleshy, smooth, with a wide, white midvein. **STEM LEAVES:** Similar to basal leaves but smaller, alternate. **FLOWER CLUSTERS:** Flat-topped, with small heads at the tips of each branch, each head with pointed, green-and-red bracts around the base. **FLOWERS:** Tiny, 7–10 per head, dark pink, with 5 tiny lobes and 2 long style branches. **FLOWERING:** Jul–Oct. **RANGE:** FL west to LA, north to NC. **NOTES:** Dried leaves smell like vanilla. **SIMILAR TO** Chaffhead (*Carphephorus corymbosus*), which occurs in wet pine flatwoods in southeast Georgia; its stems are hairy, and the flower heads have round-tipped bracts.

Hairy Chaffhead

Carphephorus paniculatus (J. F. Gmelin) Hebert
SYNONYM: *Trilisa paniculata* (J. F. Gmelin) Cassini
Aster Family (Asteraceae)

PERENNIAL HERB found in pinelands and savannas in the Coastal Plain. **STEMS:** 1–4 feet tall, covered with long, spreading hairs. **BASAL LEAVES:** 2–14 inches long and ½–1¼ inches wide, spatula-shaped, mostly entire, smooth, not gland-dotted. **STEM LEAVES:** Similar to basal leaves but much smaller, alternate, pressed upward. **FLOWER CLUSTERS:** Narrow, erect, cylindrical, with small heads at the tips of short, reddish branches, each head with tiny, purple bracts around the base. **FLOWERS:** Tiny, 12–35 per head, dark pinkish-purple, with 5 lobes and 2 long style branches. **FLOWERING:** Aug–Oct. **RANGE:** FL west to AL, north to NC. **SIMILAR TO** Sticky Chaffhead (*Carphephorus tomentosus*), which occurs in similar habitats in southeast Georgia; its leaves and stems are covered with sticky, shiny resin dots, and its flower heads are in widely branched clusters.

Bachelor's Buttons, Cornflower

Centaurea cyanus Linnaeus
Aster Family (Asteraceae)

ANNUAL HERB found on roadsides and in fields and pastures throughout Georgia. **STEMS:** 1–3 feet tall, covered with silky, white hairs. **LEAVES:** Up to 6 inches long, grasslike, alternate, covered with silky, white hairs, sometimes lobed and toothed. **FLOWER HEADS:** Up to 2 inches across, with 25–35 deep blue, spreading, petal-like disk flowers (sometimes white, pink, or purple), the larger ones deeply toothed at the tip (there are no ray flowers); several series of green, purple-fringed bracts surround the base of the flower head. **FLOWERING:** Apr–Jun. **RANGE:** Native to the Mediterranean region, now spread throughout North America. **SIMILAR TO** Chicory (*Cichorium intybus*), another European species that occurs in disturbed areas in north Georgia; its ray flowers are light lavender-blue, oblong, and tipped with 5 tiny teeth; its stems and leaves have milky sap; it flowers Jun–Oct.

Sun-Bonnets

Chaptalia tomentosa Ventenat
Aster Family (Asteraceae)

PERENNIAL HERB found in wet pine flatwoods and savannas, seepage slopes, and bogs in the Coastal Plain. **LEAVES:** Up to 8 inches long and 2¾ inches wide, in a basal rosette, elliptic to oval, with tiny, widely spaced teeth; upper surface dark green and smooth, lower surface densely covered with white hairs. **FLOWER HEADS:** At the top of a leafless stalk 2–8 inches tall, head often nodding, with several whorls of narrow, pale green bracts forming a cylinder about ½ inch high around the base of the head. **FLOWERS:** Ray flowers about ½ inch long, white on top, dark pink beneath, surrounding a central disk of tiny, cream-colored flowers. **FLOWERING:** Jan–May. **RANGE:** FL west to TX, north to NC. **SIMILAR TO** no other wildflower in Coastal Plain wetlands.

Green-and-Gold

Chrysogonum virginianum Linnaeus
Aster Family (Asteraceae)

PERENNIAL HERB found in upland forests in north Georgia and several upper Coastal Plain counties. **FLOWERING STEMS:** 1–10 inches tall, hairy. **LEAVES:** 1–4 inches long and ½–1¾ inches wide, oval to triangular, with leaf tissue running down along the stalks, hairy, with toothed or scalloped margins. **FLOWER HEADS:** 1–1½ inches wide, with 5 yellow, 3-toothed ray flowers surrounding a central disk of tiny, yellow flowers; the wide, green bracts that surround the base of the head are visible between the ray flowers. **FLOWERING:** Mar–May. **RANGE:** FL west to LA, north to OH and NY. **SIMILAR TO** no other spring-blooming wildflower. The two varieties that occur in Georgia—variety *australe* and variety *brevistolon*—spread by runners and form mats; variety *virginianum*, which occurs farther north, lacks runners and does not form mats.

Cottony Golden-Aster

Chrysopsis gossypina (Michaux) Elliott
SYNONYM: *Heterotheca gossypina* (Michaux) Shinners
Aster Family (Asteraceae)

BIENNIAL OR SHORT-LIVED PERENNIAL HERB found in sandhills, sandy woodlands, and disturbed areas in the Coastal Plain. **STEMS:** Up to 3 feet long, sprawling across the ground, turning up and becoming erect near the tips, densely covered with long, white, tangled hairs. **BASAL ROSETTE LEAVES:** Up to 4 inches long and 1¼ inches wide, oblong to spatula-shaped with blunt or rounded tips, entire, densely covered with long, white, tangled hairs; stem leaves smaller and alternate. **FLOWER HEADS:** Up to 2 inches wide with 16–30 yellow ray flowers surrounding a central disk of many tiny, yellow flowers; the base of the head is surrounded by whorls of pointed, green bracts usually covered with white hairs and/or tiny, stalked glands. **FLOWERING:** Sep–Oct. **RANGE:** FL, GA, SC, NC, and VA. **SIMILAR TO** Maryland Golden-aster (*Chrysopsis mariana*), below.

Maryland Golden-Aster

Chrysopsis mariana (Linnaeus) Elliott
SYNONYM: *Heterotheca mariana* (Linnaeus) Shinners
Aster Family (Asteraceae)

PERENNIAL HERB found in dry, sandy woodlands, prairies, savannas, rock outcrops, and clearings throughout Georgia. **STEMS:** Up to 3 feet tall, usually covered with long, white, tangled hairs (mature plants may lack hairs). **BASAL ROSETTE LEAVES:** Up to 10 inches long and 1½ inches wide, oval to spatula-shaped, covered with long, white hairs; stem leaves smaller, alternate. **FLOWER HEADS:** Up to 2 inches wide, with 10–22 bright yellow ray flowers surrounding a central disk of many tiny, dark yellow flowers; several whorls of pointed, green bracts covered with tiny, stalked glands surround the base of the head (use 10× lens). **FLOWERING:** Jun–Oct. **RANGE:** FL west to TX, north to OH and CT. **SIMILAR TO** Cottony Golden-aster (*Chrysopsis gossypina*), above, but with a less hairy, erect stem.

Tall Thistle

Cirsium altissimum Michaux
SYNONYM: *Carduus altissimus* Linnaeus
Aster Family (Asteraceae)

BIENNIAL OR SHORT-LIVED PERENNIAL HERB found in bottomlands, pastures, and other disturbed areas in north Georgia. **STEMS:** Up to 13 feet tall, hairy or smooth, vertically ribbed, often branched. **BASAL LEAVES:** Up to 16 inches long and 5 inches wide, oval or elliptic, toothed or shallowly lobed, upper surface green, lower surface densely covered with white hairs, margins with weak spines; lower leaves are usually withered by flowering time. **STEM LEAVES:** Similar to basal leaves but smaller, alternate, clasping the stem. **FLOWER HEADS:** Up to 3 inches tall and 2 inches wide, crowded with pink to purple disk flowers, the base of the head surrounded by green, spine-tipped bracts. **FLOWERING:** Jun–Oct. **RANGE:** FL to TX, north to ND and CT. **NOTES:** Tall Thistle is the least spiny of Georgia's thistles. **SIMILAR TO** Bull Thistle (*Cirsium horridulum*), below, and Nodding Thistle (*Carduus nutans*), page 64.

Bull Thistle, Spiny Thistle

Cirsium horridulum Michaux
SYNONYM: *Carduus spinosissimus* Walter
Aster Family (Asteraceae)

BIENNIAL HERB found in disturbed areas throughout Georgia. **STEMS:** Up to 8 feet tall, smooth to very hairy. **BASAL LEAVES:** Up to 16 inches long, 4 inches wide, deeply lobed and toothed, green and hairy on both surfaces, with spines up to 1 inch long. **STEM LEAVES:** Smaller than basal leaves, alternate, clasping the stem, very spiny. **FLOWER HEADS:** Up to 3½ inches tall, with many yellow, pink, white, or purple disk flowers; spine-tipped bracts and narrow, spiny, erect leaves surround the base of the head. **FLOWERING:** Mar–Jun. **RANGE:** FL to TX, north to OK and ME. **NOTES:** Bull Thistle is an important larval host and nectar source for butterflies. **SIMILAR TO** Field Thistle (*Cirsium vulgare*), a European weed found in disturbed areas; its leaves are whitened on the lower surface with a dense layer of hair, and the stems have spiny wings.

Mistflower, Ageratum

Conoclinium coelestinum (Linnaeus) A. P. de Candolle
SYNONYM: *Eupatorium coelestinum* Linnaeus
Aster Family (Asteraceae)

PERENNIAL HERB found in low woods and disturbed wet areas throughout Georgia. **STEMS:** Up to 3 feet tall, very hairy; forming colonies by the spread of underground stems. **LEAVES:** Up to 4 inches long and 2 inches wide, opposite; oval, lance-shaped, or heart-shaped, with 3 main veins arising from the base of the leaf, hairy or smooth, the lower surface dotted with tiny glands (use a 10× lens), the margins toothed or scalloped. **FLOWER CLUSTERS:** With numerous small heads at the tips of short branches; heads have no ray flowers. **DISK FLOWERS:** Tiny, blue or purple (rarely white), with long style branches giving each head a fuzzy look. **FLOWERING:** Jun–Nov. **RANGE:** FL west to TX, north to NE and ON. **SIMILAR TO** no other native wildflower in Georgia.

Horseweed

Conyza canadensis (Linnaeus) Cronquist
SYNONYM: *Erigeron canadensis* Linnaeus
Aster Family (Asteraceae)

ANNUAL HERB found in disturbed areas throughout Georgia. **STEMS:** Up to 6 feet tall, occasionally twice that, hairy or smooth, branched above the middle. **LEAVES:** 1–4 inches long, up to ½ inch wide, alternate, narrowly lance-shaped, entire or toothed, with long, stiff hairs around the margins. **FLOWER HEADS:** About ¼ inch long, cylinder- or vase-shaped, with 20–45 tiny, white ray flowers surrounding a disk of yellowish flowers; several series of narrow, smooth, green bracts enclose the base of the head; mature heads explode into round puffs of white bristles. **FLOWERING:** Jul–Nov. **RANGE:** Throughout North America. **SIMILAR TO** South American Horseweed (*Conyza bonariensis*), an exotic species that occurs infrequently in disturbed areas, mostly in the outer Coastal Plain; it has 60–150 ray flowers per head and very hairy bracts surrounding the base of the head.

Large-Flowered Coreopsis

Coreopsis grandiflora Hogg ex Sweet
Aster Family (Asteraceae)

PERENNIAL HERB found on granite outcrops and in dry, sandy soils in the Piedmont. **STEMS:** Up to 3 feet tall, leafy, smooth. **UPPER LEAVES:** Up to 3 inches long, opposite, divided into several narrow segments and a broader central lobe. **LOWER LEAVES:** Entire, undivided, lance-shaped. **FLOWER HEADS:** Held at the top of a leafless stalk 2–8 inches long, each head 1½–2 inches wide, with 8 yellow, toothed ray flowers surrounding a disk of many tiny, yellow, 5-lobed flowers, the base of the head surrounded by 2 series of bracts, the inner series oval and straw-colored, the outer series green and narrow. **FLOWERING:** May–Aug. **RANGE:** FL west to TX, north to KS and NC. **SIMILAR TO** Larkspur Coreopsis (*Coreopsis delphiniifolia*), which occurs in dry woodlands; its leaves are mostly entire and undivided, and the stalks supporting the flower heads are less than 2 inches long.

Savanna Coreopsis

Coreopsis linifolia Nuttall
Aster Family (Asteraceae)

PERENNIAL HERB found in bogs, seepage slopes, and wet flatwoods in the Coastal Plain. **STEMS:** Up to 3½ feet tall, smooth, vertically ribbed. **BASAL LEAVES:** Up to 7 inches long and ¾ inch wide, lance-shaped, tapering to a long stalk, thick and stiff, entire, with minute dark dots visible on the lower surface. **STEM LEAVES:** Similar to basal leaves but smaller, mostly opposite. **FLOWER HEADS:** Usually solitary at the top of the stem, up to 2 inches wide, with 5–20 (usually 13) yellow, toothed ray flowers surrounding a central disk of tiny, yellow to reddish-orange, 4-lobed flowers; several series of small, green, pointed bracts surround the base of the head. **FLOWERING:** May–Aug. **RANGE:** FL west to TX, north to VA. **SIMILAR TO** Swamp Coreopsis (*Coreopsis gladiata*), which occurs mostly in Coastal Plain swamps; it has alternate stem leaves and lacks dark dots on the leaves.

Woodland Coreopsis

Coreopsis major Walter
Aster Family (Asteraceae)

PERENNIAL HERB found in dry woodlands and clearings in north Georgia and the upper Coastal Plain. **STEMS:** Up to 3 feet tall, hairy. **LEAVES:** Opposite, deeply divided into 3 oval or lance-shaped leaflets up to 4 inches long and 1¼ inches wide, appearing as 6 whorled leaves without stalks. **FLOWER HEADS:** 1½–3 inches wide, with 8 yellow, toothed ray flowers surrounding a central disk of many tiny, yellow flowers (aging to maroon); 2 series of bracts surround the base of the head: the outer series is green, narrow, and pointed, the inner series is shorter, wider, and downcurved at the tips. **FLOWERING:** May–Jul. **RANGE:** FL west to LA, north to IN and MA. **SIMILAR TO** Tall Coreopsis (*Coreopsis tripteris*), found in woodlands throughout Georgia; it is 3–9 feet tall, its ray flowers are not toothed, and the leaves are stalked.

Georgia Tickseed

Coreopsis nudata Nuttall
Aster Family (Asteraceae)

PERENNIAL HERB found in cypress ponds, wet pine flatwoods, and wet savannas in Georgia's Coastal Plain. **STEMS:** Up to 4 feet tall, slender, smooth. **LEAVES:** 4–16 inches long, needle-like, alternate, smooth. **FLOWER HEADS:** 1–10 per plant, with 8 pink ray flowers surrounding the central yellow disk; 2 series of bracts surround the base of the head: the outer series is erect, oval, and green, the inner series is longer, papery, and golden green. **FLOWERING:** Apr–May. **RANGE:** GA and FL, west to LA. **SIMILAR TO** no other common tickseed, all of which have yellow ray flowers. Pink Tickseed (*Coreopsis rosea*), a Georgia Special Concern species, occurs in ponds in only one southeast Georgia county; its leaves are opposite. Also compare with Bartram's Rose-gentian (*Sabatia bartramii*), page 215, which occurs in similar habitats.

Plains Tickseed

Coreopsis tinctoria Nuttall
Aster Family (Asteraceae)

ANNUAL HERB found on roadsides and in other disturbed areas in north Georgia and a few Coastal Plain counties. **STEMS:** 1–3 feet tall, smooth. **LEAVES:** Opposite, most divided into many linear segments up to 2 inches long and ¼ inch wide, smooth. **FLOWER HEADS:** About 1½ inches wide, with 8 yellow ray flowers, reddish-brown at the base and 3-toothed at the tip, surrounding a central disk of tiny, maroon or yellow, 4-lobed flowers; 2 series of bracts surround the base of the head: the outer bracts are small, green, and oval, the inner bracts are larger, triangular, and reddish-brown. **FLOWERING:** Jun–Sep. **RANGE:** Native to the Great Plains, now spread throughout North America. **SIMILAR TO** Texas Tickseed (*Coreopsis basalis*), a western species now spread throughout North America in disturbed areas; its leaves are divided into lance-shaped segments, and its disk flowers have 5 lobes.

Slender Scratch-Daisy

Croptilon divaricatum (Nuttall) Rafinesque
SYNONYM: *Haplopappus divaricatus* (Nuttall) A. Gray
Aster Family (Asteraceae)

ANNUAL HERB found in openings in dry woodlands, pinelands, and sandy, disturbed areas throughout Georgia. **STEMS:** 1–4½ feet tall, slender, widely branched in the upper half, covered with a mix of glandular and nonglandular hairs. **LEAVES:** Up to 4 inches long and ½ inch wide, alternate, linear, rough-hairy, with a few tiny teeth on the margins. **FLOWER HEADS:** Held at the tips of widely spreading branches, each head about ¾ inch wide, with 7–11 yellow ray flowers around a central disk of tiny, yellow flowers. **FLOWERING:** Jul–Nov. **RANGE:** FL west to TX, north to OK and VA. The heads and widely branched stems are **SIMILAR TO** Sticky Golden-aster (*Pityopsis aspera*), page 106, which has long, silky-hairy basal leaves.

Appalachian Flat-Topped White Aster

Doellingeria infirma (Michaux) E. Greene
SYNONYM: *Aster infirmus* Michaux
Aster Family (Asteraceae)

PERENNIAL HERB found in dry to moist forests in north Georgia. **STEMS:** 1½–4 feet tall, arising from a woody crown, slightly zigzag, smooth. **LEAVES:** Up to 5 inches long and 2 inches wide, alternate, elliptic, entire, mostly smooth, with conspicuous veins and short or no stalks. **FLOWER HEADS:** Few to 45 per plant, in a widely branched cluster, each head about 1 inch wide, with 5–9 white ray flowers around a disk of many tiny, yellowish-white flowers. **FLOWERING:** Late Jun–Sep. **RANGE:** FL west to AL, north to OH and MA. **SIMILAR TO** Tall Flat-topped White Aster (*Doellingeria umbellata*), which occurs in low, moist to wet woods in the mountains; it forms patches, the stems arising from rhizomes; it is 3–6 feet tall and has 20–100 flower heads per plant.

Eastern Purple Coneflower

Echinacea purpurea (Linnaeus) Moench
Aster Family (Asteraceae)

PERENNIAL HERB found in open woodlands, limestone cedar glades, prairies, or disturbed areas in north Georgia. **STEMS:** 1½–4 feet tall, green with purplish-brown streaks, hairy. **LEAVES:** 2–12 inches long and 2–4 inches wide, mostly basal with a few smaller, alternate leaves on the stem, oval or lance-shaped, toothed, hairy. **FLOWER HEAD:** Up to 4 inches across, with 8–21 dark pink or purple ray flowers and a domed or cone-shaped disk with many tiny, maroon flowers and orange-brown spines. **FLOWERING:** May–Oct. **RANGE:** FL west to TX, north to WI and OH. **SIMILAR TO** three other coneflower species that are rare in Georgia: Smooth Purple Coneflower (*E. laevigata*) has smooth lower leaf surfaces; Pale Purple Coneflower (*E. pallida*) and Prairie Purple Coneflower (*E. simulata*) have pale pink ray flowers.

False Daisy, Yerba de Tajo

Eclipta prostrata (Linnaeus) Linnaeus
SYNONYM: *Eclipta alba* (Linnaeus) Hasskart
Aster Family (Asteraceae)

ANNUAL HERB found in disturbed wetlands throughout Georgia (except the mountains). **STEMS:** Up to 2 feet long, sprawling and rooting at the nodes, often purple, covered with stiff, upwardly pointing hairs. **LEAVES:** Up to 4 inches long and 1¼ inches wide, opposite, lance-shaped, toothed, covered with stiff hairs. **FLOWER HEADS:** Held in the leaf axils, about ⅓ inch wide, with many narrow, white ray flowers surrounding a center of tiny, white disk flowers bearing dark purple anthers. **FLOWERING:** Jun–Nov. **RANGE:** FL west to CA, north to SD, ON, and MA; globally widespread. **SIMILAR TO** Peruvian Daisy (*Galinsoga quadriradiata*), a weedy exotic species found in upland disturbed areas; its heads have 4–6 white ray flowers and yellow disk flowers.

Southern Elephant's Foot

Elephantopus elatus Bertoloni
Aster Family (Asteraceae)

PERENNIAL HERB found in pine flatwoods and sandhills in the Coastal Plain. **STEMS:** 1–4 feet tall, hairy, with few or no leaves. **LEAVES:** In a basal rosette pressed flat against the ground, each leaf up to 12 inches long and 3 inches wide, widest above the middle with a blunt tip and tapering base, toothed, hairy on both surfaces. **FLOWER CLUSTERS:** Held at the top of the leafless stem, with 3 green, very hairy, heart-shaped bracts about ½ inch long. **FLOWERS:** About ⅓ inch long, pink, deeply divided into 5 narrow segments, the base of each flower enclosed by narrow, very hairy bracts. **FLOWERING:** Aug–Sep. **RANGE:** FL west to LA, north to SC. **SIMILAR TO** Coastal Plain Elephant's-foot (*Elephantopus nudatus*), which occurs in dry woodlands; the tiny bracts enclosing the base of each flower have few hairs and clearly visible resin dots.

Elephant's Foot

Elephantopus tomentosus Linnaeus
Aster Family (Asteraceae)

PERENNIAL HERB found in woodlands and forests throughout Georgia, mostly in the Piedmont. **STEMS:** Up to 2 feet tall, hairy, with few or no leaves. **LEAVES:** In a basal rosette pressed flat against the ground, 3½–14 inches long and 2–4 inches wide, widest above the middle with a blunt tip and tapering base, hairy on both surfaces. **FLOWER CLUSTERS:** Held at the top of the stem, with 2 or 3 green, triangular, ½-inch bracts and several pink or lavender flowers, each about ⅓ inch long and deeply divided into 5 narrow lobes. **FLOWERING:** Aug–Nov. **RANGE:** FL west to TX, north to OK and MD; Mexico. **SIMILAR TO** Leafy Elephant's-foot (*Elephantopus carolinianus*), mostly found in north Georgia, which has similar flower heads but leafy stems and no basal leaf rosette.

Fireweed

Erechtites hieraciifolius (Linnaeus) Rafinesque ex de Candolle
Aster Family (Asteraceae)

ANNUAL HERB found in disturbed areas throughout Georgia. **STEMS:** Up to 8 feet tall, hairy or smooth. **LEAVES:** Up to 8 inches long and 3 inches wide, alternate, sharply toothed and sometimes irregularly lobed, hairy or smooth, lower leaves with leaf stalks, upper leaves clasping the stem; lower leaves often withered by flowering time. **FLOWER HEADS:** Cylinder-shaped, with a swollen base; many narrow, erect, green bracts about ¾ inch high surround a flat disk of tiny, white, yellow, or pink flowers; mature heads explode into round puffs of white bristles. **FLOWERING:** Jul–Nov. **RANGE:** Most of North America except the Rocky Mountain states. **NOTES:** Fireweed is not related to the disturbance-following Fireweed (*Epilobium angustifolium*) found in western and northern states. **SIMILAR TO** Horseweed (*Conyza canadensis*), page 69.

Philadelphia Fleabane

Erigeron philadelphicus Linnaeus
Aster Family (Asteraceae)

ANNUAL OR SHORT-LIVED PERENNIAL HERB found in moist to wet disturbed areas in north Georgia and a few Coastal Plain counties. **STEMS:** Up to 2½ feet tall, ribbed, leafy, shaggy-hairy. **BASAL LEAVES:** Up to 4½ inches long and 1 inch wide, spoon- to lance-shaped, toothed, shaggy-hairy. **STEM LEAVES:** Smaller than basal leaves, alternate, heart-shaped, entire, hairy, clasping the stem. **FLOWER HEADS:** Up to 1 inch across, with 150 or more very narrow, white or pale pink ray flowers surrounding a central disk of tiny, yellow flowers; 2 or 3 series of narrow, pointed, green bracts surround the base of the head. **FLOWERING:** Apr–Aug. **RANGE:** North America. **SIMILAR TO** Whitetop Fleabane (*Erigeron vernus*), which occurs in wet savannas, flatwoods, and ditches in the Coastal Plain; its stems are usually leafless and smooth, the leaves are smooth and fleshy, and there are only 20–30 ray flowers per head.

Robin's Plantain

Erigeron pulchellus Michaux
Aster Family (Asteraceae)

BIENNIAL OR SHORT-LIVED PERENNIAL HERB found in open woodlands and on stream banks and roadsides in north Georgia. **STEMS:** 6–24 inches tall, densely hairy. **BASAL LEAVES:** Up to 7 inches long and 2 inches wide, round to spatula-shaped, densely hairy. **STEM LEAVES:** Gradually smaller up the stem, alternate, very hairy, clasping the stem. **FLOWER HEADS:** Usually 1 per stem, about 1 inch wide, composed of 50–100 white or pink ray flowers around a central disk of tiny, yellow flowers; several whorls of narrow, hairy, pointed, green bracts surround the base of the head. **FLOWERING:** Mar–Jun. **RANGE:** FL west to TX, north to MN and QC. **SIMILAR TO** Philadelphia Fleabane (*Erigeron philadelphicus*), above, and Rough Daisy Fleabane (*E. strigosus*), page 77.

Rough Daisy Fleabane

Erigeron strigosus Muhlenberg ex Willdenow
Aster Family (Asteraceae)

ANNUAL HERB found in woodlands, glades, and disturbed areas throughout Georgia. **STEMS:** 1–5 feet tall, finely ribbed, slender, with short hairs pressed upward (except near the base of the stem, where hairs are long and spreading). **LEAVES:** 1–6 inches long, usually less than ½ inch wide, alternate, linear to spatula-shaped, mostly entire, rough-hairy; basal leaves sometimes toothed near the tip, usually withered by flowering time. **FLOWER HEADS:** About ½ inch across, with 40–100 white (rarely pink) ray flowers surrounding a central disk of tiny, yellow flowers; several series of narrow, hairy, green bracts surround the base of the head. **FLOWERING:** Apr–Oct. **RANGE:** North America except the southwestern states. **SIMILAR TO** Daisy Fleabane (*Erigeron annuus*), which occurs throughout Georgia in disturbed areas; it has spreading stem hairs and many toothed leaves wider than ½ inch.

White Thoroughwort

Eupatorium album Linnaeus
Aster Family (Asteraceae)

PERENNIAL HERB found in dry woodlands, sandhills, and disturbed areas throughout Georgia. **STEMS:** Up to 3 feet tall, very hairy. **LEAVES:** 1½–5 inches long and ½–1½ inches wide, opposite, oval to lance-shaped, gland-dotted, with toothed margins and no leaf stalks. **FLOWER CLUSTERS:** Large and flat-topped. **FLOWER HEADS:** About ⅓ inch high, with 4 or 5 tiny, white, 5-lobed flowers surrounded by narrow, green bracts with long, white, pointed tips; 2 long style branches extending from each flower give the heads a fuzzy look. **FLOWERING:** Jul–Sep. **RANGE:** FL west to LA, north to OH and CT. **NOTES:** All parts of the plant are covered with short, white, curled hairs. **SIMILAR TO** Justiceweed (*Eupatorium leucolepis*), which occurs in woodlands and savannas in the Coastal Plain; it has smaller heads (about ¼ inch high) and narrower leaves (less than ½ inch wide).

Dog-Fennel

Eupatorium capillifolium (Lamarck) Small
Aster Family (Asteraceae)

PERENNIAL HERB found in clearcuts, pastures, and other disturbed areas throughout Georgia. **STEMS:** Up to 6½ feet tall, stout, hairy, reddish, much-branched in the upper half, lower stem often clothed in dead, brown leaves. **LEAVES:** 1–4 inches long, opposite or alternate, divided into threadlike segments, bright green, smooth, dotted with glands that emit a strong, fennel-like odor when crushed. **FLOWER CLUSTERS:** Elongated, drooping at the tips of branches, with many tiny heads of white flowers. **FLOWERING:** Aug–Nov. **RANGE:** FL west to TX, north to MO and CT. **SIMILAR TO** Yankeeweed (*Eupatorium compositifolium*), which occurs throughout Georgia in sandhills and disturbed areas; its leaves are hairy, grayish-green, and divided into broader segments up to ¼ inch wide.

Hyssop-Leaf Thoroughwort

Eupatorium hyssopifolium Linnaeus
Aster Family (Asteraceae)

PERENNIAL HERB found in dry upland woods and disturbed areas throughout Georgia. **STEMS:** Up to 3 feet tall, hairy, branched near the top. **LEAVES:** Up to 3¼ inches long and less than ¼ inch wide, linear, in whorls of 3 or 4, with clusters of shorter leaves in the leaf axils. **FLOWER HEADS:** Held at the tips of branches, with 5 tiny, white or pinkish flowers surrounded by 8–10 hairy, greenish-white bracts; 2 long style branches extending from each flower give the heads a fuzzy look. **FLOWERING:** Late Jul–Oct. **RANGE:** FL west to TX, north to WI and MA. **SIMILAR TO** Narrow-leaf Bushy Thoroughwort (*Eupatorium linearifolium*), which occurs in dry areas in the Coastal Plain and Piedmont; it branches from the base, giving the plants a bushy look, and its leaves are usually more than ¼ inch wide.

Boneset

Eupatorium perfoliatum Linnaeus
Aster Family (Asteraceae)

PERENNIAL HERB found in wetlands and bottomlands in north and southwest Georgia. **STEMS:** Up to 5 feet tall, covered with long, spreading hairs. **LEAVES:** Up to 8 inches long and 1¾ inches wide, usually opposite, the bases united and surrounding the stem, the tips tapering to a point, margins toothed, upper leaf surfaces nearly smooth, and lower leaf surfaces hairy and gland-dotted. **FLOWER HEADS:** In flat-topped clusters at the tips of branches, each head with 7–14 tiny, white, 5-lobed flowers; whorls of green, hairy, gland-dotted bracts surround the base of each head; 2 long style branches extending from each flower give the heads a fuzzy look. **FLOWERING:** Aug–Oct. **RANGE:** FL west to TX, north to MB and NS. Flower heads are **SIMILAR TO** those of other *Eupatorium* species, but none of these has the fused leaf bases.

Roundleaf Thoroughwort

Eupatorium rotundifolium Linnaeus
Aster Family (Asteraceae)

PERENNIAL HERB found in savannas and flatwoods, bottomlands, bogs, and disturbed wet areas throughout Georgia. **STEMS:** 1–5 feet tall, covered with soft hairs and tiny, golden gland dots. **LEAVES:** Up to 4¾ inches long and 2½ inches wide, mostly opposite, triangular or heart-shaped, toothed, with 3 main veins arising from the leaf base and no leaf stalk. **FLOWER CLUSTERS:** Large, open, with opposite branches covered with white hairs and golden glands. **FLOWER HEADS:** With 5 tiny, white flowers and whorls of tiny, green, hairy, pointed bracts surrounding the base of the head. **FLOWERING:** Jul–Oct. **RANGE:** FL west to TX, north to OK and MA. **SIMILAR TO** Rough Boneset (*Eupatorium pilosum*), which occurs infrequently throughout Georgia in wet areas; its leaves are nearly twice as long as wide and often have a purple border; its upper leaves are mostly alternate.

Late-Flowering Thoroughwort

Eupatorium serotinum Michaux
Aster Family (Asteraceae)

PERENNIAL HERB found in open woodlands, floodplains, and disturbed areas throughout Georgia. **STEMS:** Up to 6 feet tall, hairy, often dark red. **LEAVES:** Up to 8 inches long and 4 inches wide, opposite, oval to lance-shaped, gland-dotted, smooth or short-hairy, coarsely toothed, with rounded bases and 3 prominent veins; the stalk is ½–1 inch long. **FLOWER HEADS:** About ¼ inch high, with 9–15 tiny, white, 5-lobed flowers; several whorls of hairy, green-and-white bracts surround the base of the head; 2 long style branches extend from each flower, giving the head a fuzzy look. **FLOWERING:** Aug–Oct. **RANGE:** FL west to TX, north to MN and MA. **SIMILAR TO** Tall Thoroughwort (*Eupatorium altissimum*), which occurs in northwest Georgia in dry woodlands and old fields, often over mafic or limestone bedrock; it has narrow leaves with no leaf stalks.

White Wood Aster, Heart-Leaved Aster

Eurybia divaricata (Linnaeus) G. L. Nesom
SYNONYM: *Aster divaricatus* Linnaeus
Aster Family (Asteraceae)

PERENNIAL HERB found in lower-elevation forests in the mountains and upper Piedmont of Georgia. **STEMS:** 1–3 feet tall, hairy, slightly zigzag, usually in large colonies. **LOWER LEAVES:** Up to 8 inches long and 4 inches wide, with stalks up to 2¾ inches long, heart-shaped, toothed, hairy, withered by flowering time; upper leaves smaller, alternate, lance-shaped. **FLOWER HEADS:** About 1 inch wide, with 5–12 white ray flowers surrounding a central disk of tiny, yellow flowers that turn red with age; several series of green-and-white bracts surround the base of the head. **FLOWERING:** Aug–Oct. **RANGE:** GA and AL, north to MI and QC. **SIMILAR TO** Blue Ridge Heart-leaved Aster (*Eurybia chlorolepis*), which occurs above 4,000 feet in the Blue Ridge and has 8–20 white ray flowers.

Big-Leaf Aster

Eurybia macrophylla (Linnaeus) Cassini
SYNONYM: *Aster macrophyllus* Linnaeus
Aster Family (Asteraceae)

PERENNIAL HERB found in moist to dry, high-elevation forests in the Blue Ridge. **STEMS:** Up to 3½ feet tall, smooth or hairy, covered with tiny stalked glands, usually in large colonies. **BASAL LEAVES:** Up to 12 inches long and 8 inches wide, oval to heart-shaped, toothed, with winged leaf stalks up to 3 inches long, mostly withered by flowering time; stem leaves smaller than basal leaves, alternate, oval, without stalks. **FLOWER HEADS:** Up to 1¼ inches wide, with 9–20 lavender or white ray flowers surrounding a central disk of tiny, white or pale yellow flowers (aging maroon); several series of green-and-white bracts surround the base of the head. **FLOWERING:** Aug–Oct. **RANGE:** GA north to MB and QC. **SIMILAR TO** White Wood Aster (*Eurybia divaricata*), page 80, and Heart-leaf Aster (*Symphyotrichum cordifolium*), page 120.

Creeping Aster

Eurybia surculosa (Michaux) G. L. Nesom
SYNONYM: *Aster surculosus* Michaux
Aster Family (Asteraceae)

PERENNIAL HERB found in dry, rocky woodlands, rock outcrops, and sandy soils in the mountains and upper Piedmont of Georgia. **STEMS:** Up to 3 feet tall, hairy, sometimes sticky-hairy, often dark red, spreading ("creeping") by rhizomes. **BASAL AND LOWER LEAVES:** Up to 6 inches long and 1½ inches wide, oval, elliptic, or lance-shaped, entire or few-toothed, firm, usually rough-hairy; upper leaves smaller, alternate, without stalks. **FLOWER HEADS:** About 1½ inches wide, with 13–35 pale blue-violet ray flowers surrounding a central disk of tiny, yellow flowers (aging to red); several series of green, spreading, triangular bracts surround the base of the head. **FLOWERING:** Aug–Oct. **RANGE:** GA north to KY and VA. **SIMILAR TO** Slender Aster (*Eurybia compacta*), which occurs in pine savannas in the Coastal Plain; it has smaller heads with only 5–14 ray flowers.

Carolina Goldentop

Euthamia caroliniana (Linnaeus) Greene ex Porter & Britton

SYNONYMS: *Euthamia tenuifolia* (Pursh) Nuttall, *Euthamia minor* (Michaux) Greene, *Solidago tenuifolia* Pursh

Aster Family (Asteraceae)

PERENNIAL HERB found in sandhills and other pinelands, river dunes, and open disturbed areas in the Coastal Plain. **STEMS:** 1–3 feet tall, smooth. **LEAVES:** Up to 2¾ inches long and ⅛ inch wide, alternate, linear, gland-dotted, with 1 main vein, often clustered in the axils of single leaves. **FLOWER HEADS:** In large, branched clusters at the top of the stem, each head about ⅓ inch wide, with 7–25 yellow ray flowers surrounding several tiny, yellow, 5-lobed disk flowers. **FLOWERING:** Aug–Dec. **RANGE:** FL west to LA, north to MI and NS. **SIMILAR TO** Marsh Goldentop (*Euthamia hirtipes*), which occurs in marshes and wet hammocks in the Coastal Plain; its leaves are wider and have 3 main veins, and none are held in axillary clusters.

Joe-Pye Weed

Eutrochium fistulosum

SYNONYMS: *Eupatorium fistulosum* Barratt, *Eupatoriadelphus fistulosus* (Barratt) King & H. E. Robinson

Aster Family (Asteraceae)

PERENNIAL HERB found in bottomlands, bogs, and wet roadside ditches, mostly in north Georgia. **STEMS:** Up to 11 feet tall, red or purple with a white, waxy coating, hollow. **LEAVES:** Up to 12 inches long and 6 inches wide, with a stalk up to 2 inches long, lance-shaped, toothed, in whorls of 4–7 at each node. **FLOWER CLUSTERS:** At the top of the stem, large and showy, with whorled branches and a rounded or pointed top, with many small, pinkish-purple flower heads. **FLOWERING:** Jul–Oct. **RANGE:** FL west to TX, north to MI and ME. **SIMILAR TO** Spotted Joe-Pye Weed (*Eutrochium maculatum*), which occurs in cove forests in Georgia's Blue Ridge; its stems are purple or green with purple spots, not hollow, and lack the waxy coating; its flower clusters are flat-topped or slightly rounded.

Sweet Joe-Pye Weed

Eutrochium purpureum (Linnaeus) E. E. Lamont
SYNONYMS: *Eupatorium purpureum* Linnaeus, *Eupatoriadelphus purpureus* (Linnaeus) R. M. King & H. Robinson
Aster Family (Asteraceae)

PERENNIAL HERB found in moist, upland hardwood forests in north Georgia. **STEMS:** 2–6½ feet tall, purple at the leaf nodes, solid (hollow only at the base), smooth. **LEAVES:** Up to 12 inches long and 7 inches wide, lance-shaped, tapering to a winged stalk about ¾ inch long, sharply toothed, mostly smooth, in whorls of 3–5 at each node. **FLOWER CLUSTERS:** Large and showy, with whorled branches and a rounded top, with many small, white or pinkish-purple, sweet-scented flower heads, each with 4–7 tiny flowers. **FLOWERING:** Jul–Oct. **RANGE:** FL west to LA, north to MN and ON. **SIMILAR TO** Joe-Pye Weed (*Eutrochium fistulosum*), page 82.

Sandhill Blanket-Flower

Gaillardia aestivalis (Walter) H. Rock
Aster Family (Asteraceae)

ANNUAL OR SHORT-LIVED PERENNIAL HERB found in dry, sandy uplands and disturbed areas in the Coastal Plain. **STEMS:** Up to 2 feet tall, usually branched, hairy, mostly leafless. **LEAVES:** Up to 2½ inches long and ½ inch wide, alternate, linear to narrowly lance-shaped, toothed or entire, hairy. **FLOWER HEADS:** Up to 2½ inches wide, with 6–15 widely spaced, yellow (rarely white, red, or purple), 3-toothed ray flowers surrounding the central disk of tiny, maroon flowers; 2 series of green, pointed, spreading or downcurved bracts surround the base of the head. **FLOWERING:** Jun–Oct. **RANGE:** FL west to TX, north to KS and NC. **SIMILAR TO** Purple Honeycomb Head (*Balduina atropurpurea*), a Georgia Special Concern species found in wet savannas and flatwoods; it has narrow, strap-shaped ray flowers and somewhat fleshy, spoon-shaped leaves.

Blanket-Flower, Fire-Wheel

Gaillardia pulchella Fougeroux
Aster Family (Asteraceae)

ANNUAL OR SHORT-LIVED PERENNIAL HERB found behind sand dunes on barrier islands and in sandy, disturbed areas elsewhere in the Coastal Plain. **STEMS:** Up to 2 feet tall, usually branched, hairy. **LEAVES:** Up to 3 inches long and 1 inch wide, alternate, hairy, linear to spatula-shaped; toothed, lobed, or entire. **FLOWER HEADS:** Up to 3 inches wide, with 8–14 ray flowers surrounding a central disk of tiny, maroon flowers; ray flowers are red with yellow, 3-toothed tips. **FLOWERING:** Apr–Dec. **RANGE:** FL west to TX, north to NC; widespread as a cultivated plant. **SIMILAR TO** no other wildflower in Georgia.

Bitter-Weed

Helenium amarum (Rafinesque) H. Rock
Aster Family (Asteraceae)

ANNUAL HERB found in overgrazed pastures, roadsides, and other disturbed areas throughout Georgia. **STEMS:** Up to 2 feet tall, smooth, green or red, very leafy, branched only in the upper half, giving the plants a top-heavy look. **LEAVES:** Up to 3 inches long, very narrow, alternate, often with clusters of smaller leaves in the axils, smooth, gland-dotted, bitter-tasting; lower leaves withered by flowering time. **FLOWER HEADS:** About ¾ inch wide, with 8–10 yellow, drooping, 3-toothed ray flowers and a dome-shaped central disk of tiny, yellow flowers; several green, drooping, pointed bracts surround the base of each head. **FLOWERING:** Spring–summer. **RANGE:** FL west to TX, north to KS and NY. **NOTES:** Cows that graze on Bitter-weed produce bitter milk; in large quantities, the foliage is toxic to horses and cows. **SIMILAR TO** Yellow Buttons (*Balduina angustifolia*), page 59.

Autumn Sneezeweed

Helenium autumnale Linnaeus
Aster Family (Asteraceae)

PERENNIAL HERB found in moist to wet forests and disturbed areas throughout Georgia. **STEMS:** 2–5 feet tall, winged, hairy. **STEM LEAVES:** Up to 6 inches long and 1½ inches wide, alternate, elliptic to lance-shaped, usually toothed, hairy, with leaf tissue forming wings on the stem; basal leaves withered by flowering time. **FLOWER HEADS:** About 2 inches wide, with 8–21 yellow, spreading or downcurved, toothed ray flowers, surrounding a ball-shaped center about ½ inch high, covered with many tiny, yellow flowers. **FLOWERING:** Aug–Oct. **RANGE:** Most of North America. **SIMILAR TO** Savannah Sneezeweed (*Helenium vernale*), which occurs in wet savannas in the Coastal Plain and blooms Apr–May; its stems are usually smooth and nearly leafless, the basal leaves persist till flowering time, and the central disk is a half-dome.

Southern Sneezeweed

Helenium flexuosum Rafinesque
Aster Family (Asteraceae)

PERENNIAL HERB found in moist to wet forests, savannas, flatwoods, ditches, and other disturbed areas throughout Georgia. **STEMS:** Up to 3 feet tall, rough-hairy. **LEAVES:** Up to 6 inches long and 1 inch wide, smaller upward on the stem, alternate, narrowly lance-shaped to oval, usually entire, rough-hairy, with leaf tissue forming wings on the stem; basal leaves withered by flowering time. **FLOWER CLUSTERS:** Large, leafy, much-branched, with a flower head at the tip of each branch. **FLOWER HEADS:** Up to 2 inches wide, with 8–13 drooping, yellow, triangular, 3-toothed ray flowers surrounding a ball-shaped center about ½ inch high and tightly packed with tiny, maroon flowers. **FLOWERING:** May–Aug. **RANGE:** FL west to TX, north to WI and NS. **SIMILAR TO** Sandhill Blanket-flower (*Gaillardia aestivalis*), page 83, which occurs in dry uplands.

Narrow-Leaf Sunflower

Helianthus angustifolius Linnaeus
Aster Family (Asteraceae)

PERENNIAL HERB found in moist to wet soils in savannas, marshes, fields, and ditches throughout Georgia. **STEMS:** 2–5 feet tall, rough-hairy, green or reddish. **LEAVES:** Up to 8 inches long and ½ inch wide, opposite and alternate, linear, dark green and rough-hairy on the upper surface, with very short or no stalks, a deeply inset midvein, and tightly curled margins nearly covering the pale green lower surface. **FLOWER HEADS:** 1½–3 inches wide, with 8–21 yellow ray flowers and a central disk of tiny, maroon flowers; several series of narrow, pointed, green bracts surround the base of the head. **FLOWERING:** Jul–Oct. **RANGE:** FL west to TX, north to IA and NY. **SIMILAR TO** Florida Sunflower (*Helianthus floridanus*), which occurs in the Coastal Plain in similar habitats; its leaves are up to 1 inch wide with wavy margins.

Appalachian Sunflower

Helianthus atrorubens Linnaeus
Aster Family (Asteraceae)

PERENNIAL HERB found in dry pine-oak woodlands, glades, and on roadsides in north Georgia. **STEMS:** Up to 6½ feet tall, very hairy and rough, reddish. **BASAL LEAVES:** Up to 10 inches long and 4 inches wide, oval to lance-shaped with long, broadly winged stalks, roughly hairy on both surfaces, margins toothed or scalloped. **STEM LEAVES:** Few, similar to basal leaves but much smaller, opposite. **FLOWER HEADS:** Up to 3 inches wide, with 10–15 yellow ray flowers around a central disk of tiny, maroon flowers; several series of oval or oblong, ¼-inch-wide, green bracts surround the base of the head. **FLOWERING:** Jul–Oct. **RANGE:** FL west to LA, north to KY and NJ. With its winged leaf stalks, rough-hairy stems, and maroon disk flowers, Appalachian Sunflower is **SIMILAR TO** no other north Georgia sunflower.

Hairy Sunflower

Helianthus hirsutus Rafinesque
Aster Family (Asteraceae)

PERENNIAL HERB found in dry, open woodlands, prairies, and roadsides in north and central Georgia. **STEMS:** 2½–6½ feet tall, with spreading hairs. **LEAVES:** Up to 7 inches long and 3 inches wide, in pairs set at right angles to the next pair, lance-shaped with rounded bases, hairy on both surfaces, often sandpapery on the upper surface, gland-dotted on the lower surface, with 3 conspicuous veins and short or no stalk. **FLOWER HEADS:** 1–3 inches wide, with 8–15 yellow ray flowers surrounding a central disk of tiny, yellow flowers; several series of pointed, green bracts form a cup under the head. **FLOWERING:** Jul–Oct. **RANGE:** FL west to TX, north to MN, ON, and CT. **SIMILAR TO** other sunflowers but distinguished by its hairy stems, right-angle leaf arrangement, and leaves with a rounded or heart-shaped base.

Long-Leaf Sunflower

Helianthus longifolius Pursh
Aster Family (Asteraceae)

PERENNIAL HERB found around sandstone and granite outcrops in northwest and west Georgia and in Longleaf Pine sandhills in the Fall Line region. **STEMS:** 1½–4 feet tall, smooth, green or purple, forming colonies. **BASAL LEAVES:** 5–12 inches long and ⅓–1 inch wide, linear to narrowly spatula-shaped, mostly entire, smooth. **STEM LEAVES:** Few, much smaller than basal leaves, opposite (alternate near the top). **FLOWER HEADS:** Up to 2½ inches wide, with 8–13 yellow ray flowers and a central ½-inch-wide disk of tiny, yellow flowers; several series of narrow, pointed, green bracts form a cup around the base of the head. **FLOWERING:** Aug–Oct. **RANGE:** AL, GA, introduced into NC. **SIMILAR TO** Barrens Sunflower (*Helianthus occidentalis*), a Georgia Special Concern species that occurs in limestone cedar glades in northwest Georgia; it has hairy basal leaves up to 3 inches wide.

Small-Headed Sunflower

Helianthus microcephalus Torrey & A. Gray
Aster Family (Asteraceae)

PERENNIAL HERB found in dry woodlands and on roadsides in north Georgia and the Fall Line region. **STEMS:** Up to 6½ feet tall, smooth, dark greenish-purple, often with a whitish waxy coating. **LEAVES:** Up to 6 inches long and 1½ inches wide, with a stalk less than 1 inch long, opposite or alternate, lance-shaped, toothed or entire, upper surface rough-hairy, lower surface gland-dotted. **FLOWER HEADS:** 1–1½ inches wide, with 5–8 yellow ray flowers surrounding a central disk of tiny, yellow flowers; pointed, green bracts surround the base of the head. **FLOWERING:** Jul–Oct. **RANGE:** FL west to LA, north to MN and CT. **SIMILAR TO** Forest Sunflower (*Helianthus decapetalus*), which occurs in moist north Georgia forests; it has sharply toothed, long-stalked leaves and heads 1½–3½ inches wide with 10–15 ray flowers.

Confederate Daisy

Helianthus porteri (A. Gray) Pruski
SYNONYM: *Viguiera porteri* (A. Gray) Blake
Aster Family (Asteraceae)

ANNUAL HERB found on and around granite outcrops in the Piedmont, forming masses of brilliantly flowering plants in the fall. **STEMS:** Up to 3 feet tall, rough-hairy. **LEAVES:** Up to 4½ inches long and ½ inch wide, opposite near the base of the stem, alternate near the top, linear but tapering at both ends, roughly hairy and gland-dotted on the upper surface, with a fringe of long hairs at the base. **FLOWER HEADS:** 1½–2 inches wide, with 7 or 8 yellow ray flowers and a central disk of tiny, yellow flowers; 2 whorls of narrow, pointed, green bracts surround the base of the head. **FLOWERING:** Aug–Oct. **RANGE:** GA, AL, SC, and NC. **SIMILAR TO** Large-flowered Coreopsis (*Coreopsis grandiflora*), page 70, a granite outcrop plant that blooms in late spring and early summer.

Round-Leaf Sunflower, Rayless Sunflower

Helianthus radula (Pursh) Torrey & A. Gray
Aster Family (Asteraceae)

PERENNIAL HERB found in pine savannas and flatwoods in the Coastal Plain. **STEMS:** 2–4 feet tall, strongly curved at the base then becoming erect, rough-hairy especially near the base, with few or no leaves. **BASAL LEAVES:** Up to 4¾ inches long and 4 inches wide, opposite, forming a rosette closely pressed to the ground, oval to round, very rough-hairy. **STEM LEAVES:** Few, opposite, much smaller than basal leaves. **FLOWER HEADS:** Usually 1 per stem, about 1 inch wide, with many tiny, dark purple disk flowers surrounded by several whorls of pointed, dark purple bracts; ray flowers absent or few and very small. **FLOWERING:** Aug–Oct. **RANGE:** FL west to LA, north to SC. **SIMILAR TO** no other wildflower in Georgia.

Resinous Sunflower

Helianthus resinosus Small
Aster Family (Asteraceae)

PERENNIAL HERB found in woodlands, thickets, and roadsides in north and southwest Georgia. **STEMS:** 3–10 feet tall, often reddish-purple, with coarse, spreading hairs. **LEAVES:** Up to 10 inches long and 3 inches wide, alternate and opposite, oval to lance-shaped, entire or toothed, hairy on both surfaces, sandpapery on the upper surface, gland-dotted on the lower surface, with no or very short, winged stalks. **FLOWER HEADS:** Up to 3½ inches wide, with 10–20 spreading, yellow ray flowers surrounding a central disk of tiny, yellow flowers; several series of hairy, long-pointed, outwardly curved bracts surround the base of the head. **FLOWERING:** Jun–Oct. **RANGE:** FL west to MS, north to NC. **SIMILAR TO** Jerusalem Artichoke (*Helianthus tuberosus*), which occurs in moist bottomlands; its leaves have 3 prominent veins and a stalk longer than ¾ inch. It is native to the western United States.

Rough-Leaf Sunflower

Helianthus strumosus Linnaeus
Aster Family (Asteraceae)

PERENNIAL HERB found in dry woodlands, prairies, and roadsides in north Georgia and a few Coastal Plain counties. **STEMS:** 3–6½ feet tall, smooth, often waxy. **LEAVES:** 3–8 inches long and 1–4 inches wide with a ¾-inch stalk, mostly opposite, lance-shaped, sometimes toothed, green and rough-hairy on the upper surface, pale green and gland-dotted on the lower surface. **FLOWER HEADS:** Up to 2½ inches wide, with 10–20 yellow ray flowers and a central disk of tiny yellow flowers; several series of loose, pointed, green bracts surround the base of the head. **FLOWERING:** Jul–Sep. **RANGE:** FL west to TX, north to ND and ME. **SIMILAR TO** Woodland Sunflower (*Helianthus divaricatus*), which occurs in north Georgia forests; its leaves are conspicuously 3-veined, rough-hairy on both surfaces, and lack stalks.

Common Camphorweed

Heterotheca latifolia Buckley
SYNONYMS: *Heterotheca subaxillaris* (Lamarck), Britton & Rusby ssp. *latifolia* (Buckley) Semple
Aster Family (Asteraceae)

ANNUAL OR BIENNIAL HERB found on roadsides and in other disturbed areas throughout Georgia. **STEMS:** Up to 6½ feet tall, usually covered with long, spreading hairs and stalked glands. **LEAVES:** Up to 2¾ inches long and 1¼ inches wide, alternate, oval to lance-shaped, veiny, hairy and glandular on both surfaces, with wavy margins. **FLOWER HEADS:** About 1 inch wide, with many yellow ray flowers surrounding a central disk of tiny, yellow flowers; green, hairy, glandular bracts surround the base of the flower head, none tipped with white hairs. **FLOWERING:** Aug–Oct. **RANGE:** Native to the south-central United States and Mexico, now widely spread throughout the United States. **NOTES:** All parts of the plant smell like camphor. **SIMILAR TO** Dune Camphorweed (*Heterotheca subaxillaris*), a short, sprawling plant that occurs on beach dunes; the bracts of its flower head have a tuft of white hairs at the tip.

Hairy Hawkweed, Beaked Hawkweed

Hieracium gronovii Linnaeus
Aster Family (Asteraceae)

PERENNIAL HERB found in sandhills, dry woodlands, glades, and dry, disturbed areas throughout Georgia. **STEMS:** Up to 1 foot tall, very hairy, especially at the base. **LEAVES:** Up to 4 inches long and 2 inches wide, elliptic to lance-shaped, tapering to the base, very hairy, without conspicuous red veins. **FLOWER CLUSTERS:** Tall and narrow, with alternate, glandular-hairy branches. **FLOWER HEADS:** Up to ¾ inch wide, with 12–20 yellow, 5-toothed ray flowers and no disk flowers; narrow, green or purplish bracts form a cylinder about ⅓ inch high around the base of the head. **FLOWERING:** Jul–Nov. **RANGE:** FL west to TX, north to MN, ON, and ME. **NOTES:** All hawkweeds have milky sap in their leaves and stems. **SIMILAR TO** Veiny Hawkweed (*Hieracium venosum*), below.

Veiny Hawkweed

Hieracium venosum Linnaeus
Aster Family (Asteraceae)

PERENNIAL HERB found in dry forests and forest openings in north Georgia. **STEMS:** Up to 32 inches tall, green or purple, hairy only at the very base, with only a few small leaves. **BASAL LEAVES:** Up to 6 inches long and 2 inches wide, with broadly rounded tips, conspicuous reddish-purple veins, long hairs, and (often) black dots on the upper surface. **FLOWER HEADS:** About ½ inch wide, with many yellow ray flowers and no disk flowers; green, glandular-hairy bracts form a cylinder about ⅓ inch high around the base of the head. **FLOWERING:** May–Jul. **RANGE:** GA west to OK, north to MI and ME. **SIMILAR TO** Leafy Hawkweed (*Hieracium paniculatum*), which occurs in upland forests in the Blue Ridge; it has leafy stems and no basal leaves; its leaves are solid green and smooth.

Woolly-White

Hymenopappus scabiosaeus L'Héritier
Aster Family (Asteraceae)

BIENNIAL HERB found in sandhills and other dry, open areas in the Coastal Plain. **STEMS:** 1–5 feet tall, vertically ribbed, densely hairy (older plants may be smooth). **BASAL LEAVES:** Up to 10 inches long and 4¾ inches wide, mostly deeply lobed and divided, very hairy; stem leaves smaller, alternate. **FLOWER HEADS:** In flat-topped clusters at the tips of branches and stems, each head nearly round, with many small, white, 5-lobed disk flowers with pink-tipped stamens arising from the center; 5–13 petal-like bracts, white with green bases, surround the head (the bracts resemble ray flowers, which are absent in this species). **FLOWERING:** Apr–Jun. **RANGE:** FL west to TX, north to NE, IN, and SC. **SIMILAR TO** Frostweed (*Verbesina virginica*), page 124.

Hairy Cat's-Ear

Hypochaeris radicata Linnaeus
Aster Family (Asteraceae)

PERENNIAL HERB found in roadsides, pastures, and other disturbed areas throughout Georgia. **STEMS:** Up to 2 feet tall, hairy at the base, leafless, often branched in the upper half. **BASAL LEAVES:** 2–14 inches long and 2¾ inches wide, forming rosettes, coarsely toothed and lobed, covered with stiff hairs. **FLOWER HEADS:** About 1 inch wide with many bright yellow ray flowers (no disk flowers); the tips of the ray flowers are squared-off and toothed, and the lower surfaces of the outer rays are often maroon. **FRUIT:** In dandelion-like heads. **FLOWERING:** Aug–Oct. **RANGE:** Native to Eurasia, now widespread and invasive throughout North America. **SIMILAR TO** Brazilian Cat's-ear (*Hypochaeris chillensis*, synonym *H. brasiliensis*), a South American weed found infrequently throughout Georgia; it has leafy stems up to 3 feet tall.

Stiff-Leaved Aster

Ionactis linariifolia (Linnaeus) Greene
SYNONYM: *Aster linariifolius* Linnaeus
Aster Family (Asteraceae)

PERENNIAL HERB found in sandhills, flatwoods, woodlands, rocky glades, and other open, dry areas throughout Georgia. **STEMS:** Up to 2 feet tall, unbranched, hairy; green, yellow, or reddish. **LEAVES:** Up to 1½ inches long and ⅛ inch wide, alternate, linear, stiffly spreading, rough-hairy, shiny dark green on the upper surface, paler green below. **FLOWER HEADS:** About 1 inch across, with 10–20 purple ray flowers, a central disk of tiny, yellow flowers (aging red), and several series of green-and-white bracts forming a cylinder around the base of the head. **FLOWERING:** Aug–Nov. **RANGE:** FL west to TX, north to WI, QC, and ME. **SIMILAR TO** other fall-flowering purple asters, but Stiff-leaved Aster is the only one with very narrow, stiff, rough-hairy leaves.

Dune Marsh-Elder

Iva imbricata Walter
Aster Family (Asteraceae)

SHRUBLIKE PERENNIAL HERB found on dunes and sandy flats on barrier islands. **STEMS:** 1–3 feet tall, succulent, reddish-purple, smooth, erect or sprawling, the bases often buried in sand. **LEAVES:** Up to 2½ inches long and ¾ inch wide, upper leaves alternate, lower leaves opposite; elliptic, occasionally toothed, smooth, thick and succulent, with 3 veins. **FLOWER HEADS:** Look like green berries held in the axils of the upper leaves, about ¼ inch long, nodding, enclosed by fleshy green bracts, flowers barely or not visible. **FLOWERING:** Jul–Nov. **RANGE:** FL west to TX, north to VA. **NOTES:** Dune Marsh-elder aids in dune creation and stabilization. **SIMILAR TO** Maritime Marsh-elder (*Iva frutescens*), which grows in and around salt marshes; it has hairy stems and leaves, and all its leaves are opposite except for those at the tips of branches.

Orange Dwarf-Dandelion

Krigia biflora (Walter) S. F. Blake
Aster Family (Asteraceae)

PERENNIAL HERB found in moist to dry forests and disturbed areas in north Georgia. **STEMS:** Up to 2 feet tall, branched in the upper half, mostly smooth, with milky juice. **BASAL LEAVES:** 1½–10 inches long, including the winged leaf stalk, and ½–2 inches wide, smooth, entire or irregularly and shallowly toothed. **FLOWER HEADS:** 1–1½ inches across, with many yellow or orange, toothed ray flowers (there are no disk flowers); a whorl of green or red-tinged bracts surrounds the base of the head. **FLOWERING:** Apr–Oct. **RANGE:** GA west to AR, north to MB and MA. **NOTES:** Leaves and stems contain milky latex. **SIMILAR TO** Mountain Dwarf-dandelion (*Krigia montana*), which occurs on cliffs and rock outcrops in the Blue Ridge; it branches near the base, and its leaves are less than ½ inch wide.

Dwarf-Dandelion

Krigia virginica (Linnaeus) Willdenow
Aster Family (Asteraceae)

ANNUAL HERB found in dry woodlands, rock outcrops, sandhills, prairies, and disturbed areas throughout Georgia. **FLOWERING STEMS:** 1–12 inches tall, unbranched, glandular-hairy, leafy only near the base, with milky juice. **BASAL LEAVES:** Up to 5 inches long and ½ inch wide, sharply and irregularly lobed and toothed, usually hairy. **FLOWER HEADS:** 1 per flowering stem, about ½ inch wide, with many yellow, 5-toothed ray flowers and no disk flowers; a whorl of green or red-tinged bracts surrounds the base of the head. **FLOWERING:** Feb–Nov. **RANGE:** FL west to TX, north to MN and ME. **NOTES:** Leaves and stems contain milky latex. **SIMILAR TO** Potato Dandelion (*Krigia dandelion*), which occurs in woodlands and disturbed areas throughout Georgia (except mountains); its stems and leaves are smooth, and it has small, potato-like tubers.

Wild Lettuce

Lactuca canadensis Linnaeus
Aster Family (Asteraceae)

BIENNIAL HERB found in disturbed areas throughout Georgia. **STEMS:** Up to 15 feet tall, stout, waxy, often purple-spotted. **LEAVES:** Up to 14 inches long and 5 inches wide, alternate, variable in shape and size, entire or the largest deeply toothed and lobed, smooth except for hairy veins. **FLOWER HEADS:** Up to 100 per plant in a large, open cluster at the top of the stem, each head dandelion-like, up to ½ inch wide, with 15–20 yellow ray flowers (no disk flowers); the base of the head is surrounded by a ½-inch-high cylinder of narrow, green, purple-tinted bracts. **FLOWERING:** Jun–Nov. **RANGE:** Throughout North America. **NOTES:** Leaves and stems contain a bitter, beige latex. **SIMILAR TO** Prickly Lettuce (*Lactuca serriola*), a European native that occurs in disturbed areas. Its leaves are prickly on the midveins and margins and are twisted sideways.

Woodland Lettuce

Lactuca floridana (Linnaeus) Gaertner
Aster Family (Asteraceae)

ANNUAL OR BIENNIAL HERB found in forests, thickets, and clearings in north Georgia and a few Coastal Plain counties. **STEMS:** Up to 6½ feet tall, stout, purplish green, smooth. **LEAVES:** Up to 8 inches long and 4 inches wide, alternate, toothed and deeply lobed, smooth except for hairy veins. **FLOWER HEADS:** Less than ½ inch wide, with 11–17 pale blue or lavender ray flowers and no disk flowers; the base of the head is surrounded by an inch-high cylinder of narrow, green, purple-tipped bracts. **FLOWERING:** Aug–Nov. **RANGE:** FL west to TX, north to MB and MA. **NOTES:** Leaves and stems contain a bitter, white latex. **SIMILAR TO** Coastal Plain Lettuce (*Lactuca graminifolia*), which occurs in sandhills and woodlands throughout Georgia except for the mountains. Most of its leaves are narrow and in a basal rosette; its flowers are blue, sometimes yellow.

Ox-Eye Daisy

Leucanthemum vulgare Lamarck
SYNONYM: *Chrysanthemum leucanthemum* Linnaeus
Aster Family (Asteraceae)

PERENNIAL HERB found on roadsides and in other disturbed areas in north Georgia and a few Coastal Plain counties. **STEMS:** Usually 1–3 feet tall, smooth or slightly hairy. **BASAL LEAVES:** 1½–6 inches long and up to 1¼ inches wide, spatula-shaped, irregularly lobed and toothed, smooth. **STEM LEAVES:** Alternate, smaller than basal leaves and less lobed. **FLOWER HEADS:** Up to 2 inches wide, with 13–35 white ray flowers surrounding a central disk about ½ inch wide that is composed of many tiny, golden yellow flowers. **FLOWERING:** Apr–Oct. **RANGE:** Native to Eurasia, now spread throughout North America. Although considered invasive in some western states, Ox-eye Daisy is not yet a problem in Georgia's natural areas. **SIMILAR TO** Robin's Plantain (*Erigeron pulchellus*), page 76.

Rough Blazing-Star

Liatris aspera Michaux
Aster Family (Asteraceae)

PERENNIAL HERB found in dry open woodlands, prairies, and glades in western Georgia both north and south of the Fall Line. **STEMS:** 1–6 feet tall, hairy. **LEAVES:** Up to 16 inches long and 1¾ inches wide at the base of the stem, much smaller up the stem, alternate, smooth or hairy, and dotted with tiny glands. **FLOWER CLUSTERS:** Elongated spike held at the top of the stem, with large, well-spaced flower heads. **FLOWER HEADS:** About 1 inch high and wide, with up to 30 small, pink flowers per head; 2 style branches extend from each flower; the bracts surrounding the base of each head have broad, ragged, upturned, pink or white margins. **FLOWERING:** Aug–Oct. **RANGE:** FL west to TX, north to ND, ON, and VA. **SIMILAR TO** other blazing-stars, but this is the only species in Georgia with such unusual bracts surrounding large flower heads.

Elegant Blazing-Star

Liatris elegans (Walter) Michaux
Aster Family (Asteraceae)

PERENNIAL HERB found in sandhills and savannas in the Coastal Plain. **STEMS:** 1–4 feet tall, hairy. **LEAVES:** Up to 12 inches long and ½ inch wide near the base of the stem, much smaller up the stem, alternate, nearly hairless, dotted with tiny glands. **FLOWER CLUSTERS:** Elongated spike held at the top of the stem, crowded with flower heads. **FLOWER HEADS:** About 1 inch high, with 4 or 5 tiny, white or pale pink flowers; 2 style branches extend from each flower; the inner bracts surrounding the base of the head are pink and petal-like, the outer bracts green and narrow. **FLOWERING:** Aug–Oct. **RANGE:** FL west to TX, north to OK and SC. With its distinctive white flowers and pink, petal-like bracts, Elegant Blazing-star is not **SIMILAR TO** any of the other blazing-stars.

Greene's Grass-Leaved Blazing-Star

Liatris elegantula (Greene) K. Schumann
SYNONYM: *Liatris graminifolia* Willdenow var. *elegantula* (Greene) Gaiser
Aster Family (Asteraceae)

PERENNIAL HERB found in sandhills, flatwoods, and savannas in the Coastal Plain. **STEMS:** 2–3½ feet tall, smooth. **LEAVES:** Up to 8 inches long and ¼ inch wide, alternate, smooth, dotted with tiny glands. **FLOWER CLUSTERS:** Elongated spike held at the top of the stem, with flower heads spaced about ⅓ inch apart. **FLOWER HEADS:** About 1 inch high, with 7–13 small pink flowers; 2 style branches extend from each flower; the bracts surrounding the base of each head are purple, gland-dotted, and round-tipped, with translucent borders. **FLOWERING:** Aug–Nov. **RANGE:** GA, FL, AL, and MS. **SIMILAR TO** Wand Blazing-star (*Liatris virgata*), a north Georgia species, page 100.

Small-Head Blazing-Star

Liatris microcephala (Small) K. Schumann
Aster Family (Asteraceae)

PERENNIAL HERB found on sandstone, granite, and Altamaha Grit outcrops throughout Georgia. STEMS: 1–2½ feet tall, smooth, forming clumps of several leafy stems. LEAVES: Up to 8 inches long and ¼ inch wide near the base of the stem, much smaller up the stem, alternate, linear, smooth, glossy. FLOWER CLUSTERS: Elongated spike held at the top of the stem, with small, well-spaced flower heads. FLOWER HEADS: Less than ¾ inch high, most on ½-inch stalks, each with 4–6 small, pink flowers; 2 style branches extend from each flower; the bracts surrounding the base of each head are green or purplish and blunt-tipped with narrow translucent margins. FLOWERING: Aug–Oct. RANGE: GA west to AL, north to KY and NC. SIMILAR TO other blazing-stars but with exceptionally small, narrow flower heads borne on conspicuous stalks.

Florist's Blazing-Star

Liatris spicata (Linnaeus) Willdenow
Aster Family (Asteraceae)

PERENNIAL HERB found in seepages, flatwoods, wet prairies, bogs, and ditches throughout Georgia. STEMS: 1–4 feet tall, smooth. LEAVES: Up to 14 inches long and ½ inch wide near the base of the stem, much smaller up the stem, alternate, 3–5-veined, mostly smooth. FLOWER CLUSTERS: Elongated spikes up to 12 inches long at the top of the stem, crowded with heads. FLOWER HEADS: Less than ½ inch across, with no stalk; each head with 4–14 small, pink flowers surrounded by purple bristles; 2 style branches extend from each flower; the bracts surrounding the base of each head are green or purplish, with rounded tips and narrow, pink or white, translucent margins. FLOWERING: Jul–Nov. RANGE: FL west to LA, north to WI, QC, and MA. SIMILAR TO other blazing-stars but distinguished by the stalkless flower heads and leaves with 3–5 obvious veins.

Scaly Blazing-Star

Liatris squarrosa (Linnaeus) Michaux
Aster Family (Asteraceae)

PERENNIAL HERB found in dry or rocky woodlands, prairies, and openings throughout Georgia (except for the Blue Ridge). **STEMS:** 1–2½ feet tall, hairy. **LEAVES:** Up to 10 inches long and ½ inch wide, alternate, broadly linear, nearly hairless, with 3–5 parallel veins. **FLOWER CLUSTERS:** Elongated spike held at the top of the stem, with widely spaced flower heads. **FLOWER HEADS:** ½–¾ inch across, with 10–60 small, pink flowers surrounded by red, feathery bristles; 2 long style branches extend from each flower; the bracts surrounding the base of each head are green, spreading or erect, smooth or hairy, with pointed tips and hairy margins. **FLOWERING:** May–Sep. **RANGE:** FL west to LA, north to MO, MI, and NJ. **SIMILAR TO** Appalachian Blazing-star (*Liatris squarrulosa*), below, which has spatula-shaped leaves with 1 main vein.

Appalachian Blazing-Star

Liatris squarrulosa Michaux
Aster Family (Asteraceae)

PERENNIAL HERB found in dry or rocky woodlands, glades, and prairies in north Georgia. **STEMS:** Up to 4 feet tall, hairy. **LEAVES:** Up to 12 inches long and 1¾ inches wide, alternate, narrowly spatula-shaped, smooth or hairy, with 1 main vein. **FLOWER CLUSTERS:** Elongated spike held at the top of the stem, with widely spaced flower heads. **FLOWER HEADS:** About 1 inch across, with 14–24 small, pink flowers surrounded by red, threadlike bristles; 2 long style branches extend from each flower; the bracts surrounding the base of each head are spreading, smooth or hairy, with hairless margins and broad, purple tips. **FLOWERING:** Jul–Nov. **RANGE:** FL west to TX, north to MO and WV. **SIMILAR TO** Scaly Blazing-star (*Liatris squarrosa*), above, which has narrower leaves with 3–5 veins.

Narrow-Leaved Blazing-Star

Liatris tenuifolia Nuttall
Aster Family (Asteraceae)

PERENNIAL HERB found in sandhills in the Coastal Plain. **STEMS:** 1–5 feet tall, smooth or nearly so. **BASAL LEAVES:** 4–12 inches long and very narrow, resembling Wire Grass leaves; stem leaves much shorter and pressed upward against the stem. **FLOWER CLUSTERS:** Elongated spike held at the top of the stem, crowded with flower heads. **FLOWER HEADS:** About ½ inch high, with 3–6 small, pink flowers surrounded by dark red bristles; 2 style branches extend from each flower; the bracts surrounding the base of each head are reddish-purple, the inner bracts are rounded, the outer bracts pointed. **FLOWERING:** Aug–Nov. **RANGE:** GA, AL, FL, and SC. **SIMILAR TO** Slender Blazing-star (*Liatris gracilis*), which occurs in sandhills and dry pine flatwoods in the Coastal Plain; it has hairy stems and wider basal leaves (⅛–½ inch wide), the stem leaves are widely spreading, and its flower heads are more widely spaced.

Wand Blazing-Star

Liatris virgata Nuttall
SYNONYM: *Liatris graminifolia* Willdenow var. *virgata* (Nuttall) Fernald
Aster Family (Asteraceae)

PERENNIAL HERB found in open pine-oak woods and on rocky slopes and road banks in north Georgia. **STEMS:** 1–2½ feet tall, smooth. **LEAVES:** Up to 6 inches long and ½ inch wide, alternate, smooth or slightly hairy, dotted with tiny glands. **FLOWER CLUSTERS:** Elongated spike held at the top of the stem, slightly zigzag, with flower heads spaced about ½ inch apart. **FLOWER HEADS:** About 1 inch high, with 7–12 small, pink flowers; 2 style branches extend from each flower; the bracts surrounding the base of each head are covered with gland dots and have thickened, pointed tips and translucent margins. **FLOWERING:** Aug–Oct. **RANGE:** GA, SC, NC, and VA. **SIMILAR TO** Greene's Grass-leaved Blazing-star (*Liatris elegantula*), a Coastal Plain species, page 97.

Rose-Rush

Lygodesmia aphylla (Nuttall) Torrey & A. Gray
Aster Family (Asteraceae)

PERENNIAL HERB found in sandhills and other dry woodlands in southwest Georgia. **STEMS:** 1–2½ feet tall, usually branched, leafless, with milky sap. **LEAVES:** In a basal rosette, 4–12 inches long and less than ⅛ inch wide, usually withering before the plant flowers; stem leaves are reduced to small scales. **FLOWER HEADS:** Held singly at the tips of branches, about 2–3 inches wide, with 8–10 pink or lavender, toothed ray flowers (there are no disk flowers); grayish-green bracts form a slender, ¾-inch-high cylinder around the base of the head. **FLOWERING:** Feb–May. **RANGE:** GA, FL. **NOTES:** A delicate 2-branched style about ½ inch long, its base enclosed by a sleeve of purple and white stamens, arises from the tube at the base of each ray flower. **SIMILAR TO** no other wildflower in Georgia.

Narrow-Leaf Barbara's Buttons

Marshallia tenuifolia Rafinesque
SYNONYM: *Marshallia graminifolia* (Walter) Small ssp. *tenuifolia* (Rafinesque) L. E. Watson
Aster Family (Asteraceae)

PERENNIAL HERB found in bogs, wet pine savannas, and flatwoods in the Coastal Plain. **STEMS:** 1–3 feet tall, branched, without dried leaves at the base. **BASAL LEAVES:** Up to 4 inches long and ½ inch wide, tapering at both ends. **STEM LEAVES:** Small, linear, pressed upward. **FLOWER HEADS:** About 1 inch wide, with many white, pink, or lavender disk flowers opening from the outside of the head to the center (there are no ray flowers). **FLOWERS:** Up to ½ inch long, with 5 twisted lobes and a purple tube of stamens. **FLOWERING:** Aug–Oct. **RANGE:** GA and FL, west to TX. **SIMILAR TO** Spoon-leaved Barbara's Buttons (*Marshallia obovata*), which occurs in dry woodlands throughout Georgia; its leaves are spatula-shaped, and its unbranched stems are topped with a single, white-flowered head.

Salt-and-Pepper, Snowy Square-Stem

Melanthera nivea (Linnaeus) Small
Aster Family (Asteraceae)

PERENNIAL HERB found in sandy woodlands, dunes, and dry, disturbed areas in Georgia's Coastal Plain. **STEMS:** 1½–7 feet tall, erect or sprawling, much-branched and bushy, 4-angled, rough-hairy. **LEAVES:** Up to 6 inches long and 4 inches wide, opposite, oval or triangular, often with 3 lobes, toothed, rough-hairy, conspicuously 3-veined. **FLOWER HEADS:** About 1 inch wide, dome-shaped, with many small, white, tubular, 5-lobed disk flowers, each flower with black anthers in the center (there are no ray flowers); 2 or 3 whorls of pointed, green bracts form a cup under the head. **FLOWERING:** Jun–Oct. **RANGE:** FL west to LA, north to IL, KY, and SC; West Indies, Central and South America. The flower heads are **SIMILAR TO** those of Barbara's Buttons, page 101, which have round, not 4-angled, stems.

Climbing Hempweed

Mikania scandens (Linnaeus) Willdenow
Aster Family (Asteraceae)

PERENNIAL HERBACEOUS VINE found in marshes and swamps throughout Georgia, except for the Blue Ridge. **STEMS:** Up to 50 feet long, hairy or smooth, twining and sprawling over other plants. **LEAVES:** Up to 6 inches long and 4½ inches wide, opposite, hairy or smooth, heart-shaped with wavy, sometimes toothed, margins. **FLOWER CLUSTERS:** Up to 6 inches wide, held on long stalks in the leaf axils, containing many small, white or pink flower heads. **FLOWER HEADS:** Less than ½ inch high, with 4 tiny, tubular flowers; 2 style branches extend from each flower, giving the heads a fuzzy look. **FLOWERING:** Jun–Dec. **RANGE:** FL west to TX, north to WI and ME. **SIMILAR TO** Heartleaf Climbing Hempweed (*Mikania cordifolia*), which is common in Florida but known only from Bryan County in Georgia; it has very hairy leaves and stems and larger flower heads.

Whorled Wood Aster

Oclemena acuminata (Michaux) Greene
SYNONYM: *Aster acuminatus* Michaux
Aster Family (Asteraceae)

PERENNIAL HERB found in moist, usually high-elevation forests in the Blue Ridge. **STEMS:** Up to 2½ feet tall, hairy, slightly zigzag, often in large colonies. **LEAVES:** Up to 6½ inches long and 2 inches wide, alternate, lance-shaped, coarsely toothed, usually smooth but sometimes rough on the upper surface, with no or a very short stalk; upper leaves are often crowded and appear whorled. **FLOWER HEADS:** About 2 inches wide, with 9–20 white ray flowers surrounding a central disk of tiny, yellow flowers (aging to red); several series of narrow, pointed, pale green or purple-tinged bracts surround the base of the head. **FLOWERING:** Jul–Sep. **RANGE:** GA north to ON and NF. **SIMILAR TO** White Wood Aster (*Eurybia divaricata*), page 80.

Piney Woods Aster

Oclemena reticulata (Pursh) G. L. Nesom
SYNONYM: *Aster reticulatus* Pursh
Aster Family (Asteraceae)

PERENNIAL HERB found in moist to wet pine flatwoods in the Coastal Plain. **STEMS:** Up to 3 feet tall, very hairy, usually in clumps or large colonies. **LEAVES:** Up to 4¼ inches long and 1½ inches wide, alternate, oval or elliptic, entire or few-toothed, very hairy, conspicuously veined, with no or very short stalks. **FLOWER HEADS:** About 1½ inches wide, with 5–20 white ray flowers surrounding a central disk of tiny, yellow flowers (aging to red); several series of narrow, pointed, hairy bracts surround the base of the head. **FLOWERING:** Apr–Jun. **RANGE:** GA, FL, AL, and SC. **SIMILAR TO** Southern White-top Aster (*Doellingeria sericocarpoides, Aster umbellatus*), which occurs in pine flatwoods in the Coastal Plain; its stems and leaves are nearly hairless, it blooms in the fall, and the heads have only 2–7 ray flowers.

Small's Ragwort

Packera anonyma (Wood) W. A. Weber & Á. Löve

SYNONYMS: *Senecio anonymus* Wood, *Senecio smallii* Britton

Aster Family (Asteraceae)

PERENNIAL HERB found in rocky woodlands, and on rock outcrops and roadsides throughout Georgia. **STEMS:** Up to 2½ feet tall, smooth except at the base and leaf axils. **BASAL LEAVES:** Up to 3½ inches long and ¾ inch wide, elliptic to lance-shaped, finely toothed, smooth. **STEM LEAVES:** Smaller, alternate, deeply lobed and toothed. **FLOWER HEADS:** About ¾ inch wide, with 8–13 yellow ray flowers surrounding a central disk of tiny, yellow flowers; narrow, pointed, green bracts form a cup around the base. **FLOWERING:** Apr–May. **RANGE:** FL west to LA, north to AR and NY. **SIMILAR TO** Round-leaf Ragwort (*Packera obovata*), which occurs in rich woods and on limestone cedar glades and cliffs in north Georgia, the Fall Line, and the southwestern Coastal Plain; its basal leaves are oval or round.

Golden Ragwort

Packera aurea (Linnaeus) Á. & D. Löve

SYNONYM: *Senecio aureus* Linnaeus

Aster Family (Asteraceae)

PERENNIAL HERB found in wetlands in north Georgia. **STEMS:** 1–2 feet tall, cobwebby when young, smooth when mature. **BASAL LEAVES:** Up to 2½ inches long and wide, nearly round or heart-shaped, not lobed, hairy on the lower surface, with scallop-toothed margins and long stalks. **STEM LEAVES:** Smaller, lobed, toothed, smooth, without stalks. **FLOWER HEADS:** Up to 1½ inches wide, with 8–13 yellow ray flowers surrounding a disk of tiny, yellow flowers; small, green or purplish bracts form a cup around the base of the head. **FLOWERING:** Feb–May. **RANGE:** FL west to OK, north to MB and NL. **NOTES:** Unopened flower heads are purple and cobwebby. **SIMILAR TO** Butterweed (*Packera glabella*), which occurs in bottomlands and disturbed wetlands throughout Georgia; it has red-tinged stems, and all of its leaves are toothed and deeply lobed.

Woolly Ragwort

Packera tomentosa (Michauxii) C. Jeffrey
SYNONYM: *Senecio tomentosus* Michauxii
Aster Family (Asteraceae)

PERENNIAL HERB found on granite and Altamaha Grit outcrops, roadsides, and dry woodlands throughout Georgia. **STEMS:** 1–2 feet tall, covered with long, white, cobwebby hairs. **BASAL LEAVES:** Up to 5 inches long and 2 inches wide, held nearly erect, oval or lance-shaped, abruptly narrowed to the stalk, toothed or entire, covered with cobwebby hairs (the upper surface becomes smooth as the leaf ages). **STEM LEAVES:** Similar to basal leaves but much smaller and toothed or lobed. **FLOWER HEADS:** About 1 inch wide, with 10–13 yellow ray flowers surrounding a central disk of tiny, yellow flowers; narrow, pointed, green bracts form a cup around the base. **FLOWERING:** Apr–early Jun. **RANGE:** FL west to TX, north to OK and NJ. **SIMILAR TO** other ragworts, but this is the only species covered with cobwebby hairs.

Wild Quinine

Parthenium integrifolium Linnaeus
Aster Family (Asteraceae)

PERENNIAL HERB found in dry woodlands and disturbed areas in north Georgia. **STEMS:** 1–3 feet tall, smooth or hairy. **LOWER LEAVES:** Up to 10 inches long and 5 inches wide, alternate, oval to lance-shaped, toothed and sometimes slightly lobed, very rough, with winged stalks. **UPPER LEAVES:** Smaller than lower leaves, without stalks. **FLOWER HEADS:** Held in flat-topped clusters at the top of the stem, each head about ⅓ inch wide, with 5 or 6 small, white ray flowers surrounding a central disk of many tiny, white, tubular flowers. **FLOWERING:** May–Oct. **RANGE:** GA west to TX, north to MN and MA. **NOTES:** Native Americans and pioneers used Wild Quinine to treat fevers and inflammation. Wild Quinine heads are **SIMILAR TO** those of Frostweed (*Verbesina virginica*), page 124.

Sticky Golden-Aster, Pineland Silk-Grass

Pityopsis aspera (A. Gray) Small
SYNONYM: *Heterotheca adenolepis* (Fernald) H. E. Ahles
Aster Family (Asteraceae)

PERENNIAL HERB found in dry upland woods, sandhills, and dry, disturbed areas throughout Georgia. **STEMS:** 8–20 inches tall, covered with a mix of long, silky hairs and gland-tipped hairs. **LEAVES:** Up to 14 inches long and ½ inch wide near the base of the stem, smaller and alternate up the stem, all leaves (variety *adenolepis*) or lower leaves only (variety *aspera*) covered with silky hairs. **FLOWER HEADS:** About ½ inch wide, with 6–10 yellow ray flowers around a central disk of tiny, yellow flowers; green, gland-covered bracts form a cylinder around the base of the head; the stalks supporting the heads are also covered with gland-tipped hairs. **FLOWERING:** Jul–Oct. **RANGE:** FL west to LA, north to VA. **SIMILAR TO** Grass-leaved Golden-aster (*Pityopsis graminifolia*), below.

Grass-Leaved Golden-Aster, Silk-Grass

Pityopsis graminifolia (Michaux) Nuttall
SYNONYMS: *Chrysopsis graminifolia* (Michaux) Elliott, *Heterotheca graminifolia* (Michaux) Shinners
Aster Family (Asteraceae)

PERENNIAL HERB found in pine-oak woodlands, sandhills, rocky woodlands, and dry, disturbed areas throughout Georgia. **STEMS:** Up to 2½ feet tall, often growing in large patches. **LEAVES:** Up to 16 inches long and ¾ inch wide near the base of the stem, grasslike, smaller and alternate up the stem. **FLOWER HEADS:** About 1 inch wide, with 9–13 yellow ray flowers around a central disk of tiny, yellow flowers; narrow, green bracts form a cylinder around the base of the head. **FLOWERING:** Jun–Oct. **RANGE:** FL west to TX and OK, north to OH and DE. **NOTES:** All parts of the plant are covered with long, white, silky hairs. **SIMILAR TO** Sticky Golden-aster (*Pityopsis aspera* variety *adenolepis*), above.

Camphorweed, Marsh Fleabane

Pluchea camphorata (Linnaeus) A. P. de Candolle

Aster Family (Asteraceae)

ANNUAL OR PERENNIAL HERB found in wetlands throughout Georgia. **STEMS:** 2–6½ feet tall, hairy. **LEAVES:** Up to 6 inches long and 2¾ inches wide, with ½-inch long stalks, elliptic, toothed, and covered with tiny, grainy, glistening glands that exude a strong odor. **FLOWER CLUSTERS:** Held at the top of the stem and ends of branches, large, round-topped. **FLOWER HEADS:** About ¼ inch high, with many tiny, pink disk flowers surrounded by whorls of pink, gland-dotted bracts (there are no ray flowers). **FLOWERING:** Aug–Oct. **RANGE:** FL west to TX, north to KS and DE. **SIMILAR TO** Salt Marsh Camphorweed (*Pluchea odorata*), which has smaller, flat-topped flower clusters held at the tips of branches; it occurs in salt marshes on the Georgia coast.

Stinking Camphorweed

Pluchea foetida (Linnaeus) A. P. de Candolle

Aster Family (Asteraceae)

ANNUAL OR PERENNIAL HERB found in swamps and other freshwater wetlands in the Coastal Plain. **STEMS:** Up to 3 feet tall, purplish, very hairy, and gland-dotted. **LEAVES:** Up to 5 inches long and 1½ inches wide, alternate; oblong, oval, or lance-shaped; blunt-tipped, finely toothed, very hairy, gland-dotted, thick, clasping the stem; with a strong, unpleasant odor. **FLOWER CLUSTERS:** Held at the top of the stem and ends of branches, large, rounded or flat-topped. **FLOWER HEADS:** About ¼ inch high and wide, with many tiny, creamy white disk flowers surrounded by whorls of pink or white, hairy, gland-dotted bracts. **FLOWERING:** Jul–Oct. **RANGE:** FL west to TX, north to MO and NJ. **SIMILAR TO** Rosy Camphorweed (*Pluchea baccharis*, synonym *Pluchea rosea*), which occurs in wet flatwoods and marshes in the Coastal Plain; its flower heads are pink.

Tall Rattlesnake-Root, Gall-of-the-Earth

Prenanthes altissima Linnaeus
SYNONYM: *Nabalus altissimus* (Linnaeus) Hooker
Aster Family (Asteraceae)

PERENNIAL HERB found in upland forests in north Georgia. **STEMS:** 1½–6 feet tall, mostly smooth, often zigzag at the top. **LEAVES:** Up to 6 inches long and wide, smaller up the stem, alternate, smooth, with milky latex; leaf shape varies from triangular with few teeth to deeply lobed; young, triangular leaves often dot the forest floor near mature plants. **FLOWER HEADS:** Drooping from the tips of short branches, with 5 pale yellow, petal-like ray flowers emerging from the end of a ½-inch cylinder formed by 5 smooth, pale green bracts. **FLOWERING:** Aug–Nov. **RANGE:** GA west to TX, north to MI and NL. **SIMILAR TO** Lion's Paw (*Prenanthes serpentaria*), which occurs in similar habitats; it has bell-shaped (not cylindrical) heads, with 8–14 ray flowers and 8 spreading, hairy bracts.

Rabbit-Tobacco

Pseudognaphalium obtusifolium (Linnaeus) Hilliard & Burtt
SYNONYM: *Gnaphalium obtusifolium* Linnaeus
Aster Family (Asteraceae)

ANNUAL HERB found in dry, open woodlands and disturbed areas throughout Georgia. **STEMS:** Up to 3 feet tall, matted with white, woolly hairs. **LEAVES:** Up to 4 inches long and ⅜ inch wide, alternate, linear and oblong, with no stalk; upper surface green, lower surface covered with white, woolly hairs. **FLOWER HEADS:** Held at the tips of woolly branches, each head about ¼ inch long and ⅛ inch wide, pointed or flat-topped, enclosed by several series of white, papery, woolly bracts; tiny yellow flowers are just visible at the top of the head. **FLOWERING:** Aug–Nov. **RANGE:** FL west to TX, north to MN and NL. **SIMILAR TO** Heller's Rabbit Tobacco (*Pseudognaphalium helleri*), which occurs in similar habitats in north Georgia; it has green, spreading-hairy stems covered with tiny, sticky, stalked glands.

Blackroot

Pterocaulon pycnostachyum (Michaux) Elliott
Aster Family (Asteraceae)

PERENNIAL HERB found in sandhills and other pinelands in the Coastal Plain. **STEMS:** Up to 2½ feet tall, winged and densely covered with white or rusty hairs. **LEAVES:** Up to 4⅓ inches long and 1½ inches wide, alternate, elliptic or lance-shaped, dark green and nearly hairless on the upper surface, whitened by a dense layer of hairs on the lower surface, leaf bases continuing down along the stem as wings. **FLOWER CLUSTERS:** A spike of densely packed flower heads at the top of the stem, 1–4 inches long, white tinged with pink, nodding at the tip before the flowers open; individual flower heads have tiny, yellowish flowers. **FLOWERING:** May–Jun. **RANGE:** FL west to MS, north to NC. With its winged stems, dense covering of white hairs, and thick, white flower spikes, Blackroot is **SIMILAR TO** no other Georgia wildflower.

False Dandelion

Pyrrhopappus carolinianus (Walter) A. P. de Candolle
Aster Family (Asteraceae)

ANNUAL HERB found in woodland edges, roadsides, pastures, and meadows throughout Georgia. **STEMS:** Up to 3 feet tall, mostly smooth, with milky sap. **BASAL LEAVES:** Up to 10 inches long and 2½ inches wide, toothed or irregularly lobed, mostly smooth; stem leaves few, similar to basal leaves but smaller, alternate. **FLOWER HEADS:** Up to 2½ inches wide, with many toothed, yellow ray flowers and 2 series of narrow bracts surrounding the base of the head; there are no disk flowers, but many slender yellow styles, their bases enclosed by a sleeve of dark stamens, curve up from the bases of the ray flowers in the center. **FLOWERING:** Mar–Jun. **RANGE:** FL west to TX, north to NE and PA. **SIMILAR TO** Common Dandelion (*Taraxacum officinale*), a much shorter plant with no stem and a leafless flower stalk. Also see Hairy Cat's-ear (*Hypochaeris radicata*), page 92.

Gray-Headed Coneflower

Ratibida pinnata (Ventenat) Barnhart
Aster Family (Asteraceae)

PERENNIAL HERB found in glades and prairies over limestone and chalky soils in northwest and central Georgia. **STEMS:** Up to 4 feet tall, vertically ribbed, hairy. **LEAVES:** 2–16 inches long and 1–6 inches wide, alternate, divided into many pointed, toothed segments (upper leaves may be entire), rough-hairy, gland-dotted. **FLOWER HEADS:** 6–15 drooping, yellow ray flowers, each up to 2⅓ inches long, surround a grayish-green, thimble-shaped center up to ¾ inch high, eventually covered with tiny, maroon disk flowers that open from the bottom up. **FLOWERING:** May–Oct. **RANGE:** FL west to TX, north to MN, QC, and VT. **SIMILAR TO** Cutleaf Coneflower (*Rudbeckia laciniata*), page 111.

Eastern Coneflower

Rudbeckia fulgida Aiton
Aster Family (Asteraceae)

PERENNIAL HERB found in dry to wet woods, prairies, stream banks, and roadsides in north Georgia. **STEMS:** Up to 4 feet tall, smooth or hairy. **BASAL LEAVES:** 2–12 inches long and ½–3 inches wide with long stalks, oval or lance-shaped, toothed or entire, rough-hairy, brittle. **STEM LEAVES:** Similar to basal leaves but gradually reduced in size up the stem, alternate, upper leaves without stalks. **FLOWER HEADS:** 1½–2½ inches wide, with 10–15 dark yellow ray flowers around a dome-shaped disk of tiny, brownish-purple flowers; 2 or 3 whorls of green, spreading or downcurved bracts surround the base of the head. **FLOWERING:** Aug–Oct. **RANGE:** FL west to TX, north to WI, ON, and MA. **SIMILAR TO** Mohr's Coneflower (*Rudbeckia mohrii*), which occurs in wet pine savannas in the Coastal Plain; it has smooth stems and long, smooth, grasslike leaves.

Black-Eyed Susan

Rudbeckia hirta Linnaeus
Aster Family (Asteraceae)

BIENNIAL OR SHORT-LIVED PERENNIAL HERB found in dry woodlands, pastures, prairies, and roadsides throughout Georgia. **STEMS:** Up to 3 feet tall, with long, stiff hairs. **BASAL LEAVES:** 3–12 inches long and ¼–3 inches wide, oval or lance-shaped, toothed or entire, rough-hairy, grayish-green, with long stalks. **STEM LEAVES:** Gradually reduced in size up the stem, narrowly lance-shaped, alternate. **FLOWER HEADS:** 1½–4½ inches wide, with 8–16 dark yellow ray flowers around a dome-shaped disk of tiny, brownish-purple flowers; the ray flowers often have a maroon splotch at the base; 2 or 3 whorls of green, spreading or downcurved bracts surround the base of the head. **FLOWERING:** May–Jul. **RANGE:** Throughout North America. **SIMILAR TO** Woolly Coneflower (*Rudbeckia mollis*), which occurs in southeast Georgia sandhills; it flowers Aug–Oct and has softly hairy or woolly stems and leaves.

Cutleaf Coneflower

Rudbeckia laciniata Linnaeus
Aster Family (Asteraceae)

PERENNIAL HERB found in bottomlands, seepages, and ditches in north Georgia and a few Coastal Plain counties. **STEMS:** 2–10 feet tall, smooth, with a white, waxy coating. **LEAVES:** Up to 18 inches long and 10 inches wide near the base of the stem, reducing in size up the stem, alternate, midstem leaves with 3–7 large lobes, upper leaves with fewer or no lobes, mostly smooth, drooping on long leaf stalks. **FLOWER HEADS:** Up to 1½ inches wide, with 5–13 yellow ray flowers around a central disk of tubular, yellowish-green flowers, the disk about 1 inch high and wide. **FLOWERING:** Jul–Oct. **RANGE:** FL west to TX, north to MB and NB. The sparsely flowered heads are **SIMILAR TO** those of Wing-stem (*Verbesina alternifolia*), page 124. Other *Rudbeckia* species are "black-eyed Susans" and have dark brown disk flowers.

Toothed White-Topped Aster

Sericocarpus asteroides (Linnaeus) B.S.P.
SYNONYM: *Aster paternus* Cronquist
Aster Family (Asteraceae)

PERENNIAL HERB found in dry woodlands and sandhills, on roadsides, and around rock outcrops in north Georgia and a few Coastal Plain counties. **STEMS:** Up to 2 feet tall, hairy. **BASAL LEAVES:** Up to 6 inches long and 1¾ inches wide, oval to lance-shaped with winged stalks, toothed near the tip, usually hairy; stem leaves alternate, fewer and smaller than basal leaves, and without stalks. **FLOWER HEADS:** About ½ inch wide, with 3–7 white ray flowers surrounding a central disk of tiny white, cream, or pink-tinged flowers; several series of whitish bracts with spreading green tips surround the base of the head. **FLOWERING:** Jun–Aug. **RANGE:** FL west to LA, north to OH and ME. **SIMILAR TO** Narrow-leaf White-topped Aster (*Sericocarpus linifolius*), below, and Twisted Leaf White-topped Aster (*S. tortifolius*), page 113.

Narrow-Leaf White-Topped Aster

Sericocarpus linifolius (Linnaeus) B.S.P.
SYNONYM: *Aster solidagineus* Michaux
Aster Family (Asteraceae)

PERENNIAL HERB found in dry woodlands, prairies, and clearings in north Georgia. **STEMS:** Up to 2½ feet tall, mostly smooth. **STEM LEAVES:** Up to 3 inches long and ⅓ inch wide, alternate, linear, entire, smooth, not twisted, with very short or no stalks; lower leaves withered by flowering time. **FLOWER HEADS:** In a flat-topped cluster, each head about 1 inch wide, with 2–6 white ray flowers around a central disk of tiny, whitish flowers; 3 or 4 series of green-and-white bracts surround the base of the head. **FLOWERING:** Jun–Jul. **RANGE:** GA west to LA, north to IN and NH. **SIMILAR TO** Toothed White-topped Aster (*Sericocarpus asteroides*), above, and Twisted-leaf White-topped Aster (*S. tortifolius*), page 113.

Twisted-Leaf White-Topped Aster

Sericocarpus tortifolius (Michaux) Nees
SYNONYM: *Aster tortifolius* Michaux
Aster Family (Asteraceae)

PERENNIAL HERB found in dry woodlands and sandhills and on roadsides in the Coastal Plain and a few north Georgia counties. **STEMS:** Up to 3 feet tall, hairy. **LEAVES:** Up to 1½ inches long and ½ inch wide, alternate, oval with broad, rounded tips, entire, rough-hairy, slightly sticky, twisted at the base so they are held perpendicular to the ground. **FLOWER HEADS:** In large, flat-topped clusters at the top of the stem, each head about ½ inch wide, with 2–7 white ray flowers surrounding a central disk of tiny, dull white flowers; several series of tiny, whitish, green-tipped bracts surround the base of the head. **FLOWERING:** Aug–Oct. **RANGE:** FL west to LA, north to NC. **SIMILAR TO** Toothed White-topped Aster (*Sericocarpus asteroides*) and Narrow-leaf White-topped Aster (*S. linifolius*), page 112.

Starry Rosinweed

Silphium asteriscus Linnaeus
Aster Family (Asteraceae)

PERENNIAL HERB found in open woodlands, clearings, and prairies, and on roadsides throughout Georgia. **STEMS:** 1–5 feet tall. **LEAVES:** Up to 6 inches long and 2 inches wide, alternate or opposite, smooth or hairy, coarsely toothed or entire, tapering or rounded at the base. **FLOWER HEADS:** About 2½ inches wide, with 8–20 yellow ray flowers surrounding a central disk of tiny, yellow flowers and tiny, green bracts; 2 or 3 series of spreading, triangular bracts surround the base of the head. **FLOWERING:** Jun–Sep. **RANGE:** FL west to TX, north to MO and VA. **NOTES:** A single yellow, unbranched style, its base enclosed in a tube of purple stamens, extends beyond the top of each disk flower. **SIMILAR TO** the sunflowers, pages 86–90, which have 2-branched styles and narrow, pointed bracts around the base of the head.

Kidney-Leaf Rosinweed

Silphium compositum Michaux
Aster Family (Asteraceae)

PERENNIAL HERB found in dry forests and woodlands, sandhills, and roadsides throughout Georgia. **STEMS:** 1½–8 feet tall, smooth, often with a white, waxy coating. **LEAVES:** Up to 14 inches long and 20 inches wide with long stalks, in a cluster at the base of the stem, toothed or deeply lobed, smooth or hairy, often with a red midvein. **FLOWER HEADS:** Up to 2 inches wide, with 6–12 yellow ray flowers around a central disk of tiny, yellow flowers; 2 or 3 series of oval, green bracts encircle the base of the head. **FLOWERING:** May–Sep. **RANGE:** FL west to AL, north to TN, WV, and VA. **SIMILAR TO** Prairie-dock (*Silphium terebinthinaceum*), which occurs in glades and prairies in northwest Georgia; its heads are larger, up to 3½ inches wide, with 13–21 ray flowers.

Bear's-Foot, Yellow Leaf-Cup

Smallanthus uvedalius (Linnaeus) Mackenzie ex Small
SYNONYM: *Polymnia uvedalia* (Linnaeus) Linnaeus
Aster Family (Asteraceae)

PERENNIAL HERB found in bottomlands, moist forests, and disturbed areas in north Georgia and a few Coastal Plain counties. **STEMS:** 3–10 feet tall, mostly smooth. **LEAVES:** Up to 14 inches long and wide, opposite, oval to triangular, toothed and lobed, hairy, with broadly winged stalks up to 5 inches long, often with a "cup" at the base. **FLOWER HEADS:** Up to 3 inches wide, with 7–25 yellow ray flowers around a central disk of tiny, yellow flowers; 4–6 oval, green bracts surround the base of the head and are visible between the ray flowers. **FLOWERING:** Jul–Oct. **RANGE:** FL west to TX, north to IL and NY. The combination of yellow flower heads and winged stalks with "leaf cups" is **SIMILAR TO** no other wildflower in Georgia.

Tall Goldenrod

Solidago altissima Linnaeus
SYNONYM: *Solidago canadensis* Linnaeus var. *scabra* Torrey & Gray
Aster Family (Asteraceae)

PERENNIAL HERB found in fields, roadsides, and rights-of-way throughout Georgia. **STEMS:** 2–6½ feet tall, hairy. **LEAVES:** Up to 6 inches long and 1¼ inches wide at midstem, alternate, entire or slightly toothed, hairy, with no or very short stalks. **FLOWER CLUSTERS:** A large, more or less pyramid-shaped cluster at the top of the stem, with many arching branches, the flower heads held in rows on the upper sides of the branches. **FLOWER HEADS:** About ⅓ inch wide, with 5–17 yellow ray flowers surrounding a center of tiny, yellow disk flowers. **FLOWERING:** Aug–Oct. **RANGE:** Most of North America. **NOTES:** Tall Goldenrod spreads by underground stems and forms large patches in disturbed areas. **SIMILAR TO** Smooth Goldenrod (*Solidago gigantea*), which occurs throughout Georgia in disturbed habitats and bottomlands; its stems are smooth and often have a white, waxy coating.

Blue-Stem Goldenrod, Wreath Goldenrod

Solidago caesia Linnaeus
Aster Family (Asteraceae)

PERENNIAL HERB found in moist forests in north Georgia and a few Coastal Plain counties. **STEMS:** Up to 3 feet tall, usually solitary, not angled, gray-green, blue-green, or purplish. **LEAVES:** Up to 4 inches long and ¾ inch wide, becoming smaller up the stem, alternate, lance-shaped, toothed, sparsely hairy. **FLOWER CLUSTERS:** Held in the axils of the midstem and upper leaves, with several flower heads per cluster. **FLOWER HEADS:** About ¼ inch across, with 1–6 yellow ray flowers surrounding a center of tiny, yellow disk flowers. **FLOWERING:** Aug–Oct. **RANGE:** FL west to TX, north to OK, ON, and ME. **SIMILAR TO** Curtis's Goldenrod (*Solidago curtisii*), which occurs in similar habitats in north Georgia; it has bright green, angled, grooved stems.

Erect Goldenrod

Solidago erecta Pursh
Aster Family (Asteraceae)

PERENNIAL HERB found in dry woodlands and dry, open disturbed areas in north Georgia and a few Coastal Plain counties. **STEMS:** 1–4 feet tall, smooth below the flower cluster. **LEAVES:** Basal leaves up to 12 inches long and 2 inches wide near the base of the stem, oval to lance-shaped, toothed, hairy only on the margins and on the winged, tapering stalks; stem leaves smaller, alternate. **FLOWER CLUSTERS:** An elongated spike with short, widely spaced branches. **FLOWER HEADS:** 1–10 per branch, each head about ¼ inch high and wide, with 5–9 yellow ray flowers surrounding a center of tiny, yellow disk flowers. **FLOWERING:** Aug–Oct. **RANGE:** GA west to MS, north to IN and CT. **SIMILAR TO** Hairy Goldenrod (*Solidago hispida*), which occurs in similar habitats in the north Georgia mountains; it has hairy stems and leaves.

Gray Goldenrod

Solidago nemoralis Aiton
Aster Family (Asteraceae)

PERENNIAL HERB found in prairies and disturbed areas in north Georgia and the upper Coastal Plain. **STEMS:** Up to 3 feet tall, covered with short, white hairs. **BASAL LEAVES:** Up to 4 inches long and 1½ inches wide, spoon-shaped, toothed, hairy, tapering to winged stalks. **STEM LEAVES:** Up to 2 inches long, alternate, hairy, with no stalks and clusters of small leaves in the axils. **FLOWER CLUSTERS:** About as wide as long, with nodding branches bearing many flower heads. **FLOWER HEADS:** About ¼ inch wide, with 5–11 yellow ray flowers surrounding a center of tiny, yellow disk flowers. **FLOWERING:** Jun–Nov. **RANGE:** FL west to TX, north to MB and NS. **SIMILAR TO** Forest Goldenrod (*Solidago arguta*), which occurs in woodlands throughout Georgia; its stems are smooth, its leaf bases are rounded, and there are few or no leaf clusters in the leaf axils.

Sweet Goldenrod, Licorice Goldenrod

Solidago odora Aiton
Aster Family (Asteraceae)

PERENNIAL HERB found in dry woodlands, prairies, and clearings throughout Georgia. **STEMS:** 2–4 feet tall, with lines of hairs running between the bases of the leaves. **LEAVES:** Up to 4⅓ inches long and ¾ inch wide, elliptic or lance-shaped, entire, mostly smooth, with a single main vein and translucent gland dots visible when backlit; smells like licorice when crushed. **FLOWER CLUSTERS:** Roughly pyramid-shaped, with many short, arching branches bearing heads on the upper sides. **FLOWER HEADS:** About ¼ inch wide, with 3–6 yellow ray flowers surrounding a center of tiny, yellow disk flowers. **FLOWERING:** Jul–Oct. **RANGE:** FL west to TX, north to OK and NH. **SIMILAR TO** Wrinkle-leaf Goldenrod (*Solidago rugosa*), which grows in forests and disturbed areas in north and southwest Georgia; its leaves have a network of deeply etched veins and are toothed.

Seaside Goldenrod

Solidago sempervirens Linnaeus
Aster Family (Asteraceae)

PERENNIAL HERB found on dunes, edges of salt marshes, coastal grasslands, and other coastal habitats. **STEMS:** 1½–6½ feet tall, succulent, smooth, arising from a woody, branching base. **BASAL LEAVES:** Up to 16 inches long and 2½ inches wide, narrowly oval or lance-shaped, tapering to a long, winged stalk; entire, waxy, succulent, smooth. **STEM LEAVES:** Similar to basal leaves but smaller. **FLOWER CLUSTERS:** Pyramid- or club-shaped, with downcurved branches bearing dozens of flower heads on their upper sides. **FLOWER HEADS:** About ½ inch wide, with 8–17 yellow ray flowers surrounding a center of tiny, yellow disk flowers. **FLOWERING:** Aug–Nov. **RANGE:** FL west to TX, north to ON and NL, mostly coastal. With its succulent, salt-resistant leaves and stems, Seaside Goldenrod is not **SIMILAR TO** any other goldenrod. Hybrids with Wand Goldenrod (p. 118) may be intermediate in appearance.

Wand Goldenrod

Solidago stricta Aiton
Aster Family (Asteraceae)

PERENNIAL HERB found in fresh and brackish marshes, bogs, pine flatwoods, and moist Longleaf Pine forests. **STEMS:** 1–6½ feet tall, smooth, branched only if damaged. **BASAL LEAVES:** Up to 12 inches long and 1 inch wide, thick, firm, smooth except for roughened margins, with a rounded tip and a winged stalk that sheaths the stem; stem leaves smaller, reduced to small bracts near the top. **FLOWER CLUSTERS:** An elongated spike up to 3 feet long. **FLOWER HEADS:** About ¼ inch high and wide, with 3–7 yellow ray flowers surrounding a center of tiny, yellow disk flowers. **FLOWERING:** Sep–Nov. **RANGE:** FL west to TX, north to DE. **SIMILAR TO** Southern Bog Goldenrod (*Solidago gracillima*), which has short-branched flower clusters and toothed basal leaves. Hybrids with Seaside Goldenrod (p. 117) may be intermediate in appearance.

Sow Thistle

Sonchus oleraceus Linnaeus
Aster Family (Asteraceae)

ANNUAL HERB found in gardens, roadsides, and other disturbed areas throughout Georgia. **STEMS:** ½–5 feet tall, usually smooth, with milky juice. **LEAVES:** Up to 14 inches long and 6 inches wide, alternate, toothed or divided into several toothed segments, smooth except for a few prickles on the margins, with pointed lobes strongly clasping the stem. **FLOWER HEADS:** Dandelion-like, ½–1 inch wide, with many yellow ray flowers (no disk flowers). **FLOWERING:** Jul–Oct. **RANGE:** Native to Europe and spread throughout North America. **SIMILAR TO** Prickly Sowthistle (*Sonchus asper*), also a European weed; it has very prickly leaves, and the lobes clasping the stem are rounded at the tips; it flowers Mar–Jul.

Where Have All the Asters Gone?

You won't find a single plant in this field guide with the current scientific name *Aster*. The name appears many times as a synonym, but our old friend *Aster* is no longer used for any native Georgia asters, all of which have new scientific names. What are these modern botanists thinking? As it turns out, the naming of asters has tormented plant taxonomists for at least 150 years. With the advent of molecular research, we now know that North American asters are not closely related to the *Aster*s in Europe and Asia. North American asters are themselves a diverse bunch—diverse enough, in fact, to warrant separation into several different genera. Most of the "new" genus names have been around for a while but never quite made it into the big leagues, until now. What does this mean for the average wildflower lover in Georgia? Not much. Most of Georgia's asters wound up in the genus *Symphyotrichum* (pronounced sym-fee-OT-ree-cum), so you could get by learning just one new genus. If you don't care about the complexities of plant taxonomy, you're in good company. The plants themselves haven't changed, and the common name aster is still just fine.

Ampelaster (1 species)
Doellingeria (3 species)
Eurybia (10 species)
Ionactis (1 species)
Oclemena (2 species)
Sericocarpus (3 species)
Symphyotrichum (27 species)

Eastern Silvery Aster

Symphyotrichum concolor (Linnaeus) G. L. Nesom
SYNONYM: *Aster concolor* Linnaeus
Aster Family (Asteraceae)

PERENNIAL HERB found in sandhills, dry woodlands, and prairies throughout Georgia. **STEMS:** Up to 3 feet tall, smooth or hairy. **LEAVES:** Up to 1½ inches long and ½ inch wide, oblong or lance-shaped, entire, silvery-hairy on both sides, pressed upward against the stem, without stalks. **FLOWER CLUSTERS:** Up to 8 inches long, narrow and unbranched, with stalked flower heads. **FLOWER HEADS:** About 1 inch wide, with 7–12 blue to purple ray flowers surrounding a disk of tiny, yellow (aging to red) flowers; small, silky-hairy bracts form a cup around the base. **FLOWERING:** Aug–Nov. **RANGE:** FL west to LA, north to KY and MA. **SIMILAR TO** Savannah Grass-leaved Aster (*Eurybia paludosa*), which occurs in wet savannas in the Coastal Plain; it has green, grasslike leaves up to 4 inches long and 15–35 ray flowers per head.

Heart-Leaf Aster, Blue Wood Aster

Symphyotrichum cordifolium (Linnaeus) G. L. Nesom

SYNONYM: *Aster cordifolius* Linnaeus

Aster Family (Asteraceae)

PERENNIAL HERB found in moist, upland forests in north Georgia, mostly in the mountains. **STEMS:** Up to 4 feet tall, straight or zigzag. **LEAVES:** Up to 6 inches long and 3 inches wide, alternate, toothed, hairy or smooth; midstem leaves are heart-shaped with long, slender stalks, upper leaves are oval with short, winged stalks. **FLOWER HEADS:** About ½ inch wide, with 8–20 pale purple ray flowers surrounding a central disk of tiny, yellow flowers (aging to red); narrow bracts form a cup around the base of the head, each with a dark green midrib and purplish tip. **FLOWERING:** Aug–Oct. **RANGE:** GA west to OK, north to MB and PEI. **SIMILAR TO** White Arrow-leaf Aster (*Symphyotrichum urophyllum*, synonym *Aster sagittifolius*), which occurs in similar habitats throughout Georgia; its ray flowers are white or pink.

Confusing White-Flowered Fall Asters

Birders have their confusing fall warblers, botanists have our confusing fall asters. There are several white-flowered asters in the genus *Symphyotrichum* that are conspicuous in the fall in dry to wet woodlands, roadsides, old fields, ditches, and other disturbed areas throughout Georgia. What they have in common:

- A much-branched, bushy growth form, erect or sprawling.
- Narrow, almost needle-like leaves on the flowering branches and much larger leaves on their main stems; sometimes several leaf sizes are found together on a single branch.
- Leaves of the main stems are usually withered by flowering time.
- The flower heads have white (rarely lavender) ray flowers and yellow disk flowers that turn red with age. Narrow, whitish bracts form a small cup around the base of the head, each bract with a green midrib or patch.

No single trait distinguishes any particular species, but taking several traits into consideration can usually lead to a sure identification. Here is a summary of distinguishing traits for four of the most common of the confusing white-flowered, fall-blooming asters.

BUSHY ASTER (*Symphyotrichum dumosum*, synonym *Aster dumosus*). Flower heads about ¾ inch wide, with 15–33 rays, bracts blunt-tipped. **Leaves of branches and branchlets thickish and stiff**, blunt-tipped, sometimes bent downward, roughly hairy.

CALICO ASTER, Starved Aster, Goblet Aster, Star Aster, One-sided Aster (*Symphyotrichum lateriflorum*, synonym *Aster lateriflorus*). Flower heads are about ⅓ inch wide, with **relatively few (8–12) rays** and blunt-tipped bracts. **Heads are often held on one side of the branches.** The main stem and flowering branches may be smooth or shaggy-hairy.

FROSTED ASTER, Hairy Aster, Old Field Aster, White Heath Aster, Awl Aster (*Symphyotrichum pilosum*, synonym *Aster pilosus*). Flower heads are about ¾ inch wide, usually with **30 (6–35) rays** and sharply pointed bracts. Tiny white hairs cover the stems, flowering branches, and leaves, giving the plant a slightly "frosted" look (sometimes the "frost" is very sparse).

SMALL WHITE ASTER (*Symphyotrichum racemosum*, synonym *Aster vimineus*). This aster is found in the **wettest habitats** of all these. **Flower heads about ¾ inch wide,** with 7–20 rays. Bracts pointed, but not sharply so. Stems covered with tiny hairs.

Bushy Aster
(*Symphyotrichum dumosum*)

Frosted Aster
(*Symphyotrichum pilosum*)

Calico Aster
(*Symphyotrichum lateriflorum*)

Small White Aster
(*Symphyotrichum racemosum*)

Clasping Aster

Symphyotrichum patens (Aiton) G. L. Nesom
SYNONYM: *Aster patens* Aiton
Aster Family (Asteraceae)

PERENNIAL HERB found in dry, open woodlands in north Georgia and a few Coastal Plain counties. **STEMS:** Up to 4 feet tall, erect or sprawling, hairy. **LEAVES:** Up to 2½ inches long and ¾ inch wide, much smaller on the flowering branches, alternate, thick and firm, rough-hairy, oval or lance-shaped, leaf bases nearly encircling the stem. **FLOWER HEADS:** Up to 1¾ inches wide, with 12–24 blue or purple ray flowers around a central disk of tiny, yellow flowers (turning red with age); glandular-hairy, red-tipped bracts surround the base of the head. **FLOWERING:** Aug–Oct. **RANGE:** FL west to TX, north to KS and ME. **SIMILAR TO** Appalachian Clasping Aster (*Symphyotrichum phlogifolium*), which occurs in moist forests in the north-central mountains of Georgia; its leaves are thin and pliable, and the disk flowers are white with purple tips.

Purple-Stem Aster

Symphyotrichum puniceum (Linnaeus) Löve & Löve
Aster Family (Asteraceae)

PERENNIAL HERB found in wetlands in north Georgia. **STEMS:** 1–8 feet tall, stout, purple, usually much-branched, upper stem very hairy, lower stem less so. **UPPER STEM LEAVES:** 2–8 inches long and ½–1½ inches wide, alternate, oblong to lance-shaped, entire to slightly toothed, hairy to smooth, clasping the stem; lower leaves withered by flowering time. **FLOWER HEADS:** About 1 inch wide, with 20–50 bluish-purple ray flowers surrounding a central disk of tiny, yellow or red flowers; narrow, green, spreading bracts form a cup under the head. **FLOWERING:** Sep–Oct. **RANGE:** GA west to TX, north to ND and ME; Canada. **SIMILAR TO** Wavy-leaf Aster (*Symphyotrichum undulatum*), which occurs in dry woodlands and on roadsides mostly in north Georgia; its heads have 10–25 rays and pale, erect bracts with an elongated, green patch.

Walter's Aster

Symphyotrichum walteri (Alexander) G. L. Nesom
SYNONYM: *Aster walteri* Alexander
Aster Family (Asteraceae)

PERENNIAL HERB found in sandhills, Longleaf Pine woodlands, and pine flatwoods in the Coastal Plain. **STEMS:** Up to 3 feet tall, erect or sprawling, smooth or slightly hairy. **STEM LEAVES:** Up to 1¼ inches long and ½ inch wide, lance- or triangle-shaped, spreading or turned sharply downward at the tip, thick, rough-hairy. **FLOWER HEADS:** About ½ inch wide, with 11–26 blue or pale purple ray flowers surrounding a central disk of tiny, yellow flowers; several series of narrow, pointed bracts form a cup at the base of the head; the bracts are whitish with a diamond-shaped, green patch. **FLOWERING:** Oct–Dec. **RANGE:** GA, FL, SC, and NC. **SIMILAR TO** Scale-leaf Aster (*Symphyotrichum adnatum*), which occurs in the same habitats; all of its leaves turn upward, and its stems are very hairy.

Squarehead

Tetragonotheca helianthoides Linnaeus
Aster Family (Asteraceae)

PERENNIAL HERB found in sandhills, pine flatwoods, and other dry woodlands in the Coastal Plain. **STEMS:** 1–3 feet tall, sticky-hairy, usually unbranched but often in clumps of several stems. **LEAVES:** Up to 8 inches long and 4 inches wide, opposite, oval, lance-shaped, or triangular, with leaf tissue narrowing to a winged stalk, hairy and gland-dotted. **FLOWER HEADS:** About 3 inches wide, with 6–12 toothed, yellow ray flowers surrounding a central disk of many small, yellow flowers; 4 large, green bracts surround the base of the head and form a square that shows between the widely spaced rays. **FLOWERING:** Apr–Jul. **RANGE:** FL west to MS, north to TN and VA. **NOTES:** Unopened flower heads look like hairy green pyramids formed by the 4 large bracts. With its "square heads," this plant is **SIMILAR TO** no other wildflower in Georgia.

Wing-Stem

Verbesina alternifolia (Linnaeus) Britton ex Kearney
Aster Family (Asteraceae)

PERENNIAL HERB found in floodplains, woodland edges, and disturbed areas in north Georgia. STEMS: 3–10 feet tall, smooth or hairy, winged with leaf tissue that runs between the bases of the leaf stalks. LEAVES: Up to 10 inches long and 3 inches wide, alternate, lance-shaped, toothed or entire. FLOWER CLUSTERS: Large, open, ragged-looking. FLOWER HEADS: Up to 2½ inches wide, with 2–10 drooping, yellow ray flowers surrounding a domed center of spreading, greenish-yellow disk flowers. FLOWERING: Aug–Oct. RANGE: FL west to TX, north to WI, ON, and RI. NOTES: Wing-stem is a larval host plant or nectar source for many butterflies and bees. Deer and rabbits will not eat the bitter leaves. SIMILAR TO Southern Crownbeard (*Verbesina occidentalis*), which also has winged stems and yellow flower heads but has opposite leaves.

Frostweed

Verbesina virginica Linnaeus
Aster Family (Asteraceae)

PERENNIAL HERB found in floodplains, woodland edges, and clearings in north Georgia and several Coastal Plain counties. STEMS: 1½–8 feet tall, hairy, winged with leaf tissue that runs between the bases of the leaf stalks. LEAVES: Up to 5 inches long and 2½ inches wide, alternate, oval or lance-shaped, hairy; with entire, wavy, toothed, or lobed margins. FLOWER HEADS: In dense, branched clusters at the top of the stem, each head with 5 white, oval ray flowers surrounding 8–12 tiny, white, 5-lobed disk flowers; 2 whorls of hairy, green-and-white bracts form a cup around the base of the head. FLOWERING: Jul–Dec. RANGE: FL west to TX, north to KS and MD. SIMILAR TO Woolly-white (*Hymenopappus scabiosaeus*), page 92, and Appalachian Flat-topped White Aster (*Doellingeria infirma*), page 73.

Narrow-Leaved Ironweed

Vernonia angustifolia Michaux
Aster Family (Asteraceae)

PERENNIAL HERB found in sandhills and other dry woodlands in the Coastal Plain. **STEMS:** Up to 3 feet tall, hairy. **STEM LEAVES:** Up to 4½ inches long and ¼ inch wide, alternate, linear, pointed, slightly rough-hairy, with the margins curled under; basal leaves few or none. **FLOWER HEADS:** In large, branched clusters at the top of the stem, each head about 1 inch wide with 12–30 spreading, tubular, magenta flowers; several series of narrow, pointed, purple-green bracts surround the base of the head. **FLOWERING:** Jun–Aug. **RANGE:** FL west to LA, north to NC. **SIMILAR TO** Short-stem Ironweed (*Vernonia acaulis*), which occurs in similar habitats; it has a basal rosette of oval, 4-inch-long leaves, and its stem leaves are reduced to small bracts.

Tennessee Ironweed

Vernonia flaccidifolia Small
Aster Family (Asteraceae)

PERENNIAL HERB found in upland deciduous forests and woodlands in northwest Georgia. **STEMS:** 3–6½ feet tall, waxy blue-green, smooth. **LEAVES:** 4–12 inches long and ½–2½ inches wide, alternate, lance-shaped, toothed, hairy but not gland-dotted. **FLOWER HEADS:** In large, much-branched clusters at the top of the stem, each head about 1 inch wide with 15–25 spreading, tubular, magenta flowers and several series of purple-and-green, blunt-tipped bracts surrounding the base of the head. **FLOWERING:** Jun–Sep. **RANGE:** GA, AL, and TN. **SIMILAR TO** Appalachian Ironweed (*Vernonia glauca*), which occurs in similar habitats in northeast Georgia; the bracts of its flower heads have long, pointed tips. Georgia Ironweed (*V. pulchella*) occurs in sandhills in southeast Georgia; its stems are hairy, and its leaves clasp the stem. Also see Tall Ironweed (*Vernonia gigantea*), page 126.

Tall Ironweed

Vernonia gigantea (Walter) Trelease
Aster Family (Asteraceae)

PERENNIAL HERB found in floodplains and on stream banks throughout Georgia. **STEMS:** 3–12 feet tall, stout, ribbed, hairy or smooth. **LEAVES:** Up to 12 inches long and 3 inches wide, alternate, lance-shaped, toothed or entire, lower surface usually rough-hairy. **FLOWER HEADS:** In large, branched clusters at the top of the stem, each head about ¾ inch wide with 12–30 spreading, tubular, magenta flowers; several series of blunt-tipped, purple-green bracts surround the base of the head. **FLOWERING:** Aug–Oct. **RANGE:** FL west to TX, north to NE, ON, and PA. **SIMILAR TO** New York Ironweed (*Vernonia noveboracensis*), which occurs in similar habitats; it is less than 7 feet tall and its flower heads are larger, with up to 65 magenta flowers; the bracts surrounding the base of the flower heads have long, slender tips.

Jewelweed, Orange Touch-Me-Not

Impatiens capensis Meerburgh
Balsam Family (Balsaminaceae)

ANNUAL HERB found in floodplains and other wetlands in north Georgia and a few Coastal Plain counties. **STEMS:** Up to 5 feet tall, smooth, glossy, hollow. **LEAVES:** Up to 5 inches long and 2 inches wide, alternate, oval, soft, dull greenish-gray, with scalloped margins and long stalks. **FLOWERS:** About 1 inch long, yellow and orange with dark spots and streaks, dangling on slender stalks, 2-lipped, with the large lower lip extending back into a curved spur. **FRUIT:** A fleshy, green capsule about 1 inch long that opens explosively, flinging seeds several feet. **FLOWERING:** Jun–Nov. **RANGE:** Most of North America. **NOTES:** The flowers are pollinated by hummingbirds and long-tongued insects that probe the spur for nectar. **SIMILAR TO** Yellow Touch-me-not (*Impatiens pallida*), which occurs in moist coves, stream banks, and seepages in the mountains; it has yellow flowers.

Saltwort

Batis maritima Linnaeus
Saltwort Family (Bataceae)

PERENNIAL HERB found in tidal marshes and dune swales on Georgia's barrier islands. **STEMS:** Up to 3 feet tall, trailing or arching, succulent, pale green or yellow-green; where tips of arching branches touch the ground, they root and send up new shoots, forming large colonies. **LEAVES:** 2–8 inches long, less than ¼ inch wide, opposite, succulent, curved, round or half-round in cross-section. **FLOWER CLUSTERS:** Fleshy spikes less than ½ inch long held in the leaf axils. **FLOWERS:** Female and male flowers are on separate plants; female flowers lack petals; male flowers have 4 or 5 minute, white petals. **FRUIT:** Less than ½ inch long, knobby, spongy, green turning purplish-black; fruits and seeds are buoyant and can float for weeks. **FLOWERING:** Jun–Jul. **RANGE:** FL west to TX, north to SC; West Indies, Central and South America. **SIMILAR TO** Woody Glasswort (*Sarcocornia pacifica*), page 27.

Blue Cohosh

Caulophyllum thalictroides (Linnaeus) Michaux
Barberry Family (Berberidaceae)

PERENNIAL HERB found in cove forests and other rich, deciduous forests in the north Georgia mountains. **STEMS:** 1–3 feet tall, smooth, blue-green, with a waxy coating. **LEAVES:** 1 large leaf (rarely 2) at the top of the stem, divided into many waxy, veiny, blue-green leaflets, each 1–3 inches long and irregularly toothed or lobed. **FLOWERS:** About ½ inch across, with 6 yellow, green, or purple sepals and 6 tiny, nectar-producing glands at the base of the sepals; there are no petals. **FRUITS:** Do not develop; instead, each flower produces 2 fleshy, dark blue seeds held on stout stalks. **FLOWERING:** Apr–May. **RANGE:** GA, AL, north to MB, QC, and NS. **NOTES:** Most parts of Blue Cohosh are toxic to animals. **SIMILAR TO** no other wildflower in our area.

Umbrella Leaf

Diphylleia cymosa Michaux
Barberry Family (Berberidaceae)

PERENNIAL HERB found in cove forests, boulderfields, seepages, spray cliffs, and creeks at high elevations in the Blue Ridge. **STEMS:** 1–3 feet tall. **LEAVES:** 2 per plant, each up to 20 inches long and wide, divided into 2 lobed and toothed segments; the stalk is attached near the center of the leaf between the 2 segments. **FLOWER CLUSTERS:** Up to 4 inches wide, held on a long stalk that arises in the fork between the leaves. **FLOWERS:** About 1 inch wide, with 6 white petals. **FRUITS:** Up to ½ inch wide, round, dark blue, fleshy, held on red stalks. **FLOWERING:** May–Jun. **RANGE:** GA, SC, NC, TN, and VA. **NOTES:** Nonflowering plants have only 1 leaf, which may be up to 3 feet wide. Leaves are **SIMILAR TO** those of May-apple (*Podophyllum peltatum*), below.

May-Apple

Podophyllum peltatum Linnaeus
Barberry Family (Berberidaceae)

PERENNIAL HERB found in bottomlands and on moist, upland slopes in north Georgia and a few Coastal Plain counties. **STEMS:** Up to 2 feet tall, forming colonies by spreading, underground stems. **LEAVES:** Up to 16 inches across, with 5–9 lobes; stalks up to 6 inches long, attached to the center of the leaf blade; young plants have 1 leaf, mature plants have 2. **FLOWERS:** Held in the fork between leaf stalks, 1–2 inches across, with 6 or 9 white, waxy petals, a round, green ovary, and 12 or 18 stamens. **FRUIT:** A yellowish-green, fleshy, oval berry up to 2 inches long and 1½ inches wide. **FLOWERING:** Mar–May. **RANGE:** FL west to TX, north to MN and NH. **NOTES:** Only mature, 2-leaved plants flower; it takes each plant 4 or 5 years to reach maturity. **SIMILAR TO** no other wildflower in Georgia.

Cross-Vine

Bignonia capreolata Linnaeus
SYNONYM: *Anisostichus capreolata* (Linnaeus) Bureau
Bignonia Family (Bignoniaceae)

PERENNIAL WOODY VINE found in bottomlands, moist uplands, and fencerows throughout Georgia. **STEMS:** High-climbing, up to 50 feet long; bark of older vines with lengthwise splits and many lenticels; pith cross-shaped. **LEAVES:** Opposite, each leaf with 2 lance-shaped leaflets and 1 forking, twining tendril; leaflets are up to 6 inches long and 3 inches wide. **FLOWERS:** Up to 2½ inches long, tubular, orange-red with 5 yellow, spreading lobes. **FRUIT:** A flattened, green, beanlike pod 4–8 inches long with 4 stacks of flat, winged seeds. **FLOWERING:** Mar–May. **RANGE:** FL west to TX, north to MO and MD. **NOTES:** Some horticultural varieties have solid red or orange flowers. Tips of the tendrils have tiny, adhesive disks that anchor the climbing plant. **SIMILAR TO** Trumpet-creeper (*Campsis radicans*), below, and Yellow Jessamine (*Gelsemium sempervirens*), page 251.

Trumpet-Creeper, Cow-Itch Vine

Campsis radicans (Linnaeus) Seemann
Bignonia Family (Bignoniaceae)

WOODY VINE found in bottomlands, upland thickets and woodlands, fencerows, and disturbed areas throughout Georgia. **STEMS:** High-climbing by aerial roots, up to 130 feet long and 6 inches in diameter, with shredding, pale tan bark. **LEAVES:** Opposite, each leaf with 7–15 sharply toothed leaflets up to 2¾ inches long and 1⅓ inches wide. **FLOWERS:** Up to 3¼ inches long, orange or red, tubular with 5 spreading, rounded lobes. **FRUIT:** A green, beanlike pod 4–8 inches long, splitting lengthwise when dry to release flat, winged seeds. **FLOWERING:** Jun–Jul. **RANGE:** FL west to TX, north to IA and NJ. **SIMILAR TO** Cross-vine (*Bignonia capreolata*), above.

Wild Comfrey, Wild Hound's Tongue

Cynoglossum virginianum Linnaeus
Borage Family (Boraginaceae)

PERENNIAL HERB found in moist slope forests and bottomlands in north Georgia and a few southwest Georgia counties. **STEMS:** Up to 30 inches tall, very hairy. **LOWER LEAVES:** Up to 10 inches long and 4 inches wide, entire, oval or elliptic, with long stalks. **STEM LEAVES:** Smaller than lower leaves, alternate, clasping the stem. **FLOWER CLUSTERS:** 4 coiled branches that lengthen and straighten as the flowers open and set seed. **FLOWERS:** About ¾ inch wide, pale blue or white, with a short tube and 5 overlapping lobes. **FLOWERING:** Apr–Jun. **RANGE:** FL west to TX, north to BC and NB. **NOTES:** Long, rough, bristly hairs cover all parts of the plant. **SIMILAR TO** Garden Comfrey (*Cynoglossum officinale*), a European native widely planted in gardens; its flowers are reddish-purple.

Turnsole

Heliotropium indicum Linnaeus
Borage Family (Boraginaceae)

ANNUAL HERB found in pastures, roadsides, and other disturbed areas throughout Georgia. **STEMS:** Up to 3 feet tall, with stiff, spreading hairs. **LEAVES:** Up to 4 inches long and 1½ inches wide, alternate, hairy, oval or elliptic, with long stalks, rounded or squared-off bases, wavy margins, and deeply etched veins. **FLOWER CLUSTERS:** Coiled, densely flowered, 1-sided spikes up to 6 inches long. **FLOWERS:** About ⅒ inch wide, blue or lavender, with a yellow, hairless throat. **FLOWERING:** Jul–Nov. **RANGE:** FL west to TX, north to KS and MA; native to South America. **SIMILAR TO** the native Seaside Heliotrope (*Heliotropium curassavicum*), which occurs in back dunes and salt marsh borders and has white flowers and smooth, succulent stems and leaves; and Wild Heliotrope (*Heliotropium amplexicaule*), which occurs in disturbed areas and has stalkless leaves and pink or purple flowers with hairy throats.

Hoary Puccoon

Lithospermum canescens (Michaux) Lehmann
Borage Family (Boraginaceae)

PERENNIAL HERB found in cedar glades, prairies, and dry woodlands over limestone bedrock in northwest Georgia. **STEMS:** 4–20 inches tall, often in clumps, softly hairy. **LEAVES:** Up to 2 inches long and ½ inch wide, alternate, lance-shaped or narrowly oblong, tapering at both ends, entire, softly hairy, without stalks. **FLOWER CLUSTERS:** Coiled at the tips of branches, lengthening and straightening as the flowers open. **FLOWERS:** About ½ inch across with a ¼-inch tube and 5 spreading, rounded lobes, yellow or orange. **FRUIT:** About ⅛ inch wide, hard, white, shiny. **FLOWERING:** Apr–May. **RANGE:** GA west to TX, north to SK and ON. **SIMILAR TO** Coastal Plain Puccoon (*Lithospermum caroliniense*), below.

Coastal Plain Puccoon

Lithospermum caroliniense (Walter ex J. F. Gmelin) MacMillan
Borage Family (Boraginaceae)

PERENNIAL HERB found in sandhills and other dry habitats in the Coastal Plain. **STEMS:** 1–3 feet tall, rough-hairy, branched at the top, often in clumps. **LEAVES:** Up to 3 inches long and ¾ inch wide, alternate, rough-hairy, linear, with blunt tips and no stalks. **FLOWER CLUSTERS:** Coiled at the tips of branches, uncurling as the flowers open. **FLOWERS:** About ¾ inch across, yellow-orange, with a ½-inch tube and 5 spreading, rounded lobes. **FRUIT:** About ⅛ inch wide, hard, white, shiny. **FLOWERING:** Apr–Jun. **RANGE:** FL west to TX, north to VA. **SIMILAR TO** Southern Stoneseed (*Lithospermum tuberosum*), which occurs in rich forests over limestone or mafic bedrock in widely scattered locations in north and southwest Georgia; it has a cluster of basal leaves and smaller, pale yellow flowers.

Spring Forget-Me-Not

Myosotis verna Nuttall
Borage Family (Boraginaceae)

ANNUAL HERB found in dry woodlands and disturbed areas in north Georgia. **STEMS:** Up to 1½ feet tall, covered with long, spreading hairs. **LEAVES:** Up to 2 inches long and ½ inch wide, smaller upward, alternate, oval or oblong with winged or no stalks and long, spreading hairs. **FLOWER CLUSTERS:** 2–8 inches long, coiled, uncurling at the tip as each new flower opens; flowers and fruits are on short stalks and spaced about ⅓ inch apart. **FLOWERS:** About ⅛ inch wide with 5 white, rounded lobes at the top of a short tube; 5 green, hairy sepals enclose the tube and the fruits. **FLOWERING:** Mar–Jul. **RANGE:** Most of North America. **SIMILAR TO** Big-seeded Forget-me-not (*Myosotis macrosperma*), which occurs in bottomland forests; its flowers and fruits are spaced ½–2 inches apart, and its fruits are not enclosed by sepals.

Virginia Marbleseed

Onosmodium virginianum (Linnaeus) Alphonse de Candolle
SYNONYM: *Lithospermum virginianum* Linnaeus
Borage Family (Boraginaceae)

PERENNIAL HERB found in dry woodlands and shell middens in the Coastal Plain, and in prairies, glades, and woodlands over limestone or mafic bedrock in north Georgia. **STEMS:** 8–32 inches tall, branched near the top. **LEAVES:** Up to 5 inches long and ¾ inch wide, alternate, oval to oblong, with 3 or 5 prominent veins. **FLOWER CLUSTERS:** Coiled, leafy spikes, uncurling at the tips as the flowers open. **FLOWERS:** About ⅓ inch long, pale yellow, tubular with 5 pointed, upright lobes. **FRUIT:** About 1/10 inch long, white, shiny. **FLOWERING:** Apr–Sep. **RANGE:** FL west to LA, north to MA. **NOTES:** Long, stiff, spreading hairs cover all parts of the plant. **SIMILAR TO** Western Marbleseed (*Onosmodium molle* ssp. *occidentale*), a Georgia Special Concern species that occurs on limestone in cedar glades in northwest Georgia; its flowers are greenish-white.

Mustard Family (Brassicaceae and Cruciferae)

Plants in the Mustard Family are often called crucifers—crucifer being the Latin word for "cross-bearing"—because of their 4-petaled, cross-shaped flowers. The flowers are usually white or yellow, sometimes pink. The leaves, roots, and rhizomes contain mustard oils that give members of this family their characteristic peppery "bite" and produce a pungent, sulfurous odor familiar to anyone who has overcooked broccoli. The fruits are distinctive capsules that are called siliques when they are long and narrow or silicles when they are short and rounded. About 60 species of crucifers occur in Georgia, many of them well known to southern gardeners, such as collards and cabbage (both *Brassica oleracea*), mustard greens (*Brassica juncea*), turnip greens (*Brassica rapa*), radish (*Raphanus sativus*), and kale (*Brassica napus*).

Most of the crucifers in Georgia are introductions from Europe and Asia that escaped from cultivation and flourish on roadsides. However, 24 species are native to Georgia, including the familiar toothworts (*Cardamine* spp.) found in springtime forests.

Mouse-Ear Cress

Arabidopsis thaliana (Linnaeus) Heynhold
Mustard Family (Brassicaceae)

ANNUAL HERB found in disturbed areas and floodplains throughout Georgia, mostly in the Piedmont. **STEMS:** 1–20 inches tall, lower half hairy, upper half smooth, branching from the base. **LEAVES:** Mostly in a basal rosette, each leaf up to 1½ inches long and ½ inch wide, oval or spoon-shaped, hairy, with a few shallow teeth; stem leaves few, alternate, smaller than basal, not toothed. **FLOWERS:** About ¼ inch wide, with 4 white, rounded petals. **FRUIT:** A smooth, slender, green pod about ½ inch long. **FLOWERING:** Feb–May. **RANGE:** Native to Eurasia, now spread throughout most of North America. **NOTES:** This species has a small number of chromosomes and completes its life cycle in 6 weeks, making it an ideal plant for lab experiments. **SIMILAR TO** Short-fruited Draba (*Draba brachycarpa*), page 138, Whitlow-grass (*Draba verna*), and Poor Man's Pepper (*Lepidium virginicum*), page 139.

Sicklepod, Canada Rock-Cress

Boechera canadensis (Linnaeus) Al-Shehbaz
SYNONYM: *Arabis canadensis* Linnaeus
Mustard Family (Brassicaceae)

BIENNIAL HERB found around rock outcrops and on rocky slopes, especially over mafic or limestone bedrock, in the Piedmont and upper Coastal Plain. **STEMS:** 1–4 feet tall, smooth except at the base. **BASAL LEAVES:** 2½–8 inches long and ½–2 inches wide, lance-shaped with widely spaced teeth, usually hairy, often withered by flowering time. **STEM LEAVES:** Similar to basal leaves, alternate, clasping the stem. **FLOWERS:** About ¼ inch long, with 4 white petals and straight or drooping stalks about ½ inch long. **FRUIT:** A narrow, flat, curved pod 1½–4 inches long, drooping when ripe. **FLOWERING:** Apr–Jul. **RANGE:** FL west to TX, north to MN and QC. **SIMILAR TO** Smooth Rock-cress (*Boechera laevigata*), which occurs in similar habitats; it has a waxy, white coating on its stem and leaves, and the flowering and fruiting stalks are nearly erect.

Harper's Sea-Rocket

Cakile harperi Small
SYNONYM: *Cakile edentula* (Bigelow) Hooker ssp. *harperi* (Small) Rodman
Mustard Family (Brassicaceae)

ANNUAL HERB found on beaches, dunes, and edges of salt marshes on Georgia's barrier islands. **STEMS:** Up to 2½ feet tall, succulent, slightly woody at the base, lower branches sprawling. **LEAVES:** Up to 6 inches long and 2¾ inches wide, alternate, succulent, with wavy or slightly toothed margins. **FLOWERS:** About ½ inch across, with 4 white, pink, or lavender petals. **FRUIT:** A fleshy, 2-segmented pod; the outer segment is ½–¾ inch long with 8 ribs. **FLOWERING:** Mar–Oct. **RANGE:** GA, FL, SC, and NC. **NOTES:** The outer fruit segment is corky—it breaks off and floats long distances; the lower segment remains attached, dropping its seed nearby. **SIMILAR TO** Northern Sea Rocket (*Cakile edentula*), which occurs from NC north to NL; its upper fruit segment is 4-angled and less than ½ inch long.

Shepherd's Purse

Capsella bursa-pastoris (Linnaeus) Medikus
Mustard Family (Brassicaceae)

ANNUAL OR BIENNIAL HERB found in fields, roadsides, gardens, and other disturbed areas throughout Georgia. **STEMS:** 4–20 inches tall, hairy, branched. **BASAL LEAVES:** Up to 4 inches long and 1¼ inches wide, deeply lobed. **STEM LEAVES:** Up to 2 inches long and ½ inch wide, alternate, arrowhead-shaped, toothed or entire, clasping the stem. **FLOWER CLUSTERS:** Up to 12 inches long, cylindrical, flowers opening from the bottom upward. **FLOWERS:** Less than ⅛ inch wide, with 4 white petals. **FRUIT:** A flat, heart-shaped pod about ¼ inch long and slightly wider, on a long stalk. **FLOWERING:** Feb–Oct. **RANGE:** Native to Europe, now globally widespread. **SIMILAR TO** Poor Man's Pepper (*Lepidium virginicum*), page 139.

Slender Toothwort, Two-Leaved Toothwort

Cardamine angustata O. E. Schulz
SYNONYM: *Dentaria heterophylla* Nuttall
Mustard Family (Brassicaceae)

PERENNIAL HERB found in rich, moist forests in north Georgia. **STEMS:** 4–16 inches tall, smooth or hairy, often purplish. **BASAL LEAVES:** With 3 oval, toothed leaflets up to 4 inches long and 2⅓ inches wide, often evergreen. **STEM LEAVES:** 2 at midstem, each with 3 toothed leaflets that are much narrower and differently shaped than the basal leaflets. **FLOWERS:** About 1 inch wide when fully open, white, lavender, or pale pink, with 4 oval petals. **FLOWERING:** Mar–May. **RANGE:** GA west to AR, north to IN and NJ. **SIMILAR TO** Cut-leaf Toothwort (*Cardamine concatenata*), page 136, which has 3 stem leaves. The basal leaflets of Broad-leaved Toothwort (*Cardamine diphylla*), page 136, are the same shape as its stem leaflets.

Cut-Leaf Toothwort

Cardamine concatenata (Michaux) O. Schwarz
SYNONYM: *Dentaria laciniata* Muhlenberg ex Willdenow
Mustard Family (Brassicaceae)

PERENNIAL HERB found in rich, moist forests in north and southwest Georgia. **STEMS:** 4–8 inches tall, mostly smooth (hairy near the top). **BASAL LEAVES:** 3–8 inches long, with 3 narrowly oblong, toothed leaflets; often absent or growing at a distance from the stem. **STEM LEAVES:** 3 leaves in a whorl at midstem, each divided into 3 leaflets about the same size and shape as the basal leaflets. **FLOWERS:** ½–1½ inches wide when fully open, with 4 white or pale pink petals, fragrant. **FLOWERING:** Feb–May. **RANGE:** FL west to TX, north to NE, ON, and ME. **SIMILAR TO** Dissected Toothwort (*Cardamine dissecta*, synonym *Dentaria multifida*), which occurs in rich forests in northwest Georgia; both its basal and stem leaves are divided into 9 or more narrow segments.

Broad-Leaved Toothwort, Two-Leaved Toothwort

Cardamine diphylla (Michaux) A. Wood
SYNONYM: *Dentaria diphylla* Michaux
Mustard Family (Brassicaceae)

PERENNIAL HERB found in rich, moist forests in the mountains and upper Piedmont of Georgia. **STEMS:** 5–16 inches tall, smooth. **BASAL LEAVES:** 2–10 inches long, with 3 oval, toothed leaflets with conspicuous pale veins, often evergreen. **STEM LEAVES:** 1 pair at midstem, each leaf with 3 leaflets similar in size and shape to the leaflets of the basal leaves. **FLOWERS:** About 1 inch wide when fully open, white (turning pink with age), with 4 oval petals. **FLOWERING:** Apr–May. **RANGE:** GA west to AR, north to MN and NS. **NOTES:** Although toothwort rhizomes are edible, digging them up is not recommended. Toothworts rarely produce seeds and rely largely on the spread of rhizomes for reproduction. **SIMILAR TO** Cut-leaf Toothwort (*Cardamine concatenata*), above, and Slender Toothwort (*C. angustata*), page 135.

Blue Ridge Bittercress

Cardamine flagellifera O. E. Schulz
Mustard Family (Brassicaceae)

PERENNIAL HERB found in rich cove forests and seepages in the Blue Ridge. **STEMS:** 4–10 inches tall, hairy near the base, smooth except near the top. **BASAL LEAVES:** Blade ½–1¾ inches wide, nearly round, on a stalk up to 5 inches long. **STEM LEAVES:** 2–5 leaves in a whorl at midstem, each divided into 3 or 5 oval or round, toothed or lobed leaflets; the top leaflet is ½–2 inches long. **FLOWERS:** About ½ inch wide when fully open, with 4 white, oval petals. **FLOWERING:** Feb–May. **RANGE:** GA, north to TN, WV, and VA. **SIMILAR TO** Mountain Bittercress (*Cardamine clematitis*), a Georgia Special Concern species that occurs in high-elevation boulderfields and rocky seepages; its stem is smooth throughout.

Hairy Bittercress

Cardamine hirsuta Linnaeus
Mustard Family (Brassicaceae)

ANNUAL OR BIENNIAL HERB found in disturbed areas throughout Georgia. **STEMS:** 4–16 inches tall, hairy at the base, much-branched. **BASAL LEAVES:** 1–7 inches long, with 5–9 oval or round leaflets; the base of the stalk is purplish and fringed with long hairs. **STEM LEAVES:** Few, similar to basal leaves but smaller. **FLOWERS:** Less than ¼ inch wide, with 4 white, oval petals. **FLOWERING:** Feb–May. **RANGE:** Native to Europe, now spread throughout much of North America. **SIMILAR TO** Wavy Bittercress (*Cardamine flexuosa*), a Eurasian weed that occurs infrequently in disturbed uplands throughout Georgia; it lacks basal leaves, its stem leaves are up to 4 inches long, its stems and leaf stalks are smooth, and the stem is curved or bent. Also see Pennsylvania Bittercress (*Cardamine pensylvanica*), page 138.

Pennsylvania Bittercress

Cardamine pensylvanica Muhlenberg ex Willdenow
Mustard Family (Brassicaceae)

ANNUAL OR BIENNIAL HERB found in wetlands throughout Georgia. **STEMS:** 6–22 inches tall, smooth, sometimes hairy near the base, slightly succulent, usually branched from the base. **BASAL LEAVES:** Usually withered by flowering time. **STEM LEAVES:** 1½–6 inches long, alternate, smooth, with 5–13 leaflets, each leaflet at least ¼ inch wide with 2 or 3 lobes or rounded teeth; the stalk is not fringed or hairy. **FLOWERS:** About ¼ inch wide, with 4 white, oval petals. **FLOWERING:** Mar–Jul. **RANGE:** Throughout North America. **SIMILAR TO** Sand Bittercress (*Cardamine parviflora*), a smaller plant that occurs in north Georgia and a few Coastal Plain counties in wet, shallow soils, especially on rock outcrops and glades; its stem leaves are less than 1½ inches long, and the leaflets are less than ¼ inch wide; the flowers are ⅛–¼ inch wide.

Short-Fruited Draba

Draba brachycarpa Nuttall ex Torrey & A. Gray
Mustard Family (Brassicaceae)

ANNUAL HERB found on granite outcrops and in woodlands and disturbed areas in north Georgia and a few Coastal Plain counties. **STEMS:** 1–8 inches tall, often branched from the base, hairy, purplish in strong sun. **BASAL LEAVES:** Up to ¾ inch long and ¼ inch wide, oval, entire, hairy. **STEM LEAVES:** Few, similar to basal leaves but smaller, alternate. **FLOWERS:** Less than ¼ inch wide, with 4 white, rounded or slightly notched petals. **FLOWERING:** Feb–Apr. **RANGE:** FL west to TX, north to KS and VA. **SIMILAR TO** Whitlow-grass (*Draba verna*), which occurs in similar habitats; its stems are leafless, and its petals are deeply notched.

Poor Man's Pepper, Peppergrass

Lepidium virginicum Linnaeus
Mustard Family (Brassicaceae)

ANNUAL HERB found in fields, roadsides, and other disturbed areas throughout Georgia. **STEMS:** 2–22 inches tall, hairy, much-branched. **BASAL LEAVES:** 1–4 inches long and ¼–1¼ inches wide, with many toothed lobes, usually withered by flowering time. **STEM LEAVES:** Up to 3 inches long and ½ inch wide, alternate, oval or lance-shaped, tapering to the base, toothed or entire. **FLOWER CLUSTERS:** Up to 4 inches long, cylindrical, flowers opening from the bottom upward. **FLOWERS:** Less than ⅛ inch wide, with 4 white petals. **FRUIT:** A flat, round pod up to ⅛ inch wide with a tiny notch in the top. **FLOWERING:** Mar–Sep. **RANGE:** Most of North America. **SIMILAR TO** Field Pepperwort (*Lepidium campestre*), which occurs in disturbed areas in north Georgia; its stem leaves are arrowhead-shaped and clasp the stem.

Wild Radish

Raphanus raphanistrum Linnaeus
Mustard Family (Brassicaceae)

ANNUAL HERB found in fields, roadsides, and other disturbed areas throughout Georgia. **STEMS:** 8–32 inches tall, usually very hairy. **BASAL LEAVES:** Up to 8 inches long and 2 inches wide, deeply lobed and divided. **STEM LEAVES:** Fewer and smaller than basal, alternate, often undivided. **FLOWERS:** 1–2 inches wide, with 4 pale yellow or white petals that are spoon-shaped with broad, rounded tips and darker veins. **FRUIT:** ¾–2½ inches long including the beak, less than ¼ inch wide, grooved lengthwise, strongly narrowed between the seeds. **FLOWERING:** Feb–Jun. **RANGE:** Native to Europe, now spread throughout North America. **SIMILAR TO** two other European species—Canola or Rapeseed (*Brassica napus*) and Turnip (*Brassica rapa*)—that are cultivated and widely escaped to roadsides. Both have smooth stems. Canola flowers are pure pale yellow; turnip flowers are darker golden yellow.

Pineland Cress

Warea cuneifolia (Muhlenberg ex Nuttall) Nuttall
Mustard Family (Brassicaceae)

PERENNIAL HERB found in sandhills and river dunes in Fall Line and southeast Georgia counties. **STEMS:** 1–2 feet tall, smooth, branched in the upper half. **LEAVES:** Up to 1½ inches long and ½ inch wide, alternate, oval or lance-shaped, rounded at the tip and tapered at the base, entire, smooth, with a very short stalk. **FLOWERS:** About 1 inch wide, in rounded clusters at the tips of branches, with 4 pink or white, spreading, spoon-shaped petals, 4 long, spreading stamens, and 4 downward-spreading sepals. **FRUITS:** Up to 2 inches long, slender, curving, long-stalked. **FLOWERING:** Jul–Sep. **RANGE:** GA, AL, FL, SC, and NC. **NOTES:** The long stamens and fruits give the flower clusters a "spidery" look. **SIMILAR TO** Slender-leaf Clammy-weed (*Polanisia tenuifolia*), which occurs in similar habitats; its flowers are white, and its narrow leaves are almost round in cross-section.

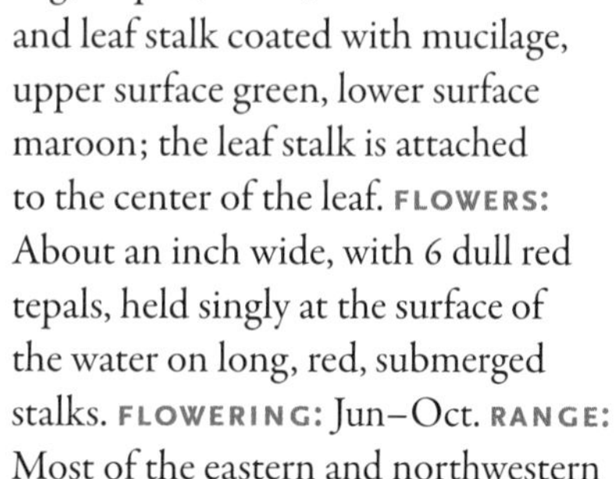

Water-Shield

Brasenia schreberi J. F. Gmelin
Water-Shield Family (Cabombaceae)

PERENNIAL HERB floating on the surface of ponds, lakes, and slow-moving streams, mostly in the Coastal Plain. **STEMS:** Usually submerged, coated with a thick layer of clear, jelly-like mucilage. **LEAVES:** Up to 5 inches long and 3 inches wide, floating, elliptic, entire, the lower surface and leaf stalk coated with mucilage, upper surface green, lower surface maroon; the leaf stalk is attached to the center of the leaf. **FLOWERS:** About an inch wide, with 6 dull red tepals, held singly at the surface of the water on long, red, submerged stalks. **FLOWERING:** Jun–Oct. **RANGE:** Most of the eastern and northwestern United States and adjacent Canada. **NOTES:** Water-shield is known in some circles as Snot-lily. **SIMILAR TO** water lilies (p. 262), but Water-shield has elliptic (not round), mucilage-coated leaves.

Eastern Prickly-Pear

Opuntia humifusa (Rafinesque) Rafinesque
SYNONYM: *Opuntia compressa* (Salisbury) J. F. Macbride
Cactus Family (Cactaceae)

SHRUB found in dry or rocky woodlands, granite outcrops, and beach dunes in the Piedmont and Coastal Plain. **STEMS:** Sprawling and forming mats or clumps, consisting of flat, oval, succulent pads up to 7 inches long and 5 inches wide; tufts of bristles, usually with a long, stiff spine rising from the center, are scattered across the pads; the spines are white, gray, or brown. **LEAVES:** Absent. **FLOWERS:** Up to 4 inches across, with many satiny, yellow tepals (often orange-red at the base) surrounding a cluster of yellow stamens; each flower lasts only a day. **FRUIT:** Up to 2 inches long and 1 inch wide, green to red, fleshy, with tufts of bristles. **FLOWERING:** Apr–Jul. **RANGE:** FL west to TX, north to SD and MA. **SIMILAR TO** Jumping Cactus (*Opuntia pusilla*), below.

Jumping Cactus, Creeping Cactus

Opuntia pusilla (Haworth) Nuttall
SYNONYM: *Opuntia drummondii* Graham
Cactus Family (Cactaceae)

SHRUB found in dunes on barrier islands. **STEMS:** Sprawling and forming mats of flat or cylindrical, succulent pads (joints) up to 5 inches long and 3 inches wide; tufts of bristles, with 1–4 spines rising from the center, are scattered across the pads; the spines are barbed, stout, straight, and reddish-brown (aging to gray). **LEAVES:** Absent. **FLOWERS:** Up to 3 inches wide, with satiny, yellow tepals and a cluster of yellow stamens. **FRUIT:** Up to 1¼ inches long and ¾ inch wide, barrel-shaped, green to red, fleshy, smooth, without bristles. **FLOWERING:** Apr–Jun. **RANGE:** FL west to TX, north to NC. **NOTES:** Joints easily break off and painfully attach themselves to skin and clothing. **SIMILAR TO** Erect Prickly-pear (*Opuntia stricta*), which occurs on beach dunes and in maritime forests; it is an erect shrub with yellow spines.

Sweet Shrub, Carolina Allspice

Calycanthus floridus Linnaeus
Sweetshrub Family (Calycanthaceae)

SHRUB found in moist to dry upland forests throughout Georgia except for the southeastern counties. **STEMS:** Up to 10 feet tall, reddish-brown, forming colonies. **LEAVES:** 2–5 inches long and ¾–3 inches wide, opposite, oval, entire, upper surface glossy green, lower surface pale green and sometimes hairy, spicy-fragrant when crushed. **FLOWERS:** About 1 inch long, with many maroon, strap-shaped tepals curving over the center of the flower; before they are pollinated they smell like strawberries or bananas. **FLOWERING:** Apr–May. **RANGE:** GA west to MS, north to OH and PA. **SIMILAR TO** Georgia Sweet Shrub (*Calycanthus brockianus*), known only from Georgia; it has yellowish-green flowers and may be only a color variation rather than a separate species. Georgia Sweet Shrub is sold in nurseries as the cultivar 'Athens' and is especially fragrant.

Beetles: Unsung Heroes of Pollination

More than 30,000 beetle species occur in the United States. Many are important pollinators, but we usually overlook them in favor of the colorful butterflies and hummingbirds. But a whole group of ancient flowering plants depends on beetles for pollination. When flowering plants first evolved—about 200 million years ago—bees, flies, and butterflies had not yet appeared. Beetles were already numerous, though, and early flowering plants such as Magnolia, Sweet Shrub, and Pawpaw evolved flowers to exploit their pollinator potential. Typical beetle-pollinated flowers have strong, fruity or spicy odors—appealing to beetles that use their sense of smell to find food and egg-laying sites.

When Sweet Shrub flowers, the outer tepals curve attractively outward and the inner petals curve inward, forming a one-way passage down which the beetle crawls in pursuit of nectar and tiny food bodies attached to the tips of the inner tepals and stamens. As the beetle bumps around inside the flower, first feeding and then trying to escape, the pollen that attached to its body during visits to other flowers brushes off onto waiting stigmas, and voila! Pollination. At that point, the anthers open, showering the beetle with fresh pollen. The tepals then open outward, releasing the pollen-dusted beetle to visit another flower. The tepals then turn brown and the fragrance disappears, signaling that the flower has been pollinated and is no longer open for business. Humans don't much care for beetles and their flower-munching ways (sometimes called "mess and spoil pollination"), but without them we would lose some of our most beloved trees and shrubs.

American Bell-Flower

Campanula americana Linnaeus
SYNONYM: *Campanulastrum americana* (Linnaeus) Small
Bellflower Family (Campanulaceae)

ANNUAL OR BIENNIAL HERB found in rich, moist forests, especially over mafic bedrock, in north Georgia and a few Coastal Plain counties. **STEMS:** 2–6 feet tall, stout, smooth or slightly hairy. **LEAVES:** Up to 6 inches long and 2½ inches wide, alternate, lance-shaped, toothed, rough-hairy on upper surface. **FLOWER CLUSTERS:** An erect spike at the top of the stem, occasionally branched. **FLOWERS:** Up to 1 inch wide, blue, with 5 spreading lobes and a white ring around the throat; the anthers are coiled, and the style turns sharply upward at the tip. **FLOWERING:** Jul–Sep. **RANGE:** FL west to LA, north to SD, ON, and NY. **NOTES:** Long-tongued bees such as bumblebees are the primary pollinators. **SIMILAR TO** Clasping Venus's Looking-glass (*Triodanis perfoliata*), page 146.

Southern Harebell, Appalachian Bell-Flower

Campanula divaricata Michaux
Bellflower Family (Campanulaceae)

PERENNIAL HERB found on rocky, wooded slopes, trailsides, and cliffs in north Georgia and a few counties in the upper Coastal Plain. **STEMS:** 8–20 inches tall, with many delicate, spreading or drooping branches. **LEAVES:** Up to 3 inches long and 1 inch wide, alternate, lance-shaped, toothed, smooth. **FLOWERS:** Dangling at the tips of long stalks, each flower nearly ⅓ inch long, pale blue or lavender, bell-shaped with 5 upturned lobes and a long, blue style; calyx green, cup-shaped, with 5 narrow, spreading lobes. **FRUIT:** A top-shaped capsule less than ¼ inch long. **FLOWERING:** Aug–Oct. **RANGE:** GA, AL, north to KY and MD. **SIMILAR TO** no other wildflower in Georgia.

Cardinal Flower

Lobelia cardinalis Linnaeus
Bellflower Family (Campanulaceae)

PERENNIAL HERB found in wetlands throughout Georgia. **STEMS:** 1½–6½ feet tall, stout, usually unbranched, mostly smooth. **LEAVES:** Up to 7 inches long and 2 inches wide, alternate, elliptic or lance-shaped, mostly smooth, with toothed margins, each tooth tipped with a tiny knob. **FLOWER CLUSTERS:** An erect spike, 4–20 inches long, at the top of the stem. **FLOWERS:** About 1½ inches long, bright red, with an erect tube, a 3-lobed lower lip, 2 narrow, spreading upper lobes, and a filament tube topped by gray, fuzzy anthers and white style branches. **FLOWERING:** Jul–Oct. **RANGE:** FL west to TX, north to MN, ON, and NB; Central and South America. **NOTES:** Hummingbirds and long-tongued butterflies such as swallowtails and Cloudless Sulphurs are the primary pollinators. **SIMILAR TO** no other wildflower in Georgia.

Glade Lobelia

Lobelia glandulosa Walter
Bellflower Family (Campanulaceae)

PERENNIAL HERB found in wetlands in the Coastal Plain. **STEMS:** 1½–3 feet tall, smooth or slightly hairy. **LEAVES:** 3–6 inches long and about ¼ inch wide, linear or narrowly lance-shaped, tapering at the base with no stalk, thick, entire or with knob-tipped teeth on the margins. **FLOWER CLUSTERS:** An erect spike with few to 20 flowers held on one side of the stem. **FLOWERS:** About 1 inch long, blue-purple, with 2 spreading lips, the upper lip with 2 narrow, upright segments, the lower lip 3-lobed with white markings and a patch of hair at the base. The calyx is cup-shaped with 5 narrow, toothed (sometimes entire) segments. **FLOWERING:** Sep–Oct. **RANGE:** FL west to MS, north to MD. **SIMILAR TO** Longleaf Lobelia (*Lobelia elongata*), which occurs in similar habitats; its flowers lack the patch of hairs, and the calyx segments lack teeth.

Nuttall's Lobelia

Lobelia nuttallii J. A. Schultes
Bellflower Family (Campanulaceae)

ANNUAL HERB found in pine savannas and flatwoods, seepages, and bogs in the Coastal Plain and two mountain counties. **STEMS:** 8–30 inches tall, slender, hairy near the base, often with long, upright branches. **LEAVES:** Less than 1½ inches long and very narrow, linear to elliptic. **FLOWER CLUSTERS:** An erect spike with fewer than 20 widely spaced flowers. **FLOWERS:** Less than 1 inch long, pale blue-purple with a white eye, tubular with 2 spreading lips, the upper lip with 2 narrow, upright segments, the lower lip 3-lobed; the calyx has 5 narrow, pointed segments. **RANGE:** FL west to LA, north to KY and NY. **FLOWERING:** May–Nov. **SIMILAR TO** Fold-ear Lobelia (*Lobelia flaccidifolia*), which occurs in Coastal Plain wetlands; most of its leaves are 2–3 inches long and more than ⅓ inch wide; the calyx segments are arrowhead-shaped.

Downy Lobelia

Lobelia puberula Michaux
Bellflower Family (Campanulaceae)

PERENNIAL HERB found in moist forests, floodplains, and roadsides throughout Georgia except for the southeastern counties. **STEMS:** 1–5 feet tall, purple-tinged, hairy. **LEAVES:** Up to 4¾ inches long and 1½ inches wide, alternate, oval to lance-shaped, hairy on both surfaces, with no stalk; the margins have tiny, callus-tipped teeth. **FLOWERS:** In an elongated spike, each flower ½–1 inch long, purple, tubular with 2 spreading lips, the upper lip with 2 narrow, upright segments, the lower lip 3-lobed with white markings; the calyx has 5 narrow, pointed segments about half the length of the flower. **RANGE:** FL west to LA, north to IL and NJ. **FLOWERING:** Late Jul–Oct. **SIMILAR TO** Southern Lobelia (*Lobelia amoena*), which occurs in wetlands, mostly in the Blue Ridge and Piedmont; its stems are smooth, and its larger leaves taper to slightly winged stalks.

Pale-Spike Lobelia

Lobelia spicata Lamarck
Bellflower Family (Campanulaceae)

ANNUAL HERB found in woodlands, prairies, and clearings in north Georgia. **STEMS:** 1–4 feet tall, slender, angular, rough-hairy only near the base. **LEAVES:** Up to 4½ inches long and 1½ inches wide, oval to spoon-shaped with a tapered base, usually without a stalk, sometimes toothed and hairy. **FLOWERS:** In a densely flowered spike, each flower about ½ inch long, pale blue to white, tubular with 2 spreading lips, the upper lip with 2 narrow, upright segments separated by a knob of gray anthers, the lower lip 3-lobed with a patch of hair and 2 yellowish bumps. **RANGE:** GA west to TX, north to AB and NS. **FLOWERING:** Late May–Aug. **SIMILAR TO** Indian-Tobacco (*Lobelia inflata*), which occurs in woodlands in north Georgia; it has ¼-inch, pale blue or white flowers, and green, inflated calyxes surrounding a round or oval ovary.

Clasping Venus's Looking-Glass

Triodanis perfoliata (Linnaeus) Nieuwland
SYNONYM: *Specularia perfoliata* (Linnaeus) Alphonse de Candolle
Bellflower Family (Campanulaceae)

ANNUAL HERB found in dry woodland edges, prairies, meadows, and disturbed areas throughout Georgia. **STEMS:** Up to 1 foot tall, slender, angular with a line of hairs on each ridge. **LEAVES:** Up to 1 inch long and wide, alternate, heart-shaped and clasping the stem, with toothed margins. **FLOWERS:** Held in the leaf axils, about ½ inch wide, blue or purple, with 5 spreading lobes, several flowers open at the same time. **RANGE:** Most of North America. **FLOWERING:** Apr–Jun. **NOTES:** Stems and leaves have milky latex. **SIMILAR TO** Venus's Looking-glass (*Triodanis biflora*), which occurs in similar habitats, often with Clasping Venus's Looking-glass (but not in the mountains). Its leaves also lack stalks but are only slightly or not at all clasping; only a single flower is open at a time.

Wahlenbergia

Wahlenbergia marginata (Thunberg) Alphonse de Candolle
Bellflower Family (Campanulaceae)

PERENNIAL HERB found in sandy disturbed areas in the Coastal Plain and a few Piedmont counties. **STEMS:** Up to 2 feet tall, branched, smooth or slightly hairy, in clumps of several stems. **LEAVES:** Up to 1½ inches long and ⅓ inch wide, alternate, mostly near the base of the stem, linear to lance-shaped, with wavy, sometimes toothed margins. **FLOWERS:** Up to ½ inch wide, blue, tubular with 5 spreading, pointed lobes and a cup-shaped calyx with 5 pointed lobes. **FLOWERING:** Feb–Dec. **RANGE:** FL west to TX, north to NC; native to Asia and Oceania. **SIMILAR TO** Florida Bluehearts (*Buchnera floridana*), page 272.

Southern Bush-Honeysuckle

Diervilla sessilifolia Buckley
Honeysuckle Family (Caprifoliaceae)

SHRUB found on rocky mountaintops, bluffs, rock outcrops, and dry woodlands in the Blue Ridge. **STEMS:** 2–6 feet tall, woody, twigs smooth, 4-angled. **LEAVES:** 2–7 inches long and ¾–2½ inches wide, opposite, lance-shaped, toothed, smooth, often red-veined, with no stalks. **FLOWERS:** In clusters at the tips of twigs, each flower about ¾ inch long, 5-lobed, pale yellow (turning red with age), with 5 protruding stamens. **FLOWERING:** Jun–Aug. **RANGE:** GA, AL, TN, SC, and NC. **SIMILAR TO** Northern Bush-honeysuckle (*Diervilla lonicera*), a Georgia Special Concern species that occurs in similar habitats; its twigs are round, and its leaves have short stalks and bristly margins. Hairy Southern Bush-honeysuckle (*Diervilla rivularis*) occurs in northwest Georgia on dry bluffs and canyon ledges; its twigs and lower leaf surfaces are hairy.

Yellow Honeysuckle

Lonicera flava Sims
Honeysuckle Family (Caprifoliaceae)

PERENNIAL VINE found in rocky upland forests in the Blue Ridge and Piedmont. **STEMS:** Up to 16 feet long, twining, smooth. **LEAVES:** Up to 2½ inches long and 2 inches wide, opposite, entire, broadly oval to oblong, gray (but not waxy white) on the lower surface, deciduous; the last 2 pairs of leaves on each stem are fused at their bases and appear to be pierced by the stem. **FLOWERS:** In clusters at the tips of stems, each flower about ¾ inch long, golden yellow to orange, tubular with 2 spreading lips, the upper lip with 4 lobes. **FLOWERING:** Apr–May. **RANGE:** GA west to AR, north to MO and NC. **SIMILAR TO** Limber Honeysuckle (*Lonicera dioica*), which occurs in high-elevation rocky woods and seepages in the Blue Ridge; its flowers are yellow to purplish-red, and the leaves are heavily waxy-whitened on the lower surface.

Coral Honeysuckle

Lonicera sempervirens Linnaeus
Honeysuckle Family (Caprifoliaceae)

PERENNIAL VINE found in upland forests throughout Georgia. **STEMS:** Up to 16 feet long, twining or trailing, new growth with a whitish, waxy coating. **LEAVES:** Up to 2¾ inches long, opposite, entire, oval or elliptic, upper surface dark green, often tinged with purple, lower surface pale green, both surfaces smooth and waxy, evergreen; the last pair of leaves at the tip of the stem are fused together and appear to be pierced by the stem. **FLOWERS:** Clustered at the tip of the stem, up to 2 inches long, coral-red, tubular with 5 tiny, spreading lobes, yellow inside the tube. **FLOWERING:** Apr–Jul. **RANGE:** FL west to TX, north to IA, QC, and ME. Leaves are **SIMILAR TO** those of the exotic pest plant Japanese Honeysuckle (*Lonicera japonica*), but none of its leaves are fused, and the lower leaves are deeply toothed or lobed.

Pink Family (Caryophyllaceae)

The pinks get their common name from a distinctive feature of many of their flowers—the 5 petals are often deeply fringed on the tips as though trimmed with pinking shears. Sometimes the fringe is so deep that the flower seems to have 10 or more petals. Pinks have many other features in common that make theirs one of the easiest families to recognize. The leaves are opposite, and pairs of leaves are often set at right angles to each other. The leaves are connected by a line that crosses the leaf node from one leaf base to the other. The stem is swollen at the leaf nodes (think carnations). The flowers of Georgia's native species always have 5 petals, and the sepals are fused together into a tube or an inflated cylinder. Many pinks are popular garden or florist plants, such as Sweet William, Campion, and Baby's Breath. Some plants in this family contain saponin, a bitter-tasting compound that discourages animals from eating the leaves and that foams when mashed with water to form a soap substitute. The Pink Family is large, with more than 2,200 species, most of them found in the Mediterranean region. Georgia has 30 native pinks and 20 that are exotic, most of these escaped from gardens.

Sticky Mouse-Ear Chickweed

Cerastium glomeratum Thuillier
Pink Family (Caryophyllaceae)

ANNUAL HERB found in disturbed areas throughout Georgia. **STEMS:** 2–18 inches tall, hairy, forming clumps of semi-erect to sprawling stems covered with white hairs, most hairs with glandular tips. **LEAVES:** Up to 1 inch long and ½ inch wide, opposite, clasping the stem, broadly oval, covered with long, white hairs. **FLOWERS:** Held in compact clusters at the tips of branches, each flower about ⅓ inch wide, with 5 deeply notched, white petals, 5 green, pointed sepals, and 10 white stamens. **FLOWERING:** Mar–May. **RANGE:** Native to Europe, now spread through most of North America. **SIMILAR TO** Common Mouse-ear (*Cerastium fontanum* ssp. *vulgare*, synonym *C. holosteoides*), another European weed; its stems are matted at the base and rooted at the leaf nodes. Also see Starry Chickweed (*Stellaria pubera*) and the similar Common Chickweed (*Stellaria media*), page 154.

Deptford Pink

Dianthus armeria Linnaeus
Pink Family (Caryophyllaceae)

PERENNIAL HERB found in fields, pastures, rights-of-way, and roadsides in north Georgia and a few Coastal Plain counties. **STEMS:** Up to 3 feet tall, hairy, knobby where the paired leaves meet. **LEAVES:** Up to 4 inches long and ¼ inch wide, opposite, linear, hairy, joined at the base by a stem-encircling sheath. **FLOWERS:** About ½ inch wide, with 5 toothed, spreading, pink petals dotted with white and dark pink spots, surrounded by 3 narrow bracts. **FLOWERING:** May–Sep. **RANGE:** Native to Europe but introduced nearly worldwide; it is named for a village in England. **SIMILAR TO** Sleepy Catchfly (*Silene antirrhina*), which occurs in disturbed areas throughout Georgia. Its oval petals are pink, lavender, or white and have a single notch at the tip; the calyx is inflated with 5 tiny, pointed teeth at the top.

Carolina Sandwort, Pine Barren Stitchwort

Minuartia caroliniana (Walter) Mattfeld
SYNONYM: *Arenaria caroliniana* Walter
Pink Family (Caryophyllaceae)

PERENNIAL HERB found in very dry sandhills and Turkey Oak scrub in the Coastal Plain, especially the Fall Line counties. **STEMS:** Up to 1 foot tall, erect or sprawling, smooth, very leafy, stiff but not woody. **LEAVES:** Less than ½ inch long and ¼ inch wide, opposite, closely spaced and overlapping, pointed, entire, smooth, and shiny green. **FLOWERS:** Held in small clusters at the tips of branches, each flower ½–¾ inch across, with 5 white, spreading petals and a short, green, oval sepal visible between each petal. **FLOWERING:** Apr–Jun. **RANGE:** FL north to RI. The leaves and stems are **SIMILAR TO** those of Trailing Phlox (*Phlox nivalis*), page 284, which has pink flowers.

Lime-Barren Sandwort, Pitcher's Stitchwort

Minuartia patula (Michaux) Mattfeld
SYNONYM: *Arenaria patula* Michaux
Pink Family (Caryophyllaceae)

ANNUAL HERB found on limestone outcrops (cedar glades) and other dry, calcium-rich sites in northwest Georgia. **STEMS:** 2–12 inches tall, smooth or hairy, usually dark red, several spreading from the base, with forking branches. **LEAVES:** Up to ¾ inch long and very narrow, smooth, somewhat fleshy, green with purple tips, joined at the base to the opposite leaf, the leaf pairs widely spaced. **FLOWERS:** Held singly at the top of threadlike stalks; each flower up to ½ inch across, with 5 white, notched, spreading petals and 5 green, pointed sepals. **FLOWERING:** Apr–Jun. **RANGE:** GA west to TX, north to KS, WI, and PA. **SIMILAR TO** Carolina Sandwort (*Minuartia caroliniana*), page 150, and One-flowered Sandwort (*Minuartia uniflora*), below.

One-Flowered Sandwort

Minuartia uniflora (Walter) Mattfeld
SYNONYM: *Arenaria uniflora* (Walter) Muhlenberg
Pink Family (Caryophyllaceae)

ANNUAL HERB found on granite outcrops in the Piedmont and on Altamaha Grit outcrops in the Coastal Plain. **STEMS:** 2–8 inches tall, thin and wiry, smooth, unbranched. **LEAVES:** Basal leaves about ¼ inch long, lance-shaped, somewhat fleshy; stem leaves even smaller, opposite. **FLOWERS:** Held singly at the top of threadlike stalks up to 2 inches long; each flower about ⅓ inch across, with 5 white, spreading petals. **FLOWERING:** Apr–May. **RANGE:** GA, AL, SC, and NC. **SIMILAR TO** Appalachian Sandwort (*Minuartia glabra*), which occurs on granite, limestone, sandstone, and Altamaha Grit outcrops; it is a larger plant with leafy stems and 3–7 flowers per stem; its leaves are ⅓–1 inch long.

Dune Whitlow-Wort

Paronychia baldwinii (Torrey & A. Gray) Fenzl ex Walpers
SYNONYM: *Paronychia riparia* Chapman
Pink Family (Caryophyllaceae)

ANNUAL, BIENNIAL, OR PERENNIAL HERB found in dunes, sandhills, and dry, disturbed areas in the Coastal Plain. **STEMS:** Up to 2½ feet long, erect or trailing across the ground, forked, hairy or smooth. **LEAVES:** Up to 1 inch long and ¼ inch wide, opposite, elliptic, pointed, smooth except for long, spreading hairs on the margins. **FLOWERS:** Less than ⅛ inch wide, green-and-white or brownish, with 5 stiff, oval, petal-like sepals with sharp tips (there are no petals). **FLOWERING:** Jun–Oct. **RANGE:** FL west to AL, north to VA. **NOTES:** A 10× lens is recommended for identifying nailworts. **SIMILAR TO** Coastal Plain Nailwort (*Paronychia herniarioides*), below.

Coastal Plain Nailwort

Paronychia herniarioides (Michaux) Nuttall
Pink Family (Caryophyllaceae)

ANNUAL HERB found in sandhills, river dunes, and Turkey Oak scrub in the Fall Line and Coastal Plain counties. **STEMS:** Up to 8 inches long, densely and softly hairy, sprawling across the ground and forming mats with forked branches. **LEAVES:** Less than ¾ inch long and ¼ inch wide, opposite, hairy, oval or elliptic with a sharp or knobby tip. **FLOWERS:** About ⅛ inch wide, with 5 sharply pointed, stiff, green, yellow, or brownish sepals (there are no petals). **FLOWERING:** Apr–Jul. **RANGE:** GA, FL, SC, and NC. **NOTES:** A 10× lens is recommended for identifying nailworts. **SIMILAR TO** American Nailwort (*Paronychia americana*), which also occurs in sandhills and other dry sites in the Coastal Plain; its leaves are smooth or rough-hairy, and the flowers are minute, less than 1/16 inch wide, with rounded, maroon sepals.

Starry Campion

Silene stellata (Linnaeus) W. T. Aiton
Pink Family (Caryophyllaceae)

PERENNIAL HERB found in upland hardwood forests in north Georgia. **STEMS:** 1–3 feet tall, hairy. **LEAVES:** Up to 4 inches long and 1½ inches wide with no stalk; oval, elliptic, or lance-shaped; midstem leaves whorled, upper and lowermost leaves opposite. **FLOWER CLUSTERS:** Up to 1 foot long, open and loosely branched, at the top of the stems. **FLOWERS:** About 1 inch long and wide, with 5 white petals deeply divided into 8–12 segments; each petal is woolly at the base on the inside of the flower. **FLOWERING:** Jul–Sep. **RANGE:** GA west to TX, north to ND and VT. **NOTES:** The scentless flowers are pollinated primarily by moths at night and are closed during the day. **SIMILAR TO** Oval-leaved Campion (*Silene ovata*), a rare species that occurs in similar habitats, sometimes with Starry Campion; all of its leaves are opposite.

Fire Pink

Silene virginica Linnaeus
Pink Family (Caryophyllaceae)

PERENNIAL HERB found on rock outcrops, thinly wooded slopes, and in woodland borders in north Georgia and a few Coastal Plain counties. **STEMS:** 8–32 inches tall, leaning to erect, covered with sticky hairs. **STEM LEAVES:** 1–4 inches long, opposite, in 2–4 pairs, oval to lance-shaped, entire, smooth. **FLOWERS:** About 1½ inches wide and 1 inch long, with 5 deeply notched, bright red petals and a ring of small, red teeth in the center. **FLOWERING:** Apr–Jul. **RANGE:** FL west to OK, north to IA and NY. **NOTES:** Sticky hairs on the stems and calyx discourage ants from entering the flowers and stealing the nectar. **SIMILAR TO** Royal Catchfly (*Silene regia*), which has pointed or ragged petals and 10–20 pairs of stem leaves; it occurs in limestone-based woodlands in northwest Georgia, where it is rare.

Starry Chickweed, Giant Chickweed

Stellaria pubera Michaux
Pink Family (Caryophyllaceae)

PERENNIAL HERB found in rich, moist forests in north Georgia and a few Coastal Plain counties. **STEMS:** 4–16 inches tall, branched, 4-sided, with lines of soft, spreading hairs running up the stem. **LEAVES:** ¾–4 inches long and ¼–1½ inches wide, opposite, oval or lance-shaped, hairy along the margins and usually on the upper surface, with entire margins and a very short or no stalk. **FLOWERS:** Up to ½ inch wide, with 5 white petals so deeply notched they look like 10 petals, and with 10 red-tipped stamens. **FLOWERING:** Apr–May. **RANGE:** FL west to AL, north to IL and NJ. **SIMILAR TO** Common Chickweed (*Stellaria media*), a European weed found in bottomlands and disturbed areas; it has smaller flowers (less than ¼ inch across), smaller leaves (less than 1½ inches long), and weak, sprawling, mat-forming stems.

Pineland Scaly-Pink

Stipulicida setacea Michaux
Pink Family (Caryophyllaceae)

ANNUAL OR SHORT-LIVED PERENNIAL HERB found in sandhills, dry pine flatwoods, and maritime forests in the Coastal Plain. **STEMS:** 2–10 inches tall, wiry, smooth, with many forking branches. **LEAVES:** Mostly in a basal rosette, each up to ¾ inch long, spoon-shaped, often withered by flowering time; stem leaves are tiny, opposite, triangular scales, fringed at the base (use a 10× lens). **FLOWERS:** Held in small clusters at the tips of branches, each flower about ⅛ inch across, with 5 white, spreading petals and 5 pointed, reddish-brown sepals. **FLOWERING:** May–Aug. **RANGE:** FL west to LA, north to VA. **SIMILAR TO** Carolina Sandwort (*Minuartia caroliniana*), page 150.

Strawberry-Bush, Hearts-a-Bustin'-with-Love

Euonymus americanus Linnaeus
Bittersweet Family (Celastraceae)

DECIDUOUS SHRUB found in moist hardwood-pine forests throughout Georgia. **STEMS:** Up to 6½ feet tall, stiffly erect, smooth, bright green all year. **LEAVES:** Up to 4 inches long and 1½ inches wide, opposite, elliptic to lance-shaped, with tiny gland-tipped teeth on the margins. **FLOWERS:** About ⅓ inch across, with 5 rounded, greenish or reddish petals and a flat, pale green disk in the center; flowers rest on the leaves on ½-inch stalks. **FRUIT:** Round, warty, dark pink or red capsules about ¾ inch across; the showy fruit splits into 5 segments, revealing 5 scarlet, dangling seeds. **FLOWERING:** May–Jun; fruiting Sep–Oct. **RANGE:** FL west to TX, north to OK, IL, and NY. **SIMILAR TO** Running Strawberry Bush (*Euonymus obovatus*), which occurs in cove forests and boulderfields in the Blue Ridge; it is a sprawling vine, and its fruits split into 3 segments.

Gopher-Apple

Licania michauxii Prance
SYNONYM: *Chrysobalanus oblongifolius* Michaux
Coco-Plum Family (Chrysobalanaceae)

LOW SHRUB found in sandhills and other dry, sandy pinelands in the Coastal Plain. **STEMS:** 4–16 inches tall, arising from large underground stems and forming patches. **LEAVES:** Up to 4 inches long and 1½ inches wide, alternate, oblong or oval, leathery, the upper surface glossy and with a raised network of veins, the lower surface often with a feltlike covering of hairs. **FLOWERS:** In 4-inch-long clusters at the tips of stems, very small, with 5 hairy, yellowish to white petals. **FRUIT:** About 1 inch long, oval, white turning reddish when ripe, reported to taste like pink bubblegum. **FLOWERING:** May–Jun; fruiting Sep–Oct. **RANGE:** FL west to LA, north to SC. A colony of Gopher-apple is **SIMILAR TO** a patch of Running Oak (*Quercus pumila*), but Gopher-apple's glossy leaves, flowers, and fruits are distinctive.

Carolina Sun-Rose

Crocanthemum carolinianum (Walter) Spach
SYNONYM: *Helianthemum carolinianum* (Walter) Michaux
Rockrose Family (Cistaceae)

PERENNIAL HERB found in dry pine flatwoods, sandhills, and clearings in the Coastal Plain. **STEMS:** Up to 1 foot tall, very hairy. **BASAL LEAVES:** Up to 2 inches long and ¾ inch wide, elliptic or spoon-shaped, entire, hairy on both surfaces, with short stalks, often withered by flowering time. **STEM LEAVES:** Similar to basal but smaller, alternate. **FLOWERS:** Flowers about 1 inch wide, with 5 triangular, yellow petals, 6 hairy sepals (3 triangular, 3 linear), and many orange-tipped stamens; flowers drop their petals after 1 day. **FLOWERING:** Apr–May. **RANGE:** FL west to TX, north to AR and NC. **SIMILAR TO** Pine Barren Sun-rose (*Crocanthemum corymbosum*), below; and Piriqueta (*Piriqueta caroliniana*), page 339.

Pine Barren Sun-Rose, Pine Barren Frostweed

Crocanthemum corymbosum (Michaux) Britton
SYNONYM: *Helianthemum corymbosum* Michaux
Rockrose Family (Cistaceae)

PERENNIAL HERB found in coastal dunes, dry live oak hammocks, and maritime forests. **STEMS:** 4–14 inches tall, reddish, densely hairy. **LEAVES:** Up to 1¼ inches long and ½ inch wide, alternate, elliptic, very hairy on both surfaces, with short stalks and inrolled margins. **FLOWERS:** About ¾ inch wide, with 5 triangular, yellow petals, 6 very hairy sepals (3 triangular and 3 linear), and many orange-tipped stamens. Many smaller flowers without petals are mixed with the colorful flowers; these never open and are self-pollinated. **FLOWERING:** Apr–May. **RANGE:** GA, FL, AL, MS, SC, and NC. **SIMILAR TO** Rosemary Sun-rose (*Crocanthemum rosmarinifolium*), which occurs in dry, sandy woods, roadsides, and fields in the Coastal Plain; its leaves are less than ¼ inch wide, almost needle-like. Also see Piriqueta (*Piriqueta caroliniana*), page 339.

Hairy Pinweed

Lechea mucronata Rafinesque
SYNONYM: *Lechea villosa* Elliott
Rockrose Family (Cistaceae)

PERENNIAL HERB found in dry woodlands, dunes, and disturbed areas in the Coastal Plain and a few Piedmont counties. **STEMS:** Up to 3 feet tall, covered with long, spreading hairs; upper branches alternate, lower branches opposite or whorled. **LEAVES:** Up to 1¼ inches long and ½ inch wide, alternate on the upper branches, opposite or whorled below, elliptic, entire, hairy. **FLOWERS:** Held in tight clusters, each flower with 3 tiny, maroon petals and 5 slightly longer, green sepals. **FRUIT:** A tiny, round capsule; the 3 inner sepals are slightly longer than the capsule, the 2 outer sepals are slightly shorter. **FLOWERING:** Jun–Aug. **RANGE:** FL west to TX, north to NE and NH. **SIMILAR TO** Leggett's Pinweed (*Lechea pulchella*, synonym *L. leggetti*), which occurs in similar habitats; the hairs on its stems are pressed upward; its fruits are oval and longer than all 5 sepals.

Sweet-Pepperbush

Clethra alnifolia Linnaeus
Clethra Family (Clethraceae)

SHRUB found in wetlands and on stream banks in the Coastal Plain and a few Piedmont counties. **STEMS:** 3–10 feet tall with peeling, reddish-brown bark. **LEAVES:** Up to 3½ inches long and 1½ inches wide, alternate, oval or lance-shaped, wider and sharply toothed above the middle. **FLOWER CLUSTERS:** An erect, hairy spike 2½–7½ inches long. **FLOWERS:** About ½ inch wide, fragrant, with 5 white, spreading petals and 10 long, orange-tipped stamens. **FLOWERING:** Jun–Aug. **RANGE:** FL west to TX, north to NS, also TN. **SIMILAR TO** Mountain Sweet Pepperbush (*Clethra acuminata*), which occurs in Georgia's Blue Ridge in moist forests and around rock outcrops; it reaches 20 feet in height and has peeling, reddish-brown bark; its leaves are usually 4–5 inches long and smooth on both surfaces. Also see Virginia-willow (*Itea virginica*), page 228.

Hedge Bindweed

Calystegia sepium (Linnaeus) R. Brown
Morning-Glory Family (Convolvulaceae)

PERENNIAL HERBACEOUS VINE found in fields, fencerows, roadsides, and other disturbed areas throughout Georgia. **STEMS:** Up to 7 feet long, trailing over the ground or twining over other plants. **LEAVES:** Up to 4 inches long and 2⅜ inches wide, alternate, arrowhead-shaped, smooth or hairy, with a long stalk. **FLOWERS:** Funnel-shaped, 2–3 inches wide, white or pink with a white star on the face and a yellow throat; 2 green bracts, ½–1 inch long, enclose the base of the flower, including the 5-lobed calyx. **FLOWERING:** May–Aug. **RANGE:** Throughout North America and globally widespread. **SIMILAR TO** Wild Sweet Potato (*Ipomoea pandurata*), page 160, which does not have a pair of large bracts at the base of the flower.

Common Dodder, Scaldweed

Cuscuta gronovii Willdenow ex J. A. Schultes
Morning-Glory Family (Convolvulaceae)

ANNUAL HERBACEOUS VINE lacking roots and chlorophyll, and parasitizing many different host plants, including goldenrods, Jewelweed, and smartweeds, mostly in low or wet areas throughout Georgia. **STEMS:** Up to 6 feet long, orange, smooth, twining on and tightly attaching to other plants. **LEAVES:** Minute, orange scales. **FLOWERS:** In compact clusters of several flowers, each flower about ⅛ inch wide, with 5 white, rounded, spreading petals. **FLOWERING:** Aug–Oct. **RANGE:** Most of North America. **NOTES:** Very young dodder plants have roots, but after the plant attaches to its host and begins to extract nutrients, the roots wither and the plant loses all connection to the ground. **SIMILAR TO** the six other dodder species in Georgia; distinguishing among them is difficult because the flower parts are so small.

Red Morning-Glory, Scarlet Creeper

Ipomoea coccinea Linnaeus
Morning-Glory Family (Convolvulaceae)

PERENNIAL HERBACEOUS VINE found in fields, fencerows, and roadsides in north Georgia and a few Coastal Plain counties. **STEMS:** Up to 10 feet long, trailing or twining upward. **LEAVES:** Up to 4 inches long and 2½ inches wide, alternate, heart-shaped, mostly smooth, with a long stalk. **FLOWERS:** Funnel-shaped, up to 1 inch long and 1½ inches wide, with an orange tube and a scarlet face with a yellow star. **FLOWERING:** Jul–Dec. **RANGE:** FL west to TX, north to IA and MA. **NOTES:** Introduced in much of its range, Red Morning-glory is probably native to the southeastern United States. **SIMILAR TO** two other red-flowered morning-glories found in disturbed areas in the Coastal Plain. Scarlet Creeper (*Ipomoea hederifolia*) has 3-lobed leaves; Cypress-vine (*I. quamoclit*) has dissected, fernlike leaves. Both are introduced from tropical America.

Ivy-Leaf Morning-Glory

Ipomoea hederacea Jacquin
Morning-Glory Family (Convolvulaceae)

ANNUAL HERBACEOUS VINE found on fencerows and in fields, thickets, and other disturbed areas throughout Georgia. **STEMS:** Up to 6 feet long, twining around other plants. **LEAVES:** Up to 5 inches long and 4 inches wide, alternate, deeply 3-lobed or heart-shaped, sometimes both on the same plant. **FLOWERS:** Funnel-shaped, up to 1½ inches long and 2 inches wide, blue with a white throat and faint purplish star on the face; individual flowers last only a few hours in the morning and early afternoon. **FLOWERING:** Jul–Dec. **RANGE:** FL west to AZ, north to ND and ME. **SIMILAR TO** Common Morning-glory (*Ipomoea purpurea*), a tropical species that occurs in fields and other disturbed areas throughout Georgia; its flowers are purplish-blue, dark pink, or white, and its leaves are always heart-shaped.

Wild Sweet Potato, Man-Root

Ipomoea pandurata (Linnaeus) G. F. W. Meyer

Morning-Glory Family (Convolvulaceae)

PERENNIAL HERBACEOUS VINE found in disturbed areas throughout Georgia. **STEMS:** Up to 30 feet long, sprawling across the ground or twining over other plants. **LEAVES:** Up to 6 inches long and 4 inches wide, alternate, heart-shaped, smooth or hairy, with entire or wavy margins. **FLOWERS:** Funnel-shaped, up to 3 inches long and 3 inches wide, white with a maroon throat and a faint white star on the face. **FLOWERING:** May–Sep. **RANGE:** FL west to TX, north to NE, ON, and MA. **NOTES:** Wild Sweet Potato has a 20–30-pound edible tuber. **SIMILAR TO** Beach Morning-glory (*Ipomoea imperati*, synonym *Ipomoea stolonifera*), which occurs on beaches and dunes; its white flower has a yellow throat, and its leaves are oblong, sometimes lobed, and succulent.

Railroad Vine, Goat's-Foot Vine

Ipomoea pes-caprae (Linnaeus) R. Brown

SYNONYM: *Ipomoea brasiliensis* (Linnaeus) Sweet

Morning-Glory Family (Convolvulaceae)

PERENNIAL HERBACEOUS VINE found on coastal sand flats, beach dunes, and disturbed areas near the beach. **STEMS:** Up to 100 feet long, trailing and occasionally rooting at the leaf nodes. **LEAVES:** Up to 4 inches long and wide with stalks up to 6 inches long, alternate, oblong to kidney-shaped with a notched tip, leathery, smooth, glossy. **FLOWERS:** Funnel-shaped, up to 3½ inches long and 2½ inches wide, dark pink with a magenta star. **FLOWERING:** Jun–Nov. **RANGE:** FL west to TX, north to NC; widespread in the tropics. **NOTES:** Railroad Vine stabilizes dunes by rooting at the nodes and by sending its large root many feet deep into the sand. **SIMILAR TO** two other coastal morning-glories with dark pink flowers: Salt-marsh Morning-glory (*Ipomoea sagittata*) has arrowhead-shaped leaves; Coastal Morning-glory (*Ipomoea cordatotriloba*) has thin-textured, heart-shaped leaves.

Hairy Cluster-Vine

Jacquemontia tamnifolia (Linnaeus) Grisebach
Morning-Glory Family (Convolvulaceae)

ANNUAL HERBACEOUS VINE found in fields, roadsides, and other disturbed areas throughout Georgia. **STEMS:** Up to 6 feet long, trailing along the ground or forming clumps, covered with long, silky hairs. **LEAVES:** Up to 5 inches long and 4 inches wide, alternate, heart-shaped, very hairy, with stalks 1–3 inches long. **FLOWER CLUSTERS:** Held in the leaf axils, leafy, with a few open flowers and many conspicuously hairy, narrowly pointed sepals. **FLOWERS:** About ½ inch wide, funnel-shaped, blue with a white center, open only one day. **FLOWERING:** Aug–Sep. **RANGE:** FL west to TX, north to IL and PA; native to tropical America, globally widespread. **SIMILAR TO** Ivy-leaf Morning-glory (*Ipomoea hederacea*) and Common Morning-glory (*Ipomoea purpurea*) in flower color, but with much smaller, clustered flowers, page 159.

Coastal Plain Dawnflower, Sandhill Morning-Glory

Stylisma patens (Desrousseaux) Myint
SYNONYM: *Bonamia patens* (Desrousseaux) Shinners
Morning-Glory Family (Convolvulaceae)

PERENNIAL HERBACEOUS VINE found in sandhills and other dry, sandy woods in the Coastal Plain. **STEMS:** Up to 4 feet long, spreading across the ground, not climbing. **LEAVES:** Up to 2 inches long and ⅜ inch wide, alternate; linear, elliptic, or oblong, entire, hairy. **FLOWERS:** Held singly on stalks that arise from the leaf axils; funnel-shaped, up to ¾ inch wide, white with a faint star on the face; the 5 sepals are green, hairy, and pointed. **FLOWERING:** May–Aug. **RANGE:** FL west to LA, north to NC. **SIMILAR TO** two other species that occur in dry woodlands in the Coastal Plain. Both have wider leaves and flowers in clusters of 3–12. Hairy Dawnflower (*Stylisma villosa*) has very hairy sepals. Southern Dawnflower (*S. humistrata*) has hairless sepals; it also occurs in the Piedmont.

Diamorpha, Elf Orpine, Small's Stonecrop

Diamorpha smallii Britton ex Small
Stonecrop Family (Crassulaceae)

ANNUAL HERB found on granite outcrops in the Piedmont and on sandstone outcrops in the Ridge & Valley. **STEMS:** ¾–4 inches tall, bright red, succulent. **LEAVES:** Less than ¼ inch long, alternate, red, succulent, round in cross-section. **FLOWERS:** About ¼ inch across, with 4 white petals, 8 red-tipped stamens, and a white, 4-segmented pistil in the center. **FRUIT:** About ¼ inch wide, opening along the underside of its 4 segments to release minute seeds (use 10× lens). **FLOWERING:** Late Mar–early May. **RANGE:** GA west to AL, north to TN and VA. **NOTES:** Diamorpha overwinters as tiny, red-leaf rosettes; its stems elongate in early spring. **SIMILAR TO** Granite Stonecrop (*Sedum pusillum*), a Georgia Threatened species found only on Piedmont granite outcrops. It is bluish-green (rarely reddish), and its fruits open along the upper side of the segments.

Mountain Stonecrop

Sedum ternatum Michaux
Stonecrop Family (Crassulaceae)

PERENNIAL HERB found in moist forests, bottomlands, and rock outcrops in the north Georgia mountains. **NONFLOWERING STEMS:** Lying on the ground for most of their length, then turned up at the tip and topped with a rosette of whorled leaves. **FLOWERING STEMS:** Up to 8 inches tall, erect. **LEAVES:** Up to 1 inch long and ¾ inch wide, alternate or in whorls of 3, oval to spoon-shaped with rounded tips, entire, succulent, smooth. **FLOWERS:** In a 3-branched cluster, each flower about ½ inch across, with 5 white, pointed, spreading petals and 5 red-tipped stamens. **FLOWERING:** Apr–Jun. **RANGE:** GA west to MS, north to IA, ON, and ME. **SIMILAR TO** Widow's-cross (*Sedum pulchellum*), which occurs on limestone outcrops in northwest Georgia; its leaves are alternate, linear, and round in cross-section, and its flowers are usually pink, white, or purple.

Creeping Cucumber, Melonette

Melothria pendula Linnaeus
Cucumber Family (Cucurbitaceae)

ANNUAL OR PERENNIAL HERBACEOUS VINE found in bottomlands, marshes, and moist thickets throughout Georgia. **STEMS:** Up to 15 feet long, trailing or climbing over other plants with coiled tendrils. **LEAVES:** Up to 3 inches long and wide, alternate, 3–5-lobed, rough-hairy, similar to miniature cucumber or squash leaves. **FLOWERS:** About ⅓ inch wide, yellow, with 5 notched lobes; female and male flowers are on the same plant, female flowers solitary in leaf axils with tiny, melon-like immature fruit beneath the flower; male flowers smaller, in few-flowered, long-stalked clusters. **FRUIT:** Resemble tiny watermelons dangling on long stalks, each ½–1 inch long, round to elliptic, speckled green or black. **FLOWERING:** Jun–Nov. **RANGE:** FL west to TX, north to KS and PA. Leaves and stems are **SIMILAR TO** Bur-cucumber (*Sicyos angulatus*), below.

Bur-Cucumber, Star-Cucumber

Sicyos angulatus Linnaeus
Cucumber Family (Cucurbitaceae)

ANNUAL HERBACEOUS VINE found in floodplains, moist forests, and disturbed areas throughout Georgia. **STEMS:** Up to 25 feet long, light green, brittle, hairy, climbing over other plants with branched tendrils. **LEAVES:** Up to 8 inches long and wide with a stalk up to 5 inches long, alternate, 3–5-lobed, finely toothed, rough-hairy on the upper surface. **FLOWERS:** Female and male flowers are on the same plant, each about ⅓ inch across, greenish-white, with 5 pointed lobes; female flowers in round, densely flowered clusters on short stalks; male flowers in few-flowered, long-stalked clusters. **FRUIT:** In round, compact clusters, each fruit about ½ inch long, oval, covered with long, sharp bristles. **FLOWERING:** Aug–Nov. **RANGE:** FL west to TX, north to ND, QC, and ME. Leaves and stems are **SIMILAR TO** those of Creeping Cucumber (*Melothria pendula*), above.

Buckwheat Tree, Black Titi

Cliftonia monophylla (Lamarck) Britton ex Sargent
Titi Family (Cyrillaceae)

EVERGREEN SHRUB OR SMALL TREE found in swamps, bayheads, flatwoods, and wet ditches in the Coastal Plain. **STEMS:** Up to 50 feet tall, crooked, with dark, nearly black bark on older plants, much-branched from the base, often forming dense thickets. **LEAVES:** 1–4 inches long and ½–¾ inch wide, alternate, elliptic, often slightly notched at the tip, leathery, smooth, stalkless, lower surface pale gray, upper surface with a prominent midvein (other veins not visible). **FLOWERS:** In clusters at the tips of twigs, each flower about ¼ inch wide, fragrant, with 5 white (or pinkish), rounded petals and 10 petal-like stamens. **FRUIT:** A ¼-inch, yellowish-brown capsule with 2–5 papery wings. **FLOWERING:** Mar–May. **RANGE:** FL west to LA, north to SC. **SIMILAR TO** no other flowering shrub in Georgia.

Galax

Galax urceolata (Poiret) Brummitt
SYNONYM: *Galax aphylla* Linnaeus
Diapensia Family (Diapensiaceae)

PERENNIAL HERB found in oak-pine forests and acidic coves in the mountains and upper Piedmont, often under Mountain Laurel and Rosebay Rhododendron. **STEMS:** Underground, spreading and forming large colonies. **LEAVES:** 1–4 inches long and wide, on long stalks, round or broadly heart-shaped, dark green, glossy, leathery, smooth, with finely toothed margins; individual leaves live about 18 months, turning burgundy in winter. **FLOWER CLUSTERS:** A narrow, many-flowered spike, 6–10 inches long, at the top of a long, slender stalk; the flowers open from the bottom upward. **FLOWERS:** Up to ¼ inch wide, white, with 5 petals. **FLOWERING:** May–Jul. **RANGE:** GA west to AL, north to OH and MA. **NOTES:** Galax (or perhaps a soil fungus associated with it) produces a "skunky" smell in warm weather. **SIMILAR TO** Devil's Bit (*Chamaelirium luteum*), page 393.

Pink Sundew

Drosera capillaris Poiret
Sundew Family (Droseraceae)

ANNUAL OR PERENNIAL HERB found in bogs, wet pine flatwoods and savannas, and wet ditches in the Coastal Plain. **LEAVES:** In a basal rosette up to 4 inches across, red or green, the blades oval or spoon-shaped and covered with red, gland-tipped hairs; stalks slightly longer than the blade, with a few nonglandular hairs. **FLOWER CLUSTERS:** Nodding at the top of a hairless stalk 1½–12 inches tall, with 2–20 flowers. **FLOWERS:** Nearly ½ inch across, with 5 pink petals. **FLOWERING:** May–Aug. **RANGE:** FL west to TX, north to VA; tropical America. **SIMILAR TO** Dwarf Sundew (*Drosera brevifolia*), which occurs in similar habitats; its rosettes are less than 1½ inches across, the flowers are white or pink, and the flower stalks are covered with gland-tipped hairs.

Sundews: Denizens of the Damp

Glistening like a jewel in the sunshine, sundew's beauty belies its deadliness for small insects. Sundew leaves are covered with glandular hairs, each tipped with a dewlike drop of sweet, sticky mucilage that attracts and ensnares its prey. Once the insect is secured, glands secrete digestive enzymes that break down its body for absorption. Like other carnivorous plants such as pitcher plants and butterworts, sundews depend on their prey to provide nitrogen and phosphorus, nutrients that are at low levels in the saturated soils where they live. Sundew flowers are arranged in a drooping cluster, with the flowers opening one per day from the bottom of the cluster up. The flowers are held on long stalks well above the sticky leaves, so pollinators are relatively safe. Flowers usually open at midmorning and close within five hours, though sometimes a flower may linger a second day. Five of the 115 sundew species that occur worldwide grow in Georgia.

Round-Leaved Sundew

Drosera rotundifolia Linnaeus
Sundew Family (Droseraceae)

PERENNIAL HERB found in mountain bogs and seepages in Georgia's Blue Ridge. **LEAVES:** In a basal rosette about 4 inches across, red or green, the blades nearly round and covered with red, gland-tipped hairs; stalks ½–2 inches long. **FLOWER CLUSTERS:** Nodding at the top of a smooth stalk up to 12 inches tall, with 3–15 flowers. **FLOWERS:** About ½ inch across, with 5 white or pinkish petals. **FLOWERING:** Jul–Sep. **RANGE:** Eastern United States, Canada, Pacific Northwest, and northern Europe and Asia. **SIMILAR TO** Pink Sundew (*Drosera capillaris*), page 165, which has oval or spoon-shaped leaf blades.

Tracy's Dew-Threads, Tracy's Sundew

Drosera tracyi (Diels) MacFarlane
Sundew Family (Droseraceae)

PERENNIAL HERB found in wet savannas and flatwoods and wet, sandy roadsides in the Coastal Plain. **STEMS:** Up to 2 feet tall. **LEAVES:** Up to 20 inches long, narrow but not threadlike, uncurling from the base in early spring and forming clusters of erect leaves. **FLOWER CLUSTERS:** A drooping, 1-sided spike, its flowers opening from the bottom to the top, that becomes erect as the topmost flowers open. **FLOWERS:** About ¾ inch wide, with 5 oval, dark pink petals. **FLOWERING:** Jun. **RANGE:** GA, FL, west to LA. **NOTES:** The leaves and stems are covered with green or transparent, sticky, gland-tipped hairs that glisten in the sun. Insects become trapped in the sticky secretions and are digested by enzymes. **SIMILAR TO** Dew-threads (*Drosera filiformis*), known only from FL, NC, and DE; its leaves are very narrow and threadlike with red glandular hairs.

Pipsissewa, Spotted Wintergreen

Chimaphila maculata (Linnaeus) Pursh
Heath Family (Ericaceae)

DWARF SHRUB found in acidic pine-oak forests in north Georgia and the upper Coastal Plain. **STEMS:** 4–8 inches tall, somewhat woody. **LEAVES:** Up to 4 inches long and 1¼ inches wide; alternate, opposite, or whorled; leathery, toothed, oval to lance-shaped, dark green with white stripes along the veins, evergreen. **FLOWER CLUSTERS:** At the top of a reddish stalk 1½–7½ inches tall, with 1–5 flowers per stalk. **FLOWERS:** Up to ¾ inch wide, nodding, waxy, white or pink, with 5 spreading petals, a bright green ovary, and 10 stamens. **FLOWERING:** May–Jul. **RANGE:** FL west to MS, north to ON and QC; also AZ and Central America. **NOTES:** The name "wintergreen" refers to the evergreen leaves; the plant does not have a wintergreen smell. **SIMILAR TO** no other wildflower in Georgia.

Trailing Arbutus

Epigaea repens Linnaeus
Heath Family (Ericaceae)

DWARF, VINELIKE SHRUB found in acidic soils in sandy or rocky pine-oak woodlands in north Georgia and a few Coastal Plain counties. **STEMS:** Up to 20 inches long, creeping or sprawling along the ground, hidden in the leaf litter, covered with rust-colored hairs. **LEAVES:** Up to 3½ inches long and 2⅓ inches wide, alternate (or crowded and appearing opposite), evergreen, oval to oblong, entire, leathery, dull green, rough-textured, covered with rust-colored hairs when fresh. **FLOWER CLUSTERS:** Short, few-flowered spikes arising from the axils of upper leaves. **FLOWERS:** Up to ½ inch long, white or pink, waxy, tubular with 5 spreading lobes, fragrant. **FLOWERING:** Late Feb–Apr. **RANGE:** FL west to MS, north to MB and NL. **SIMILAR TO** no other wildflower in Georgia.

Pine-Sap

Hypopitys monotropa Crantz
SYNONYM: *Monotropa hypopitys* Linnaeus
Heath (Ericaceae) or Indian Pipe (Monotropaceae) Family

PERENNIAL HERB found in moist to dry forests in north Georgia; the entire plant is yellow, tan, pink, or red, and completely lacks chlorophyll. **STEMS:** 2–8 inches tall, fleshy, hairy, usually in groups of several stems. **LEAVES:** Reduced to small scales. **FLOWERS:** About ½ inch long, hairy, in a nodding cluster at the top of the stalk, each flower with 4 or 5 petals and 10 stamens. **FRUIT:** A ¼-inch-long capsule that turns upright as it matures. **FLOWERING:** May–Oct. **RANGE:** Most of North America, Europe, and Asia. **SIMILAR TO** Sweet Pine-sap (*Monotropsis odorata*), a rare species with tan, papery leaves and smooth, purplish-red stems and flowers.

A Forest Menage à Trois

Pine-sap and Indian Pipes (p. 170) totally lack chlorophyll and therefore cannot use sunlight to photosynthesize and make carbohydrates. Instead, their roots connect to an underground fungal network that is in turn attached to the roots of a green, photosynthesizing plant, usually a tree. The fungus extracts carbohydrates from the tree roots, and the roots of Pine-sap and Indian Pipes extract carbohydrates from the fungus. It's a three-way relationship—called mycoheterotrophy—among a parasitic plant, a symbiotic fungus, and a photosynthesizing plant. Until recently, mycoheterotrophs were considered saprophytes—plants that break down decaying organic matter to extract carbon. We now know that these plants lack the ability to break down organic matter and that the so-called saprophytes are either mycoheterotrophs or direct parasites, such as Bear Corn (*Conopholis americana*, p. 272). Mycoheterotrophic plants seem to acquire their nutrients for free because they do not have to make elaborate leaves and roots and chlorophyll pigments, but they do pay a price. They are completely dependent on their hosts and on specialized habitats for survival and reproduction. As a result, they sometimes end up on lists of rare and endangered plants. Worldwide, more than 400 plant species make a living as mycoheterotrophs, including Pennywort (*Obolaria virginica*, p. 214) and Northern Blue Thread (*Burmannia biflora*, p. 368).

Hairy Wicky

Kalmia hirsuta Walter
Heath Family (Ericaceae)

EVERGREEN SHRUB found in pine flatwoods and savannas in the Coastal Plain. **STEMS:** 2–3 feet tall, reddish, covered with long, sticky hairs. **LEAVES:** Up to ½ inch long and ⅓ inch wide, alternate, oval, covered with long, sticky hairs. **FLOWERS:** In clusters arising from the leaf axils, each less than ½ inch wide, cup-shaped, slightly 5-lobed, pink to white, with a ring of dark pink spots inside. **FLOWERING:** May–Jul. **RANGE:** GA, FL, AL, and SC. **NOTES:** The stamens arch downward, with each anther tucked into a tiny pocket inside the flower; when a foraging insect bumps against them, the anthers pop out and shower the insect with pollen. **SIMILAR TO** Carolina Bog Laurel (*Kalmia carolina*), listed as Threatened in Georgia, which occurs in mountain bogs and Fall Line sandhill swamps; its stems and leaves are smooth, and the leaves are in whorls of 3.

Mountain Laurel

Kalmia latifolia Linnaeus
Heath Family (Ericaceae)

EVERGREEN SHRUB found on dry to moist slopes with acid soils, stream bluffs, and ravines in north Georgia and several Coastal Plain counties. **STEMS:** Up to 30 feet tall, crooked, with reddish, shredding bark; often forming dense, impenetrable thickets (laurel "slicks" or "hells"). **LEAVES:** Up to 5 inches long and 2 inches wide, alternate, elliptic to lance-shaped, entire, smooth, leathery. **FLOWERS:** In clusters of 20–40 flowers, each about 1 inch wide, cup-shaped, slightly 5-lobed, dark pink to white, not fragrant. **FLOWERING:** Apr–Jun. **RANGE:** FL west to LA, north to IL and ME. **NOTES:** The stamens arch downward, with each anther tucked into a tiny pocket inside the flower; the anthers pop out and shower the insect with pollen. **SIMILAR TO** Piedmont Rhododendron (*Rhododendron minus*), page 173.

Shining Fetterbush

Lyonia lucida (Lamarck) K. Koch
Heath Family (Ericaceae)

EVERGREEN SHRUB found in bogs, wet pine flatwoods, and swamps in the Coastal Plain and a few north Georgia counties. **STEMS:** 1–13 feet tall, young twigs sharply angled. **LEAVES:** Up to 4 inches long and 2 inches wide, alternate, elliptic, entire, stiff and leathery, glossy dark green on the upper surface, pale green on the lower, with a raised vein running around the edge. **FLOWERS:** In dangling clusters on the previous year's growth, each flower about ⅓ inch long; red, pink, or white; cylinder-shaped with 5 tiny, upturned lobes. **FLOWERING:** Apr–early Jun. **RANGE:** FL west to LA, north to VA; Cuba. **SIMILAR TO** Dog-hobble (*Leucothoe axillaris*), which occurs in shrubby bogs and blackwater stream swamps in the Coastal Plain; its white flowers are in elongated clusters, and the leaves are lance-shaped and spiny-toothed.

Indian Pipes, Ghost Flower

Monotropa uniflora Linnaeus
Heath (Ericaceae) or Indian Pipe (Monotropaceae) Family

PERENNIAL HERB found in forests throughout Georgia; the entire plant is white or pale pink, translucent, and completely lacking chlorophyll. **STEMS:** 2–12 inches tall, waxy, smooth, usually in groups of several stems; they become shriveled and black after fruiting. **LEAVES:** Absent, replaced by a few small, pointed scales. **FLOWERS:** About ½ inch long, solitary at the top of the stem, nodding, with 3–6 petals and 10 stamens. **FRUIT:** A ½-inch capsule that turns upright as it matures. **FLOWERING:** Jun–Oct. **RANGE:** Most of North America; Mexico, Colombia, Asia. **NOTES:** Lacking chlorophyll, Indian Pipes cannot photosynthesize; instead they derive nutrients from a three-way relationship with fungus and other plants (see "A Forest Menage à Trois," p. 168). **SIMILAR TO** Pine-sap (*Hypopitys monotropa*, synonym *Monotropa hypopitys*), page 168.

Sweet Azalea, Smooth Azalea

Rhododendron arborescens (Pursh) Torrey
Heath Family (Ericaceae)

DECIDUOUS SHRUB found on stream banks and in swamps and bogs in north Georgia. **STEMS:** 3–20 feet tall, twigs smooth, sometimes waxy-whitened. **LEAVES:** Up to 4 inches long and 1½ inches wide, alternate, oval, entire, smooth with bristly margins, sweetly fragrant when crushed. **FLOWERS:** Opening after leaves expand, 1–2 inches long with a slender tube and 5 flaring lobes, white (rarely pink) without a colored patch on the upper lobe, very fragrant, glandular-hairy inside and out, with 5 red stamens and the pistil extending well beyond the lobes. **FLOWERING:** Late May–Jul. **RANGE:** GA and AL north to KY and PA. **SIMILAR TO** Swamp Azalea (*Rhododendron viscosum*), which occurs in swamps, bottomlands, and on stream banks in the Coastal Plain and several north Georgia counties; it has hairy twigs, sticky-hairy flowers and flower stalks, and white stamens and pistils.

Flame Azalea

Rhododendron calendulaceum (Michaux) Torrey
Heath Family (Ericaceae)

DECIDUOUS SHRUB found in upland forests in the mountains and upper Piedmont. **STEMS:** 4–13 feet tall. **LEAVES:** Up to 3½ inches long and 1⅓ inches wide, alternate, elliptic, entire, hairy or smooth, with bristly margins. **FLOWERS:** Opening at the same time the leaves expand, ½–1 inch long with a slender tube and 5 flaring lobes; orange, yellow, or red, with a yellow patch on the upper lobe; covered with gland-tipped hairs on the outside, 5 stamens and the pistil extending well beyond the lobes. **FLOWERING:** May–Jun. **RANGE:** GA north to OH and PA. **SIMILAR TO** Oconee Azalea (*Rhododendron flammeum*), which occurs in upland forests in the Piedmont and upper Coastal Plain; its flowers open in April and are hairy but not glandular on the outside. Cumberland Azalea (*R. cumberlandense*), a mountain species, flowers Jun–Jul.

Piedmont Azalea, Southern Pinxterbloom Azalea

Rhododendron canescens (Michaux) Sweet
Heath Family (Ericaceae)

DECIDUOUS SHRUB found in moist to dry forests, swamps, flatwoods, often on stream banks, throughout Georgia. **STEMS:** Solitary, 4–20 feet tall, new growth very hairy. **LEAVES:** Up to 5 inches long and 1½ inches wide, alternate, oval, entire, hairy on both surfaces with bristly margins. **FLOWERS:** Opening before or at the same time the leaves expand, 1–2 inches long, pink to nearly white (without yellow markings), sweetly fragrant, sticky-hairy; the tube is much longer than the 5 short, spreading lobes; 5 stamens and the pistil extend well beyond the lobes. **FLOWERING:** Mar–Apr. **RANGE:** FL west to TX, north to OK and DE. **SIMILAR TO** Wild Azalea (*Rhododendron periclymenoides*, synonym *R. nudiflorum*), which occurs in similar habitats in north Georgia; its flowers are not fragrant, and the flower tube and the lobes are about the same length.

Rosebay Rhododendron, Great Laurel

Rhododendron maximum Linnaeus
Heath Family (Ericaceae)

EVERGREEN SHRUB found on moist slopes and in bottomlands and bogs in the Blue Ridge and upper Piedmont. **STEMS:** Up to 33 feet tall, with reddish, smooth or shredding bark and crooked trunks. **LEAVES:** Up to 12 inches long and 3 inches wide, alternate, oblong, tapering to the tip and base, entire, thick, leathery. **FLOWERS:** In large clusters, each flower 1–1½ inches wide, with 10 stamens, 5 white to pale pink lobes with a patch of yellowish-green spots on the upper lobe; not fragrant. **FLOWERING:** Jun–Aug. **RANGE:** GA and AL, north to OH and ME. **SIMILAR TO** Catawba Rhododendron (*Rhododendron catawbiense*), which occurs on sandstone cliffs, high-elevation ridges, and rocky mountaintops in north Georgia; its leaves are rounded at the base and tip, and the flowers, which bloom Apr–Jun, are dark pink to purple.

Piedmont Rhododendron

Rhododendron minus Michaux
Heath Family (Ericaceae)

EVERGREEN SHRUB found on dry to moist slopes, ridges, mountaintops, and in rocky ravines in north Georgia and a few southwest Georgia counties. **STEMS:** Up to 10 feet tall, crooked. **LEAVES:** Up to 5 inches long and 2 inches wide, alternate, elliptic to oval, entire, leathery. **FLOWERS:** In clusters, each flower ½–1 inch wide, 5 pink (rarely white) lobes with a patch of yellowish spots on the upper lobe and 10 stamens; not fragrant. **FLOWERING:** Late Apr–Jun. **RANGE:** GA, AL, TN, SC, and NC. **NOTES:** Twigs, leaves, flowers, and fruits are covered with rust-brown scales. **SIMILAR TO** Rosebay Rhododendron (*Rhododendron maximum*), page 172, which has larger leaves and flowers, blooms later, and lacks the rusty scales.

Is It a Blueberry or a Huckleberry?

Blueberries and huckleberries both produce dark blue, sweet-tart berries, but only one of them makes its way into Georgia's grocery stores and produce stands. Blueberries (genus *Vaccinium*) typically have 5 to many small, soft seeds in each fruit; huckleberries (genus *Gaylussacia*) have 10 seeds in each fruit. Huckleberry seeds are relatively large and hard, making for a snack that seems to be more crunchy seed than sweet fruit. To tell these two groups of shrubs apart before the fruits appear, check the leaves. The lower surface of most huckleberry leaves is covered with many tiny, golden dots (you may need a 10× lens). The lower surface of blueberry leaves is green and often pale and waxy, and never has gland dots. Learn to tell these two apart, and you can leave the huckleberries for the birds and bears. Feast instead on the dozen or so delicious blueberry species that occur throughout Georgia.

Sparkleberry

Vaccinium arboreum Marshall
Heath Family (Ericaceae)

DECIDUOUS SHRUB OR SMALL TREE found in dry woodlands throughout Georgia. **STEMS:** Up to 30 feet tall, with thin, gray bark peeling away and exposing reddish-brown inner bark. **LEAVES:** 1–2¾ inches long and ½–1½ inches wide, alternate, elliptic to nearly round, leathery, usually entire; upper surface dark, glossy green, lower surface pale green, turning maroon in the fall. **FLOWERS:** Abundant, each about ¼ inch long, white, narrowly bell-shaped, with 5 tiny, upturned lobes and short stamens. **FRUIT:** A round, dry, black berry about ¼ inch wide. **FLOWERING:** Late Apr–Jun. **RANGE:** FL west to TX, north to MO and VA. **SIMILAR TO** Deerberry (*Vaccinium stamineum*), which also occurs in dry woods throughout Georgia; its flowers are widely bell-shaped with 5 lobes and 10 long, conspicuous stamens; the leaves are pale green, the lower surfaces often waxy white.

Highbush Blueberry

Vaccinium corymbosum Linnaeus
Heath Family (Ericaceae)

DECIDUOUS SHRUB found in mountain bogs and wet to moist forests in Georgia's mountains and adjacent Piedmont. **STEMS:** 6–12 feet tall, with a single much-branched trunk; young twigs green, smooth or with whitish hairs in lines, older twigs reddish-brown. **LEAVES:** 1¼–3 inches long and ¾–1¼ inches wide, alternate, oval or elliptic, entire or finely toothed, with white hairs on both surfaces. **FLOWERS:** About ⅓ inch long, white, cylinder-shaped with 5 tiny, curved lobes at the tip. **FRUIT:** A sweet, dark blue berry about ⅓ inch across. **FLOWERING:** May. **RANGE:** GA west to TN, north to MI and NS. **SIMILAR TO** the semi-evergreen Southern Highbush Blueberry (*Vaccinium formosum*), which occurs in wetlands in the Coastal Plain; its young twigs are smooth and waxy white, both leaf surfaces are smooth, and the lower surface is waxy white.

The BUZZZZZZZZ on Pollination

On a sunny day, you may well hear a blueberry bush before you see it. Blueberry bushes in flower are alive with bumblebees moving from bloom to bloom. Listen closely—bumblebees won't attack. When the bee lights on a blueberry flower, you'll hear another kind of buzz—a louder and deeper "Bronx cheer" kind of buzz. What you are hearing is buzz pollination (also called sonication). In most flowers, insects easily gather the pollen from the surface of the anthers, but blueberry pollen is stored *inside* tubular anthers that have a single pore at the tip. When a bumblebee grabs the tip of a blueberry anther, she vibrates her flight muscles hundreds of times per second, creating a deep, resonant buzz and shaking the pollen out of the pore. The bee tucks most of the pollen into the pollen baskets on her hind legs and carries it back to her nest. Some of the pollen left on her furry body is carried to flowers on other blueberry plants. Cross-pollination! About 20,000 species worldwide are pollinated with buzz pollination, including Wild Senna (p. 204), meadow-beauties (pp. 258–260), Eastern Shooting-star (p. 299), and species in the Nightshade Family (pp. 336–338). Honeybees, a European species introduced centuries ago to the New World, don't buzz pollinate, so, for the sake of blueberries, encourage native bees in your garden.

Mayberry

Vaccinium elliottii Chapman
Heath Family (Ericaceae)

DECIDUOUS SHRUB found in moist to dry uplands and bottomlands in the Coastal Plain and Piedmont. **STEMS:** 3–10 feet tall, much-branched, bright green. **LEAVES:** ½–1¼ inches long and ¼–½ inch wide, alternate, oval or elliptic, bright green, sometimes red-tinged, shiny, finely toothed. **FLOWERS:** In clusters of 3–6 flowers, each about ¼ inch long, white tinged with pink, cylinder-shaped with 5 tiny, curved lobes at the tip. **FRUIT:** A sweet, black berry about ¼ inch wide. **FLOWERING:** Mar–Apr, before or with the earliest leaves. **RANGE:** FL west to TX, north to AR and VA. **SIMILAR TO** other tall, shrubby *Vaccinium* species but distinguished by its bright green twigs and early spring flowering.

Evergreen Blueberry

Vaccinium myrsinites Lamarck
Heath Family (Ericaceae)

EVERGREEN SHRUB found in pine flatwoods in the Coastal Plain. **STEMS:** Up to 2 feet tall, pale green, often forming large colonies. **LEAVES:** About ½ inch long and ¼ inch wide, alternate, oval or elliptic, leathery, glossy, dark green, with tiny black, gland-tipped hairs on the lower surface and gland-tipped teeth on the margins (use 10× lens). **FLOWERS:** About ¼ inch long, white or pink, urn-shaped with 5 tiny, spreading lobes. **FRUIT:** A sweet, black berry about ¼ inch wide. **FLOWERING:** Mar–Apr, before new twig growth. **RANGE:** GA, FL, AL, and SC. **SIMILAR TO** Darrow's Blueberry (*Vaccinium darrowii*), which occurs in similar habitats, often with Evergreen Blueberry; its stems, leaves, and fruits are waxy blue-green, and the leaves lack the gland-tipped hairs and teeth.

Hillside Blueberry

Vaccinium pallidum Chapman
SYNONYM: *Vaccinium vacillans* Kalm ex Torrey
Heath Family (Ericaceae)

DECIDUOUS SHRUB found on dry slopes and ridges in north Georgia. **STEMS:** Up to 3 feet tall, green, smooth, usually forming large colonies. **LEAVES:** 1–1½ inches long and ½–1 inch wide, alternate, oval or elliptic, pale bluish-green (sometimes red-tinged), finely toothed, usually smooth, with a waxy white coating on the lower surface. **FLOWERS:** About ⅓ inch long, white tinged with pink, cylinder-shaped with 5 tiny, curved lobes at the tip. **FRUIT:** A sweet, dark blue, waxy berry about ¼ inch wide. **FLOWERING:** Mar–Apr. **RANGE:** GA west to OK and KS, north to WI and ME. **SIMILAR TO** Southern Lowbush Blueberry (*Vaccinium tenellum*), which occurs in dry woodlands in the Coastal Plain; its twigs are hairy and the lower leaf surface is covered with red, stalked glands (use a 10× lens).

Spurge Family (Euphorbiaceae)

The Spurge Family is one of the largest and most diverse of the plant families, with most of its 7,500 species occurring in the tropics, where they are famous for producing rubber, cassava, and castor oil. Georgia has about 50 native spurges and another 13 or so exotics. Some plants in the Spurge Family, including most *Euphorbia* species, ooze milky latex when they are cut or broken (break a leaf off your Christmas poinsettia, *Euphorbia pulcherrima*). *Euphorbia* species are also distinguished by their miniature inflorescences, which are called cyathia. At first glance, *Euphorbia* flowers appear to have five white or maroon petals. But the cyathium actually consists of a tiny cup that encloses a female flower (a single pistil that develops into the fruit) and several male flowers (each consisting of a single stamen). Around the rim of the cup are 5 tiny, nectar-producing glands that bear colorful, petal-like appendages that attract pollinators. Other members of the Spurge Family, such as Tread-softly and the crotons, have more "normal" flowers. This is a great family for applying your 10× hand lens. Depending on the species, look for the peculiar flowers, scaly leaves, or star-shaped hairs on stems and leaves.

Short-Stalked Copper-Leaf

Acalypha gracilens A. Gray
Spurge Family (Euphorbiaceae)

ANNUAL HERB found in woodlands, prairies, and open disturbed areas throughout Georgia. **STEMS:** Up to 2½ feet tall, very hairy, branched. **LEAVES:** Up to 2½ inches long and ¾ inch wide, alternate, oval to oblong, toothed, hairy or smooth, with 3 prominent veins. **FLOWER CLUSTERS:** A short spike with many male flowers clustered at the tip, 1–5 female flowers at the base, and a large, leafy bract with short, triangular teeth (or lobes) and tiny, stalked glands (use 10× lens). **FLOWERS:** Tiny, with 4 white, yellow, or translucent sepals; there are no petals. **FRUIT:** A tiny, hairy, 3-lobed capsule. **FLOWERING:** Jun–Nov. **RANGE:** FL west to TX, north to MI and ME. **SIMILAR TO** Rhombic Copper-leaf (*Acalypha rhomboidea*) and Virginia Copper-leaf (*A. virginica*), which occur in similar habitats; their floral bracts have elongated, lance-shaped lobes.

Tread-Softly, Spurge-Nettle

Cnidoscolus stimulosus (Michaux) Engelmann & A. Gray
Spurge Family (Euphorbiaceae)

PERENNIAL HERB found in sandhills and other dry, sandy areas in the Coastal Plain and Piedmont. **STEMS:** 20–40 inches tall. **LEAVES:** 3–12 inches wide, alternate, deeply lobed and toothed. **FLOWER CLUSTERS:** The central flower is usually female (look for a green ovary in the flower tube); the outer flowers are male (look for 8 or more stamens inside the flower tube; see photo). **FLOWERS:** About 1 inch wide, with 5 white, spreading calyx lobes and a ½-inch tube; there are no petals. **FRUIT:** A 3-lobed capsule about ½ inch long. **FLOWERING:** Mar–Aug. **RANGE:** FL west to LA, north to VA. **NOTES:** Stems, leaves, and fruits are covered with stinging hairs filled with an irritating chemical that is released when the hairs are damaged. **SIMILAR TO** no other wildflower in Georgia.

Silver Croton

Croton argyranthemus Michaux
Spurge Family (Euphorbiaceae)

PERENNIAL HERB found in sandhills and other dry pinelands in the Coastal Plain. **STEMS:** Up to 2 feet tall, usually branched, covered with reddish-brown scales. **LEAVES:** Up to 2 inches long, alternate, oval or lance-shaped, entire, long-stalked, green on the upper surface, silvery white on the lower. **FEMALE FLOWERS:** Few, small, with 5 greenish sepals and no petals. **MALE FLOWERS:** In the same cluster with female flowers, numerous, with 5 white petals and 10 stamens. **FRUIT:** A round, 3-lobed capsule about ¼ inch wide. **FLOWERING:** Spring–summer. **RANGE:** FL, GA, west to TX and OK; Mexico. **NOTES:** Minute, silvery scales cover the lower leaf surfaces, flowers, and fruits (use a 10× lens). **SIMILAR TO** Prairie-tea (*Croton monanthogynus*), which occurs on limestone outcrops in northwest Georgia and blackland prairies in central Georgia; its male flowers lack silvery scales, and it has a 2-lobed ovary.

Beach Croton

Croton punctatus Jacquin
Spurge Family (Euphorbiaceae)

ANNUAL OR SHORT-LIVED PERENNIAL HERB found on dunes and in grasslands on Georgia's barrier islands. **STEMS:** Up to 4 feet tall, with forking branches, solitary or in large, rounded clumps. **LEAVES:** Up to 2½ inches long and 1½ inches wide, alternate, oval or elliptic, entire. **FLOWERS:** Small, with a rust-colored, 5-lobed calyx and 5 tiny, white petals; female and male flowers are in the same cluster. **FRUIT:** A ¼-inch, 3-lobed capsule. **FLOWERING:** Late May–Nov. **RANGE:** FL west to TX, north to NC; Central and South America. **NOTES:** All parts of the plant are densely covered with minute, star-shaped hairs (use a 10× lens); except for those on the upper leaf surface, the hairs have a raised red dot in the center, giving the plant a rusty look. **SIMILAR TO** no other dune or beach plant.

Glade Rush-Foil

Croton willdenowii G. L. Webster
SYNONYM: *Crotonopsis elliptica* Willdenow
Spurge Family (Euphorbiaceae)

ANNUAL HERB found on Piedmont granite outcrops and other dry habitats throughout Georgia. **STEMS:** Up to 14 inches tall. **LEAVES:** Up to 1½ inches long and ¾ inch wide, alternate, linear to elliptic, entire, folded during dry weather, upper surface green and covered with branched hairs. **FEMALE FLOWERS:** About ⅛ inch wide, with a 5-lobed calyx and no petals. **MALE FLOWERS:** In the same small cluster as the female flowers, with 5 tiny, white petals. **FLOWERING:** Jun–Oct. **RANGE:** FL west to TX, north to IA and CT. **NOTES:** Except for the upper leaf surfaces, these plants are densely covered with silvery scales with a reddish-brown central dot. **SIMILAR TO** Doveweed (*Croton glandulosus* variety *septentrionalis*), which occurs in disturbed areas throughout Georgia; its leaves are coarsely toothed, and the stem is covered with long, branched hairs, not scales.

Flowering Spurge

Euphorbia corollata Linnaeus
Spurge Family (Euphorbiaceae)

PERENNIAL HERB found in dry upland woods in north Georgia and a few Coastal Plain counties. **STEMS:** Up to 3 feet tall, smooth (rarely hairy). **LEAVES:** Up to 2¾ inches long and 1½ inches wide, alternate (except for a whorl of 3 leaves below the flower cluster), angled upward, entire, oblong or elliptic, somewhat leathery, with very short or no stalks. **FLOWERS:** About ⅓ inch wide, with 5 white, petal-like structures attached to tiny, green glands in the center. **FRUIT:** About ⅛ inch wide, green, 3-lobed, held on a drooping stalk. **FLOWERING:** Jun–Sep. **RANGE:** FL west to TX, north to SD, ON, and ME. **SIMILAR TO** Southeastern Flowering Spurge (*Euphorbia pubentissima*), which occurs in similar habitats; its leaves are thin-textured, angle downward, and have short stalks; the flower is less than ¼ inch wide.

Maroon Sandhills Spurge

Euphorbia exserta (Small) Coker
SYNONYM: *Euphorbia gracilior* Cronquist
Spurge Family (Euphorbiaceae)

PERENNIAL HERB found in sandhills in southeast Georgia. **STEMS:** Up to 1 foot long, erect or sprawling. **LEAVES:** Up to 2¾ inches long and ¾ inch wide; alternate, opposite, or whorled (sometimes all three on the same plant); linear, oval, or round, green or blue-green with a narrow red margin. **FLOWERS:** About ⅛ inch wide, held at the tip of a long, red stalk, with 5 maroon, petal-like structures attached to tiny, maroon glands in the center. **FRUIT:** About ⅛ inch long and wide, dark red, 3-lobed, held on a drooping stalk that emerges from the center of the flower. **FLOWERING:** Mar–Jun. **RANGE:** FL north to VA. **SIMILAR TO** Carolina Ipecac (*Euphorbia ipecacuanhae*), which occurs in Fall Line sandhills and river dunes; it has opposite, fleshy leaves and forking branches; its flowers are yellowish-green or maroon.

Spotted Spurge, Prostrate Spurge

Euphorbia maculata Linnaeus
SYNONYMS: *Euphorbia supina* Rafinesque, *Chamaesyce maculata* (Linnaeus) Small
Spurge Family (Euphorbiaceae)

ANNUAL HERB found in sidewalk cracks, gardens, roadsides, and other disturbed areas throughout Georgia. **STEMS:** Spreading across the ground, often forming mats up to 2 feet wide, dark red, hairy, with milky latex. **LEAVES:** Up to ½ inch long and ¼ inch wide, opposite, oval or oblong, with minute teeth on the margins and usually a maroon blotch in the center. **FLOWERS:** Less than ⅛ inch across, held in the leaf axils, with 4 or 5 pink or white, petal-like glands. **FLOWERING:** All year. **RANGE:** Most of North America. **SIMILAR TO** several other widespread weedy spurges: Eyebane (*Euphorbia nutans*) has hairy, upwardly curved or nearly erect stems; Hyssopleaf Sandmat (*E. hyssopifolia*) has smooth stems and leaves with obviously toothed margins and pointed tips.

Cumberland Spurge, Mercury Spurge

Euphorbia mercurialina Michaux
Spurge Family (Euphorbiaceae)

PERENNIAL HERB found in rich, moist forests over mafic or limestone bedrock in northwest Georgia. **STEMS:** Up to 15 inches tall, branched. **LEAVES:** Up to 3 inches long and 1 inch wide, opposite or whorled, oval or elliptic, with long hairs on the margins. **FLOWERS:** Less than ¼ inch wide, green with 5 short, white gland appendages forming a narrow rim around the flower. **FRUIT:** About ⅒ inch wide, green, 3-lobed, held on a drooping stalk that emerges from the center of the flower. **FLOWERING:** Apr–Sep. **RANGE:** GA, FL, AL, TN, NC, KY, and VA. **SIMILAR TO** Summer Spurge (*Euphorbia discoidalis*), which occurs in Coastal Plain sandhills; the gland appendages are white, oval, and petal-like; its leaves are narrowly linear and less than 2½ inches wide.

Queen's Delight

Stillingia sylvatica Garden ex Linnaeus
Spurge Family (Euphorbiaceae)

PERENNIAL HERB found in sandhills and other dry woodlands in the Coastal Plain and a few Piedmont counties. **STEMS:** Up to 2½ feet tall, yellow-green, smooth, often branched from the base. **LEAVES:** Up to 3½ inches long and 1½ inches wide, alternate, elliptic, glossy, finely toothed, each tooth tipped with a tiny, red gland. **FLOWERS:** In a yellowish-green spike at the top of the stem, a few female flowers at the base and many male flowers above, with a round nectar gland on each side of the flower. **FRUIT:** About ⅓ inch long and wide, 3-lobed, green. **FLOWERING:** May–Jul. **RANGE:** FL west to NM, north to KS and VA. **SIMILAR TO** Corkwood (*Stillingia aquatica*), which occurs in ponds in the Coastal Plain; it has a single stem up to 5 feet tall, and its leaves are less than ½ inch wide.

Nettle-Leaf Noseburn

Tragia urticifolia Michaux
Spurge Family (Euphorbiaceae)

PERENNIAL HERB found in dry woodlands in the Piedmont and a few Coastal Plain counties, especially over mafic and limestone bedrock. **STEMS:** Up to 2 feet tall. **LEAVES:** Up to 2½ inches long and 1½ inches wide, alternate, coarsely toothed, lance- or heart-shaped with a rounded or straight base. **FLOWERS:** In a 1-inch spike at the top of the stem, with 3–8 greenish or purplish sepals (no petals), 1 or 2 female flowers at the base of the spike, male flowers above. **FRUIT:** About ¼ inch wide, 3-lobed, green, at the base of the spike. **FLOWERING:** May–Oct. **RANGE:** FL west to AZ, north to CO and VA. **NOTES:** Long, stinging hairs cover the stems and leaves. **SIMILAR TO** Wavyleaf Noseburn (*Tragia urens*), which occurs in sandhills and other dry woodlands; its leaves have few teeth and wedge-shaped bases.

Bean Family (Fabaceae or Leguminosae)

The Bean Family, also called the Pea or Legume Family, is the third largest plant family. Worldwide there are nearly 19,000 species, many with enormous ecological, economic, or cultural value. Georgia has about 200 species, 140 of them native to the state. Botanists have divided the Bean Family into three subfamilies, each distinguished by its own unusual flower type. Members of all three subfamilies produce bean pods (legumes), and the leaves of most species are usually divided into 3 to many leaflets. There are many differences among subfamilies, but recent molecular studies have shown that the legumes are more similar than different genetically and belong together in a single family.

The most common of the three subfamilies, the Faboideae, is the one with the typical pea-flower shape. The largest of its 5 petals is an erect banner petal (sometimes called a standard). Two wing petals spread to the side, droop downward, or project forward. The 2 remaining petals are fused into a single petal called a keel that projects forward from the center of the flower. This subfamily is the one most familiar to gardeners, farmers, and plant lovers. Most of the plants in this subfamily "fix" nitrogen. *Rhizobium* bacteria that live in nodules on the plants' roots gather nitrogen from the air and convert it to nitrates or ammonia, which enrich the soil.

Another subfamily, the Mimosoideae, is most familiar to southerners as the exotic (and invasive) Mimosa tree, which has fragrant flower clusters that look like puffs or pompoms. There are many tiny flowers in a single pompom, each with long, colorful stamens that form the ball-shaped puff. Several species in this subfamily are "sensitive plants"; their leaflets fold up when touched (see Sensitive Brier, p. 201).

Members of the Caesalpinioideae subfamily have the most "normal" flowers of the three subfamilies, with 5 nearly identical petals. This group of plants is famous for bearing extrafloral nectaries—nectar-secreting glands—on their leaves. Ants feast on the nectar and attack other insects that try to eat the leaves, a classic case of mutualism (see Wild Senna, p. 204).

Faboideae

Mimosoideae

Caesalpinioideae

Tall Indigo-Bush

Amorpha fruticosa Linnaeus
Bean Family (Fabaceae)

DECIDUOUS SHRUB found in floodplains and upland forests throughout Georgia. **STEMS:** 3–10 feet tall, upper branches light green and hairy, lower stem woody, smooth, and gray. **LEAVES:** Up to 12 inches long, alternate, with 11–35 oblong, toothed leaflets up to 2 inches long and ¾ inch wide, covered with translucent dots and tipped with a tiny bristle. **FLOWER CLUSTERS:** A densely flowered spike 4–8 inches long. **FLOWERS:** About ¼ inch long, with a single pink to dark purple petal wrapped around 10 long, yellow-tipped stamens. **FLOWERING:** Apr–Jun. **RANGE:** Most of North America. **SIMILAR TO** Dwarf Indigo-bush (*Amorpha herbacea* variety *herbacea*), which occurs in sandhills, pine flatwoods, and dry, disturbed sites, mostly in eastern Coastal Plain counties; it is usually less than 3 feet tall, its stems and leaves are very hairy, and the flowers are white or blue-violet.

Hog Peanut

Amphicarpaea bracteata (Linnaeus) Fernald
Bean Family (Fabaceae)

ANNUAL HERBACEOUS VINE found on stream banks and in moist woods and thickets in north Georgia and a few Coastal Plain counties. **STEMS:** Up to 6 feet long, twining and sprawling over other plants, hairy. **LEAVES:** 2–6 inches long, alternate, with 3 oval, entire, thin-textured leaflets up to 2½ inches long. **FLOWER CLUSTERS:** With 2–15 flowers nodding on long stalks arising from the leaf axils. **FLOWERS:** About ½ inch long; white, pale pink, or lavender, tubular with 3 petals (wings and keel) projecting forward and the upper banner petal folded backward at the tip. **FLOWERING:** Jul–Sep. **RANGE:** FL west to TX, north to MT, MB, and QC. **NOTES:** Threadlike runners spread in the leaf litter and produce small, closed, self-pollinated flowers that yield underground pods with edible seeds ("hog peanuts"). The leaves are **SIMILAR TO** those of Wild Bean (*Phaseolus polystachios*), page 202.

Ground-Nut

Apios americana Medikus
Bean Family (Fabaceae)

PERENNIAL HERBACEOUS VINE found in floodplains and other wetlands throughout Georgia. **STEMS:** Up to 10 feet long, smooth, twining over other plants. **LEAVES:** 4–8 inches long, alternate, with 5–7 oval to lance-shaped, entire leaflets, each less than 2½ inches long. **FLOWER CLUSTERS:** Up to 6 inches long, arising from the leaf axils. **FLOWERS:** About ½ inch wide, fragrant, with a folded, pinkish-brown banner petal, 2 maroon wing petals, and a twisted, maroon keel petal. **FLOWERING:** Jun–Sep. **RANGE:** FL west to TX, north to SD and NS. **NOTES:** Ground-nut's tubers, each 1–4 inches in diameter, were a popular food with Native Americans and European settlers; although they resemble small potatoes, they have more than 3 times the protein. With its folded and bicolored flowers, Ground-nut is **SIMILAR TO** no other wildflower in Georgia.

Bearded Milk-Vetch

Astragalus villosus Michaux
Bean Family (Fabaceae)

PERENNIAL HERB found in sandhills and other dry, open, sandy areas in the Coastal Plain. **STEMS:** Up to 8 inches long, sprawling across the ground and forming mats. **LEAVES:** 1½–3½ inches long, alternate, with 5–15 round or oval, opposite leaflets, each less than ½ inch long. **FLOWERS:** In a long-stalked cluster at the end of the stem, each flower about ⅓ inch long, pale yellow, with a typical pea-flower shape. **FLOWERING:** May–Jun. **RANGE:** FL west to MS, north to TN and SC. **NOTES:** Long, soft, white hairs cover stems, leaflets, flower stalks, calyxes, and seed pods. **SIMILAR TO** Canada Milk-vetch (*Astragalus canadensis*), a Georgia Special Concern species that occurs in moist forests over limestone bedrock in northwest Georgia; it is an erect plant 1½–5 feet tall.

White Wild Indigo

Baptisia alba (Linnaeus) Ventenat
SYNONYM: *Baptisia pendula* Larisey
Bean Family (Fabaceae)

PERENNIAL HERB found in dry woodlands, roadsides, and clearings throughout Georgia. **STEMS:** Up to 4 feet tall, smooth, waxy, blue-green to gray-purple. **LEAVES:** Alternate, divided into 3 bluish-green leaflets, ½–2½ inches long, with broad, rounded tips and tapering bases. **FLOWER CLUSTERS:** An erect, many-flowered spike held at the top of the stem. **FLOWERS:** About 1 inch long, white, typical pea-flower shape with an erect banner petal. **FRUIT:** A waxy, inflated pod up to 2 inches long and ½–1 inch wide, nodding, blue-green (drying black), brittle, with a short beak. **FLOWERING:** May–Jul. **RANGE:** GA, AL, FL, SC, and NC. **SIMILAR TO** Narrow-pod White Wild Indigo (*Baptisia albescens*), which occurs in similar habitats; its flowers are less than ¾ inch long, and the seed pod is about 1 inch long, erect, narrowly cylindrical, leathery, and yellowish-brown.

Blue Wild Indigo

Baptisia australis (Linnaeus) R. Brown var. *aberrans* (Larisey) M. Mendenhall
Bean Family (Fabaceae)

PERENNIAL HERB found in prairies and dry, open woodlands over mafic or limestone bedrock in north Georgia. **STEMS:** Up to 3 feet tall, with spreading branches, often with a whitish, waxy coating. **LEAVES:** Alternate, with conspicuous stipules at the base of the stalk, each leaf divided into 3 oval leaflets ½–3 inches long. **FLOWERS:** About 1 inch long, blue-violet, typical pea-flower shape. **FRUIT:** An inflated, oblong pod 1–2 inches long, leathery, drying black and brittle, with a sharply pointed tip. **FLOWERING:** Apr–May. **RANGE:** This variety occurs only in GA, TN, and NC. **SIMILAR TO** Tall Blue Wild Indigo (*Baptisia australis* variety *australis*), which occurs in northwest Georgia along streams; it grows to 5 feet tall and has nearly upright branches.

Creamy Wild Indigo

Baptisia bracteata Elliott
Bean Family (Fabaceae)

PERENNIAL HERB found in dry, open woodlands in north Georgia, especially over mafic or ultramafic bedrock. **STEMS:** 1–2 feet tall, softly hairy, branches spreading and forming low clumps. **LEAVES:** Alternate, with large stipules at the base of the stalk, each leaf divided into 3 oval or lance-shaped leaflets up to 1½ inches long and ½ inch wide. **FLOWER CLUSTERS:** Up to 8 inches long, densely flowered, drooping, the flowers usually on one side of the cluster. **FLOWERS:** Up to 1 inch long, cream-colored or pale yellow, typical pea-flower shape, with a leafy bract at the base of the flower stalk. **FRUIT:** A green (drying black), inflated pod 1–2 inches long, hairy or smooth, with a long, sharp tip. **FLOWERING:** Apr–May. **RANGE:** GA, AL, SC, and NC. **SIMILAR TO** no other wild indigo species in Georgia.

Gopherweed, Pineland Wild Indigo

Baptisia lanceolata (Walter) Elliott
Bean Family (Fabaceae)

PERENNIAL HERB found in sandhills and other dry, sandy woodlands in the Coastal Plain. **STEMS:** Up to 3 feet tall, finely hairy, much-branched and forming rounded "bushes." **LEAVES:** Alternate, divided into 3 lance-shaped leaflets, each up to 4 inches long and ½ inch wide. **FLOWERS:** About 1 inch long, yellow, solitary in the leaf axils or in small clusters at the tips of outer branches, typical pea-flower shape. **FRUIT:** An inflated pod about 1 inch long with a long beak at the tip, drying black and leathery. **FLOWERING:** Apr–May. **RANGE:** GA, AL, FL, and SC. **SIMILAR TO** Yellow Wild Indigo (*Baptisia tinctoria*), which occurs in dry woodlands in northeast Georgia and in sandhills in southeast Georgia; its flowers are about ½ inch long and wide, and the leaflets are usually less than 1 inch long with broadly rounded tips.

Cat-Bells

Baptisia perfoliata (Linnaeus) R. Brown ex Aiton
Bean Family (Fabaceae)

PERENNIAL HERB found in sandhills, river dunes, and Longleaf Pine woodlands in southeast Georgia. **STEMS:** 1–3 feet tall, yellowish-green, smooth, waxy. **LEAVES:** 2–4 inches long and wide, alternate, oval to nearly round, leathery, seemingly pierced by the stem. **FLOWERS:** About ½ inch long, yellow, held in the leaf axils, typical pea-flower shape. **FRUIT:** An inflated pod about ½ inch long and wide, with a short beak at the tip, leathery, green, drying to silvery brown. **FLOWERING:** Apr–May. **RANGE:** GA, FL, and SC. **NOTES:** Dried stems break off at ground level and form "tumbleweeds" that disperse seeds as they roll. **SIMILAR TO** the federally listed Hairy Rattleweed (*Baptisia arachnifera*), which has stems and leaves densely covered with long, white hairs. All other *Baptisia* species in Georgia have leaves with 3 leaflets.

Spurred Butterfly Pea

Centrosema virginianum (Linnaeus) Bentham
Bean Family (Fabaceae)

PERENNIAL HERBACEOUS VINE found in dry upland woods, bottomlands, and disturbed areas throughout Georgia. **STEMS:** Up to 5 feet long, trailing and twining over other plants. **LEAVES:** Alternate, with 3 leaflets up to 2¾ inches long and 1 inch wide; narrowly oval, lance-shaped, or elliptic, with conspicuous veins, especially on the lower surface. **FLOWERS:** 1–1½ inches long, pink or lavender; the banner petal is large, round, and somewhat flattened, with a central white patch and a small spur at the back; the calyx is green and cup-shaped with 5 narrow, pointed lobes about ½ inch long; the calyx is partly enclosed by 2 oval, green, ½-inch bracts. **FRUIT:** A narrow pod up to 5½ inches long. **FLOWERING:** Jun–Aug. **RANGE:** FL west to TX, north to AR and NJ. **SIMILAR TO** Pigeon Wing (*Clitoria mariana*), page 189, which has a scoop-shaped banner petal.

Partridge Pea

Chamaecrista fasciculata (Michaux) Greene
SYNONYM: *Cassia fasciculata* Michaux
Bean Family (Fabaceae)

ANNUAL HERB found in woodlands and disturbed areas throughout Georgia. **STEMS:** 1–3 feet tall, much-branched. **LEAVES:** Up to 2½ inches long, with 5–20 pairs of narrow leaflets; the stalk has a tiny, doughnut-shaped nectar gland about midway. **FLOWERS:** 1–1½ inches wide, with 5 bright yellow petals marked near the base with red, and 10 stamens with long, red or yellow anthers. **FRUIT:** A flat, narrowly oblong pod up to 3 inches long, turning dark brown as it matures. **FLOWERING:** May–Sep. **RANGE:** FL west to NM, north to SD and MA; Mexico. **NOTES:** The leaflets slowly fold up when touched. **SIMILAR TO** Sensitive Plant (*Chamaecrista nictitans*), which occurs in similar habitats; it is a smaller plant with flowers less than ½ inch wide with 5–8 stamens.

Pigeon Wing

Clitoria mariana Linnaeus
Bean Family (Fabaceae)

PERENNIAL HERB found in dry, open woodlands and forests, clearings, and roadsides throughout Georgia. **STEMS:** Up to 3 feet tall, erect or sprawling over other plants. **LEAVES:** Alternate, with 3 leaflets ½–2 inches long, oval to lance-shaped, with no veins showing on the pale lower surface. **FLOWERS:** About 2 inches long, pink or lavender, with a large, scoop-shaped banner petal with dark purple streaks in the center; the calyx is green with a ½-inch-long cylindrical tube; 2 narrow bracts clasp the base of the calyx tube. **FRUIT:** A narrow pod up to 2 inches long. **FLOWERING:** Jun–Aug. **RANGE:** FL west to TX, north to MN and NY; South America. **SIMILAR TO** Spurred Butterfly Pea (*Centrosema virginianum*), page 188, which has a flattened banner petal, obvious veins on the lower surface of the leaflets, and a short, cup-shaped calyx.

Rabbit Bells, Low Rattlebox

Crotalaria rotundifolia Walter ex J. F. Gmelin
Bean Family (Fabaceae)

PERENNIAL HERB found in dry, sandy or rocky woodlands and disturbed areas in the Coastal Plain and a few Piedmont counties. **STEMS:** Up to 2 feet long, hairy, spreading across the ground. **LEAVES:** Up to 2 inches long, alternate; oval, oblong, or linear, hairy on both surfaces (use 10× lens); stipules (winglike, leafy tissue along the stem) are usually absent. **FLOWERS:** Up to ½ inch long, yellow, typical pea-flower shape with an erect banner petal. **FRUIT:** An inflated pod up to 1 inch long and ½ inch wide, green turning black when dry. **FLOWERING:** Apr–Aug. **RANGE:** FL west to LA, north to VA; Mexico. **NOTES:** The seeds rattle around in the dried seed pods. **SIMILAR TO** Pursh's Rattlebox (*Crotalaria purshii*), which occurs in similar habitats in the Coastal Plain; it is an erect plant with winglike, arrow-shaped stipules along the stem.

Showy Rattlebox

Crotalaria spectabilis Roth
Bean Family (Fabaceae)

ANNUAL HERB found in disturbed areas throughout Georgia. **STEMS:** Up to 5 feet tall, stout, rough-hairy. **LEAVES:** Up to 6 inches long and 2 inches wide, alternate, oval or elliptic, entire, with very short stalks. **FLOWER CLUSTERS:** Up to 20 inches tall, crowded with many flowers. **FLOWERS:** About 1 inch long, yellow, typical pea-flower shape. **FRUIT:** An inflated pod up to 1¾ inches long and ½ inch wide, green, turning black when dry. **FLOWERING:** Jul–Sep. **RANGE:** Native to Asia, now widespread and invasive from FL west to TX, north to IL and VA. **SIMILAR TO** Rattleweed (*Crotalaria retusa*), another exotic species now widespread throughout the South. It is less than 3 feet tall, and its leaves are less than 3 inches long.

Summer Farewell

Dalea pinnata (J. F. Gmelin) Barneby
SYNONYM: *Petalostemum pinnatum* (J. F. Gmelin) Blake
Bean Family (Fabaceae)

PERENNIAL HERB found in sandhills, dry pine flatwoods, and Turkey Oak scrub. **STEMS:** Up to 3 feet tall, green or red. **LEAVES:** Divided into 3–15 very narrow leaflets less than ¾ inch long. **FLOWER CLUSTERS:** Held at the tips of branches, small, headlike, with 8–12 flowers and red or brown bracts forming a cup around the base. **FLOWERS:** Very different from most Bean Family flowers, with 5 white, narrow petals. **FLOWERING:** Aug–Nov. **RANGE:** FL west to LA, north to NC. **SIMILAR TO** two other species that occur in dry pinelands in the Coastal Plain; they both have elongated flower clusters. White Tassels (*Dalea albida*) has white flowers and 5 leaflets per leaf; Pink Tassels (*D. carnea*) has pink (sometimes white) flowers and 7–9 leaflets per leaf.

Small-Leaf Tick-Trefoil

Desmodium ciliare (Muhlenberg ex Willdenow) A. P. de Candolle
Bean Family (Fabaceae)

PERENNIAL HERB found in dry woodlands, prairies, and clearings throughout Georgia. **STEMS:** 1–5 feet tall, rough-hairy. **LEAVES:** Alternate, with stalks less than ¼ inch long; divided into 3 rough-hairy, oval to elliptic leaflets up to 1¼ inches long and ½ inch wide. **FLOWERS:** Less than ¼ inch long, dark pinkish-purple, typical pea-flower shape; the banner petal has 2 green or yellow patches at the base. **FRUIT:** A flat pod about ⅓ inch long with 2 or 3 oval, hairy segments that stick to fur and clothing. **FLOWERING:** Jun–Sep. **RANGE:** FL west to TX, north to MO and MA; Cuba. **SIMILAR TO** Maryland Tick-trefoil (*Desmodium marilandicum*), which occurs in similar habitats in north Georgia; its stems and leaflets are smooth or slightly hairy, and its leaf stalks are ½–1 inch long.

Panicled Tick-Trefoil, Beggar-Ticks

Desmodium paniculatum (Linnaeus) A. P. de Candolle
Bean Family (Fabaceae)

PERENNIAL HERB found along upland forest edges throughout Georgia. **STEMS:** Up to 4 feet tall, mostly smooth. **LEAVES:** Alternate, with stalks up to 2 inches long; divided into 3 narrow, sharply pointed, sparsely hairy leaflets up to 3 inches long. **FLOWER CLUSTERS:** Up to 12 inches long, the branches and flower stalks covered with rough, hooked hairs. **FLOWERS:** About ¼ inch long, dark pink, typical pea-flower shape; the banner petal has 2 green or yellow patches at the base. **FRUIT:** A flat pod with 3–5 segments covered with hooked hairs. **FLOWERING:** Jun–Sep. **RANGE:** FL west to TX, north to NE and ON. **SIMILAR TO** Smooth Tick-trefoil (*Desmodium laevigatum*), which occurs in similar habitats; it has waxy, bluish-green stems and oval, bluntly pointed leaflets that are pale and smooth on the lower surface.

Dollar-Leaf, Round-Leaf Tick-Trefoil

Desmodium rotundifolium A. P. de Candolle
Bean Family (Fabaceae)

PERENNIAL HERB found in dry, upland, hardwood forests in north Georgia and a few Coastal Plain counties. **STEMS:** 1½–6 feet long, very hairy, trailing across the ground and forming mats. **LEAVES:** Divided into 3 nearly round leaflets, each leaflet 1–2 inches wide, very hairy, lying flat on the ground. **FLOWERS:** Less than ½ inch long, dark pink to purplish, typical pea-flower shape, the banner petal with maroon markings at the base. **FRUIT:** A flat pod ½–1½ inches long, with 3–7 segments covered with hooked hairs that stick to fur and clothing. **FLOWERING:** Jun–Sep. **RANGE:** FL west to TX, north to KS and NH. **SIMILAR TO** Matted Tick-trefoil (*Desmodium lineatum*), also found in dry woodlands; its leaflets are less than 1 inch long, and its pods have 2 or 3 segments.

Coral Bean, Cherokee Bean

Erythrina herbacea Linnaeus
Bean Family (Fabaceae)

PERENNIAL SHRUBLIKE HERB found in coastal forests, sandy woodlands, and sandhills in the Coastal Plain. **STEMS:** Up to 5 feet tall, thorny, arising from a bulbous, woody base. **LEAVES:** Alternate, long-stalked, divided into 3 lobed, pointed leaflets up to 4 inches long and 3 inches wide, with curved prickles on the lower surface and stalk. **FLOWER CLUSTERS:** A foot-long spike held at the top of a long, leafless stalk. **FLOWERS:** 1½–2½ inches long, red, tubular, curved upward. **FRUIT:** A drooping pod up to 4 inches long, constricted between the seeds, drying brown and splitting to reveal several bright red seeds. **FLOWERING:** May–Jul. **RANGE:** FL west to TX and Mexico, north to NC. **NOTES:** All parts of the plant are toxic to humans, the seeds especially so. **SIMILAR TO** no other native plant in the Coastal Plain.

Elliott's Milk-Pea

Galactia elliottii Nuttall
Bean Family (Fabaceae)

PERENNIAL HERBACEOUS VINE found in moist forests, lake and marsh edges, coastal dunes, and roadsides in Georgia's coastal and near-coastal counties. **STEMS:** Up to 5 feet long, vinelike, twining around other plants, covered with short, stiff hairs. **LEAVES:** Alternate, divided into 5–9 elliptic or oblong leaflets ¾–2 inches long, slightly leathery in texture. **FLOWERS:** About ½ inch long, white with red streaks, typical pea-flower shape. **FLOWERING:** Jul–Sep. **RANGE:** GA, FL, and SC. With its twining stems, usually 7 leaflets, and white pea-flowers, Elliott's Milk-pea is **SIMILAR TO** no other wildflower in Georgia's Coastal Plain. Erect Milk-pea (*Galactia erecta*) has erect stems, white or lavender flowers, and leaves with 3 leaflets; it occurs in dry woodlands and savannas in the Coastal Plain.

Eastern Milk-Pea

Galactia regularis (Linnaeus) B.S.P.
Bean Family (Fabaceae)

PERENNIAL HERB found in dry, upland, hardwood and pine forests throughout Georgia. **STEMS:** Up to 10 feet long, usually trailing across the ground, covered with tiny, stiff hairs. **LEAVES:** Alternate, divided into 3 oval to elliptic, slightly leathery leaflets up to 2 inches long. **FLOWER CLUSTERS:** With closely spaced flowers for the entire length of the stalk. **FLOWERS:** About ½ inch long, dark pink or pink and white, typical pea-flower shape, the banner petal marked with darker pink at the base. **FLOWERING:** Jul–Oct. **RANGE:** FL west to TX, north to KS and NJ. **SIMILAR TO** Downy Milk-pea (*Galactia volubilis*), which occurs in similar habitats; its leaves and stems have long hairs, and the flowers are clustered on the upper half of the stalk.

Naked Tick-Trefoil, Beggar-Lice

Hylodesmum nudiflorum (Linnaeus) H. Ohashi & R. R. Mill
SYNONYM: *Desmodium nudiflorum* (Linnaeus) A. P. de Candolle
Bean Family (Fabaceae)

PERENNIAL HERB found in upland hardwood forests in north Georgia and a few Coastal Plain counties. **LEAFY STEMS:** 4–12 inches tall, with a cluster of leaves at the top. **FLOWERING STEMS:** Up to 3 feet tall, arising several inches from the leafy stem, leafless with a much-branched, open flower cluster at the top. **LEAVES:** Divided into 3 oval or diamond-shaped, stalked leaflets, each leaflet up to 3 inches long. **FLOWERS:** About ½ inch long, pink, typical pea-flower in shape. **FRUIT:** A flat pod about 1 inch long with 2–4 segments covered with hooked hairs that stick to clothing and fur. **FLOWERING:** Jul–Aug. **RANGE:** FL west to TX, north to ON and QC. **SIMILAR TO** other tick-trefoils but distinguished by the leafless flowering stem arising separately from the leafy stem.

Carolina Wild Indigo

Indigofera caroliniana P. Miller
Bean Family (Fabaceae)

PERENNIAL HERB found in sandhills, dry woodlands, and roadsides in the Coastal Plain. STEMS: 2–5 feet tall, much-branched and bushy, usually dark red and covered with tiny, rough, white hairs. LEAVES: Up to 4 inches long, alternate, with 7–13 paired oval leaflets up to 1 inch long and ⅓ inch wide. FLOWERS: About ⅓ inch long, pinkish-tan, yellow, or reddish-orange, the banner petal curved strongly backward and marked with green. FRUIT: A woody, oval pod about ⅓ inch long, with 1–3 seeds and a sharply pointed tip. FLOWERING: Jun–Aug. RANGE: FL west to LA, north to NC. SIMILAR TO Hairy Indigo (*Indigofera hirsuta*), a native of the Old World tropics that occurs in sandy disturbed areas in the Coastal Plain; its leaves and stems are covered with long, spreading, gray or brown hairs.

Japanese Clover

Kummerowia striata (Thunberg) Schindler
SYNONYM: *Lespedeza striata* (Thunberg) Hooker & Arnott
Bean Family (Fabaceae)

ANNUAL HERB found in fields, roadsides, and other disturbed areas throughout Georgia. STEMS: Up to 6 feet long, usually spreading across the ground, reddish-purple, hairy. LEAVES: Alternate, with 3 oval, entire leaflets up to ½ inch long, with many closely spaced parallel veins; at the base of each leaf there is a very short stalk and a pair of brown, papery stipules. FLOWERS: Up to ¼ inch long, with a pink banner petal and white wings and keel. FLOWERING: Jul–Oct. RANGE: FL west to AZ, north to IA and CT; native to Asia. NOTES: Japanese Clover is planted for forage and erosion control and has spread widely. SIMILAR TO Korean Clover (*Kummerowia stipulacea*), which has also escaped cultivation into disturbed areas; its leaves have ¼-inch-long stalks and long, bristly hairs on the margins.

Everlasting Pea, Perennial Sweet Pea

Lathyrus latifolius Linnaeus
Bean Family (Fabaceae)

PERENNIAL HERB found on roadsides, old homesites, and in other disturbed areas in north Georgia. **STEMS:** Up to 6 feet long, conspicuously winged, climbing by tendrils, with a waxy, white coating. **LEAVES:** Alternate, divided into 2 leaflets up to 4 inches long and 2 inches wide with a long, branching tendril between them; leaf stalks are winged. **FLOWERS:** In clusters of 4–10, each flower about 1 inch long, dark pink, typical pea-flower shape with a broad, notched banner petal. **FLOWERING:** May–Sep. **RANGE:** Throughout North America; native to Europe. **SIMILAR TO** Caley Pea (*Lathyrus hirsutus*), a European species found infrequently in disturbed areas throughout Georgia; its stems are narrowly winged, and the flowers are solitary.

Sericea Lespedeza, Chinese Lespedeza

Lespedeza cuneata (Dumont) G. Don
Bean Family (Fabaceae)

PERENNIAL HERB found on road banks and in other clearings throughout Georgia. **STEMS:** 1½–6 feet tall, somewhat woody, hairy, very leafy. **LEAVES:** Alternate, with 3 leaflets ¼–1 inch long and up to ¼ inch wide, wedge-shaped (tapering base, blunt tip) with a tiny bristle at the tip. **FLOWERS:** About ⅓ inch long, white, typical pea-flower shape with an erect banner petal streaked with red. **FLOWERING:** Jul–Sep. **RANGE:** Eastern and central North America; native to East Asia. **NOTES:** Sericea Lespedeza is aggressive and very difficult to eradicate, but land-management agencies continue to plant it for erosion control and wildlife food. **SIMILAR TO** other lespedezas but distinguished by its wedge-shaped leaflets.

Hairy Bush Clover

Lespedeza hirta (Linnaeus) Hornemann
Bean Family (Fabaceae)

PERENNIAL HERB found in woodland edges, sandhills, and savannas throughout Georgia. **STEMS:** 2–6 feet tall, leafy. **LEAVES:** Alternate, divided into 3 oval leaflets ½–2 inches long and ½–1 inch wide, the lateral veins parallel. **FLOWER CLUSTERS:** Held on stalks 1–2 inches long, arising from the leaf axils. **FLOWERS:** About ⅓ inch long, white or pale yellow, typical pea-flower shape with an erect banner petal streaked with red. **FRUIT:** About ¼ inch long, flat, 1-seeded, enclosed by the calyx. **FLOWERING:** Aug–Oct. **RANGE:** FL west to TX, north to MI and ME. **NOTES:** Long, white hairs cover all parts of the plant. **SIMILAR TO** Round-headed Bush Clover (*Lespedeza capitata*), whose flower clusters are held on very short stalks; and to Narrow-leaved Lespedeza (*L. angustifolia*), whose leaflets are narrowly oblong.

Smooth Trailing Lespedeza

Lespedeza repens (Linnaeus) W. Barton
Bean Family (Fabaceae)

PERENNIAL HERB found in dry upland forests and clearings throughout Georgia except for the southeastern counties. **STEMS:** Up to 3 feet long, trailing across the ground and forming loose mats, covered with straight hairs pressed flat against the stem. **LEAVES:** Alternate, divided into 3 oval leaflets up to 1 inch long, usually hairy on both surfaces. **FLOWER CLUSTERS:** Held above the leaves on a 2-inch stalk, with 2–6 flowers at the top. **FLOWERS:** About ¼ inch long, pink to violet, rarely white, typical pea-flower shape with an erect banner petal streaked with maroon. **FLOWERING:** Jul–Sep. **RANGE:** FL west to TX, north to MI and CT. **SIMILAR TO** Downy Trailing Lespedeza (*Lespedeza procumbens*), which occurs in similar habitats; it has 6–12 flowers per cluster, and its stems are covered with spreading, curved hairs.

Velvety Lespedeza

Lespedeza stuevei Nuttall
Bean Family (Fabaceae)

PERENNIAL HERB found in woodlands, prairies, savannas, and clearings throughout Georgia. **STEMS:** 2–5 feet tall, branched above the middle, hairy, grayish-green. **LEAVES:** Alternate, divided into 3 elliptic or oblong leaflets up to 1¼ inches long and up to 3 times as long as wide, hairy and grayish-green on both surfaces. **FLOWERS:** Tightly clustered in the upper leaf axils, each flower about ¼ inch long, pink to violet, typical pea-flower shape with an erect banner petal streaked with purple at the base; the keel petal is shorter than or the same length as the wings. **FLOWERING:** Jul–Sep. **RANGE:** FL west to TX, north to KS and VT. **SIMILAR TO** Slender Bush-clover (*Lespedeza virginica*), which occurs in similar habitats; its leaflets are 4–7 times longer than wide and may be smooth on the upper surface. Lespedezas hybridize with one another and are often difficult to identify.

The Ecological Importance of Legumes

Legumes play a key ecological role in several plant communities. Nodules on the roots of many species contain *Rhizobium*, a type of bacteria that converts atmospheric nitrogen into a mineral form usable by plants in a process called nitrogen fixation. The soils in frequently burned plant communities, such as Longleaf Pine–Wire Grass woodlands, are often deficient in nitrogen because fire reduces the amount of organic nitrogen in the soil. Legumes, which make up more than 10% of the plants in such communities, replace the lost nitrogen through nitrogen fixation. Other communities with nutrient-poor soils or that burned frequently in the past, such as pine-oak woodlands, prairies, glades, and barrens, also support a high number of legume species.

The legumes are widely recognized as being among the most important plant species for wildlife. The leaves and stems contain high levels of nitrogen and protein, and are an important food for browsing animals such as Gopher Tortoises, White-tailed Deer, Pocket Gophers, Wild Turkeys, and rabbits. The insects that live on the plants provide food for birds. Legumes produce abundant seeds that birds, lizards, and small mammals eat. Butterflies such as Cloudless Sulphurs, Sleepy Sulphurs, Long-tailed Skippers, and Eastern Long-tailed Blues depend on legumes as hosts for their caterpillars.

Sundial Lupine

Lupinus perennis Linnaeus
Bean Family (Fabaceae)

PERENNIAL HERB found in sandhills, dry woodlands, and roadsides, mostly in the Coastal Plain. **STEMS:** Up to 2½ feet tall, hairy or smooth, often reddish. **LEAVES:** Alternate, divided into 7–11 leaflets held in a circle at the tip of a long stalk; leaflets up to 2 inches long and ½ inch wide, upper surfaces smooth, lower surfaces and margins hairy. Leaves track the path of the sun during the day. **FLOWER CLUSTERS:** 4–10 inches high, erect. **FLOWERS:** About ½ inch long, typical pea-flower shape, blue-violet with a large, white, purple-streaked patch on the banner petal. **FLOWERING:** Apr–May. **RANGE:** FL west to TX, north to MN, ON, and NL. **NOTES:** A single colony may include pink- and white-flowering plants. With its deep blue flowers and clock-face leaflets, Sundial Lupine is **SIMILAR TO** no other native Georgia wildflowers.

Lady Lupine, Pink Sandhill Lupine

Lupinus villosus Willdenow
Bean Family (Fabaceae)

ANNUAL OR BIENNIAL HERB found in sandhills, dry woodlands, and sandy roadsides throughout Georgia's Coastal Plain. **STEMS:** Up to 2 feet tall, clustered, sprawling or erect. **LEAVES:** 2½–6 inches long and about 1 inch wide, alternate, elliptic to oblong, clustered at the base of the stem, overwintering. **FLOWER CLUSTERS:** 4–12 inches high, erect, crowded with many flowers. **FLOWERS:** About ½ inch long, typical pea-flower shape, pink or lavender with a large, maroon patch on the banner petal. **FLOWERING:** Apr–May. **RANGE:** FL west to LA, north to NC. **NOTES:** Leaves, stems, and fruits are densely covered with long, white or tawny hairs. **SIMILAR TO** Blue Sandhill Lupine (*Lupinus diffusus*), which occurs in Georgia's eastern Coastal Plain in similar habitats; its flowers are blue with a white patch on the banner petal.

Black Medic

Medicago lupulina Linnaeus
Bean Family (Fabaceae)

ANNUAL HERB found in fields, roadsides, and other disturbed areas throughout Georgia. **STEMS:** Up to 3 feet long, erect or sprawling, 4-angled, branched, usually hairy. **LEAVES:** Alternate, divided into 3 oval, finely toothed leaflets about ½ inch long; only the center leaflet is stalked. **FLOWER CLUSTERS:** Round heads about ¼ inch high and wide on long stalks. **FLOWERS:** About ⅛ inch long, yellow. **FRUIT:** In elongated clusters of many pods, each pod about ⅛ inch long, kidney-shaped, hairy, green maturing to black. **FLOWERING:** Mar–Dec. **RANGE:** Native to Europe, now globally widespread. Although a close relative of Alfalfa (*Medicago sativa*), Black Medic looks more similar to the hop clovers (*Trifolium campestre* and *T. dubium*), page 207.

White Sweetclover

Melilotus albus Medikus
Bean Family (Fabaceae)

ANNUAL OR BIENNIAL HERB found in fields, roadsides, and other disturbed areas throughout Georgia. **STEMS:** Up to 6 feet tall, branched, smooth or slightly hairy. **LEAVES:** Alternate, divided into 3 oval leaflets ½–2 inches long, grayish-green, smooth, toothed, the center leaflet stalked. **FLOWER CLUSTERS:** 1½–5 inches tall, erect, slender, with many flowers. **FLOWERS:** Tiny, white, with a nearly erect banner petal. **FLOWERING:** Apr–Oct. **RANGE:** Native to Eurasia, now globally widespread. **NOTES:** Both the flowers and the leaves are fragrant. Introduced as a forage crop, the sweetclovers are now considered invasive in much of North America. In general appearance **SIMILAR TO** Culver's Root (*Veronicastrum virginicum*), page 335, which has whorled leaves. Yellow Sweetclover (*Melilotus officinalis*), another Eurasian plant found in disturbed areas, has yellow flowers but is otherwise nearly identical.

Sensitive Brier

Mimosa microphylla Dryander
SYNONYM: *Schrankia microphylla* (Dryander) J. F. Macbride
Bean Family (Fabaceae)

PERENNIAL HERB occurring in dry woodlands and disturbed areas throughout Georgia. **STEMS:** 3–6 feet long, prickly, trailing along the ground or sprawling over other plants. **LEAVES:** Alternate, with 3–11 pairs of leaflets, each leaflet with 9–12 pairs of tiny, oblong subleaflets that fold up when touched. **FLOWER CLUSTERS:** About ¾ inch wide, round, composed of many tiny flowers with long, pink, yellow-tipped stamens. **FLOWERING:** Jun–Sep. **RANGE:** FL west to TX, north to MO and DE. **SIMILAR TO** Florida Sensitive Brier (*Mimosa quadrivalvis* variety *floridana*), which occurs in dry, sandy Coastal Plain habitats; it has 3–6 pairs of leaflets per leaf. Powder-puff Mimosa (*M. strigillosa*) occurs in floodplains and wetlands; its mat-forming stems have no prickles.

Sampson's Snakeroot

Orbexilum pedunculatum (P. Miller) Rydberg
SYNONYM: *Psoralea psoralioides* (Walter) Cory
Bean Family (Fabaceae)

PERENNIAL HERB found in open woodlands, prairies, and glades throughout Georgia. **STEMS:** 1–3 feet tall, branched from the base, ribbed, hairy. **LEAVES:** Alternate, divided into 3 leaflets ¾–2¾ inches long and up to ½ inch wide, narrowly elliptic to lance-shaped, entire, hairy or smooth. **FLOWERS:** In a crowded, long-stalked cluster, each flower about ¼ inch long, pale purple (drying to tan), typical pea-flower shape, the banner petal with a darker purple patch or midvein. **FLOWERING:** May–Jul. **RANGE:** FL west to TX, north to KS, MI, and VA. Flowers are **SIMILAR TO** those of Sandhill Scurf-pea (*Orbexilum lupinellus*), which occurs in sandhills in the Coastal Plain; it has loosely spaced flowers and 3–7 needle-like leaflets per leaf.

Buckroot, Hoary Scurf-Pea

Pediomelum canescens (Michaux) Rydberg

SYNONYM: *Psoralea canescens* Michaux

Bean Family (Fabaceae)

PERENNIAL HERB found in sandhills and other sandy woodlands in the Coastal Plain. **STEMS:** Up to 3 feet tall, much-branched and bushy. **LEAVES:** Alternate, with 3 (sometimes 1 or 2) oval to elliptic, round-tipped leaflets, up to 2 inches long. **FLOWERS:** In a long-stalked, loosely flowered cluster, each flower about ½ inch long, violet or dark blue with yellowish-green markings, drying to tan, typical pea-flower shape. **FLOWERING:** May–Jul. **RANGE:** FL west to AL, north to VA. **NOTES:** Soft hairs cover the stems, leaflets, flower stalks, and calyxes, giving the whole plant a pale, grayish look. Its bushy tumbleweed appearance is **SIMILAR TO** that of some wild indigos, pages 186–187, but none of these are covered with white hairs.

Wild Bean

Phaseolus polystachios (Linnaeus) B.S.P.

Bean Family (Fabaceae)

PERENNIAL HERBACEOUS VINE found in moist woodlands along streams, dry upland woodlands, and disturbed areas throughout Georgia. **STEMS:** 6–20 feet long, slender, hairy, new growth strongly angled, trailing and twining over other plants. **LEAVES:** Alternate, with 3 broadly oval, entire or lobed, thin-textured leaflets 1–5 inches long and ¾–3 inches wide. **FLOWERS:** About ⅓ inch long; white, pink, or purple; with a broad banner petal, 2 long, spreading wing petals, and a coiled keel petal in the center. **FLOWERING:** Jul–Sep. **RANGE:** FL west to TX, north to IA and ME. **SIMILAR TO** Sandhills Bean (*Phaseolus sinuatus*), which occurs in Coastal Plain sandhills; its stem trails across the ground, and the leaflets are leathery, 3-lobed, and less than 2 inches long. Wild Bean leaves are **SIMILAR TO** those of Hog Peanut (*Amphicarpaea bracteata*), page 184.

Dollar Weed

Rhynchosia reniformis A. P. de Candolle
Bean Family (Fabaceae)

PERENNIAL HERB found in sandhills, dry woodlands, savannas, and sandy roadsides in the Coastal Plain. **STEMS:** Up to 8 inches tall, hairy, forming small colonies. **LEAVES:** 1–2 inches wide and long, alternate, round or kidney-shaped, with long stalks, leathery with conspicuous veins and minute, amber glands on both surfaces (use 10× lens). **FLOWERS:** About ½ inch long, yellow, typical pea-flower shape with an erect banner petal. **FLOWERING:** Jun–Sep. **RANGE:** FL west to LA, north to TN and NC. **SIMILAR TO** Twining Snoutbean (*Rhynchosia difformis*), which occurs in similar habitats; it has trailing and twining stems and 3 round or oval leaflets per leaf.

Erect Snoutbean

Rhynchosia tomentosa (Linnaeus) Hooker & Arnott
Bean Family (Fabaceae)

PERENNIAL HERB found in dry woodlands, sandhills, moist or dry savannas, and sandy clearings in the Coastal Plain. **STEMS:** Up to 3 feet tall, very hairy, occasionally branched. **LEAVES:** Alternate, divided into 3 oval or elliptic leaflets up to 2¾ inches long and 1¾ inches wide, velvety-hairy. **FLOWERS:** About ⅓ inch long, yellow, typical pea-flower shape with an erect banner petal; the calyx lobes are nearly the same length as the petals. **FLOWERING:** Jun–Aug. **RANGE:** FL west to TX, north to AR and DE. **SIMILAR TO** Least Snoutbean (*Rhynchosia minima*), which occurs in pine flatwoods, beach dunes, and disturbed areas in Georgia's coastal counties; it has hairy, trailing, and twining stems and 3 oval to triangular leaflets per leaf; its petals are clearly longer than the calyx.

Wild Senna, Maryland Senna

Senna marilandica (Linnaeus) Link
SYNONYM: *Cassia marilandica* Linnaeus
Bean Family (Fabaceae)

PERENNIAL HERB found in bottomlands, woodland borders, and disturbed areas in north Georgia and a few Coastal Plain counties. **STEMS:** Up to 5 feet tall. **LEAVES:** Up to 8 inches long, alternate, smooth, with 2–11 pairs of oval or oblong leaflets up to 2½ inches long and 1 inch wide. **FLOWERS:** In clusters of 5–25 flowers, each about 1 inch wide, with 5 yellow petals. **FLOWERING:** Jul–Aug. **RANGE:** FL west to TX, north to NE and MA. **NOTES:** A small, round nectar gland near the base of the leaf stalk attracts ants that attack leaf-eating insects. **SIMILAR TO** Coffee Senna (*Senna occidentalis*), a nonnative weed with only 3–6 pairs of leaflets and 1–5 flowers per cluster. Sicklepod (*S. obtusifolia*), another widespread weed, has the nectar gland between the lowest pair of leaflets.

Orange Rattlebox

Sesbania punicea (Cavanilles) Bentham
SYNONYM: *Daubentonia punicea* (Cavanilles) A. P. de Candolle
Bean Family (Fabaceae)

SHRUB found in disturbed areas in Georgia's Coastal Plain. **STEMS:** Up to 10 feet tall, new shoots silky-hairy, older branches smooth, usually dying back in winter. **LEAVES:** 4–8 inches long, alternate, divided into 10–20 pairs of oblong leaflets up to 1 inch long and ⅓ inch wide. **FLOWERS:** In 4-inch-long clusters, each flower about ¾ inch long, orange-red, drying to purple, typical pea-flower shape. **FRUIT:** A 4-winged, reddish-brown pod 2–3 inches long and about ½ inch thick. **FLOWERING:** Jun–Oct. **RANGE:** FL west to TX, north to VA; native to South America. **SIMILAR TO** Bladderpod (*Sesbania vesicaria*; synonym *Glottidium vesicarium*), which occurs in disturbed areas in the Coastal Plain; it has yellow or orange flowers and a flat, beanlike pod.

Sand Bean

Strophostyles umbellata (Muhlenberg ex Willdenow) Britton
Bean Family (Fabaceae)

PERENNIAL HERBACEOUS VINE found in dry to moist woodlands, riverbanks, sandhills, savannas, and sandy clearings throughout Georgia. **STEMS:** Up to 6 feet long, trailing or twining over other plants, often twisted, green or purple. **LEAVES:** Alternate, with 3 oval, oblong, or linear leaflets 1–2½ inches long and ½ inch wide, entire, slightly hairy. **FLOWERS:** Held on stalks 2–8 inches long, each flower up to ¾ inch wide, dark pink, with an erect banner petal and a strongly twisted, purple-tipped, central keel petal. **FLOWERING:** Jun–Oct. **RANGE:** FL west to TX, north to MO and NY. **SIMILAR TO** Trailing Wild Bean (*Strophostyles helvola*), an annual plant that occurs on coastal dunes and in dry uplands; its leaflets are oval and usually lobed.

Pencil Flower

Stylosanthes biflora (Linnaeus) B.S.P.
Bean Family (Fabaceae)

PERENNIAL HERB found in dry, rocky or sandy woodlands throughout Georgia. **STEMS:** 4–20 inches long, spreading or erect, often with long, stiff, reddish hairs. **LEAVES:** Alternate, divided into 3 leaflets, each ½–1½ inches long and ¼ inch wide, elliptic to lance-shaped with a sharp tip, usually very hairy; the base of the leaf stalk forms a tubular sheath around the stem. **FLOWERS:** About ⅓ inch long, deep yellow with an erect banner petal sometimes streaked with red. **FRUIT:** About ¼ inch long, with 2 segments, 1 containing a seed, the other sterile, withered, and stalklike. **FLOWERING:** Jun–Aug. **RANGE:** FL west to TX, north to WI and NY. **SIMILAR TO** other yellow-flowered bean plants but distinguished by its narrow, hairy leaflets and the sheath at the base of the leaf stalk.

Spiked Goat's Rue

Tephrosia spicata (Walter) Torrey & A. Gray
Bean Family (Fabaceae)

PERENNIAL HERB found in dry to moist woodlands and clearings throughout Georgia. **STEMS:** Up to 3 feet long, erect or sprawling. **LEAVES:** Up to 4¾ inches long, the stalk less than ½ inch long, alternate, divided into 7–17 elliptic to oblong leaflets, each up to 1½ inches long. **FLOWER CLUSTERS:** Erect, with 2–20 flowers. **FLOWERS:** ½–¾ inch long, white, turning maroon with age. **FLOWERING:** Jun–Aug. **RANGE:** FL west to LA, north to KY and DE. **NOTES:** Spreading, tawny or rusty hairs cover the entire plant. **SIMILAR TO** two species that occur in Coastal Plain pinelands: Hairy Goat's Rue (*Tephrosia hispidula*) has inconspicuous gray hairs flattened against the stems and leaves; Florida Goat's Rue (*Tephrosia florida*) has short, sparse gray hairs, and its leaf stalk is usually 1–3 inches long.

Virginia Goat's Rue

Tephrosia virginiana (Linnaeus) Persoon
Bean Family (Fabaceae)

PERENNIAL HERB found in Longleaf Pine forests, oak-pine woodlands, and prairies throughout Georgia. **STEMS:** Up to 2 feet tall, often in bushy clumps. **LEAVES:** Divided into 13–23 leaflets up to 1 inch long and ¼ inch wide, oblong to elliptic. **FLOWER CLUSTERS:** Up to 3 inches tall, held erect at the top of the stem, crowded with flowers, not leafy. **FLOWERS:** About ¾ inch long and wide, typical pea-flower shape with a pale yellow banner petal and dark pink keel and wing petals. **FLOWERING:** May–Jun. **RANGE:** FL west to TX, north to MN and NH. **NOTES:** Long, gray, spreading hairs cover the entire plant. **SIMILAR TO** Dwarf Goat's Rue (*Tephrosia mohrii*), a Georgia Special Concern species found in dry pinelands in a few southwest Georgia counties; its flowers are solitary or in small, leafy clusters.

Rabbit-Foot Clover

Trifolium arvense Linnaeus
Bean Family (Fabaceae)

ANNUAL HERB found in fields, pastures, and other disturbed areas in the Piedmont and a few Coastal Plain counties. **STEMS:** 2–16 inches tall, hairy, with widely spreading branches. **LEAVES:** Alternate, divided into 3 entire or finely toothed, gray-green, narrowly oval or oblong leaflets up to ¾ inch long. **FLOWER CLUSTERS:** About 1 inch high and ½ inch wide, cylindrical, pinkish-gray, softly fuzzy, stalked. **FLOWERS:** Less than ¼ inch long, with white or pink petals and a very hairy calyx with 5 long, reddish, very narrow lobes. **FLOWERING:** Apr–Aug. **RANGE:** Introduced from the Mediterranean region, weedy and widespread throughout much of North America. **SIMILAR TO** no other clover in Georgia.

Large Hop Clover

Trifolium campestre Schreber
Bean Family (Fabaceae)

ANNUAL HERB found in lawns, roadsides, and other disturbed areas throughout Georgia. **STEMS:** 2–12 inches tall, erect or sprawling, branched, hairy. **LEAVES:** Alternate, divided into 3 oval leaflets about ½ inch long, smooth, finely toothed; the center leaflet has a longer stalk than the 2 side leaflets. **FLOWER CLUSTERS:** Round heads about ½ inch high and wide, held on hairy, 1-inch-long stalks. **FLOWERS:** About ⅛ inch long, yellow, the banner petal with conspicuous grooves (veins); flowers at first erect then spreading, drooping, and turning brown with age. **FLOWERING:** Apr–Oct. **RANGE:** Native to Eurasia, now globally widespread. **SIMILAR TO** Little Hop Clover (*Trifolium dubium*), which has smaller heads, less than ⅓ inch wide; its banner petals are inconspicuously veined and folded lengthwise; all its leaflets have very short or no stalks. Also see Black Medic (*Medicago lupulina*), page 200.

Wild White Clover

Trifolium carolinianum Michaux
Bean Family (Fabaceae)

ANNUAL HERB found in woodlands, pine savannas, rock outcrop borders, and disturbed areas in the Piedmont and a few southeast Georgia counties. **STEMS:** Up to 12 inches long, spreading but not rooting at the nodes. **LEAVES:** Alternate, divided into 3 leaflets about ½ inch long, narrow and wedge-shaped at the base, broadly rounded or notched at the tip. **FLOWER CLUSTERS:** Round heads about ½ inch wide, flowers spreading or drooping. **FLOWERS:** Less than ¼ inch long, white or pink, turning brown and drooping with age, partially enclosed by a hairy, green or red calyx. **FLOWERING:** Apr–Jul. **RANGE:** FL west to TX, north to KS and VT. **NOTES:** One of only two native clovers in Georgia. **SIMILAR TO** White Clover (*Trifolium repens*), the common clover of lawns; its stems root at the nodes, forming clumps or mats; its white flowers turn pink with age.

Crimson Clover

Trifolium incarnatum Linnaeus
Bean Family (Fabaceae)

ANNUAL HERB found in fields, roadsides, and other disturbed areas throughout Georgia. **STEMS:** 1–2 feet tall, hairy. **LEAVES:** Alternate, long-stalked, divided into 3 hairy, finely toothed, broadly oval or heart-shaped leaflets up to 1 inch long. **FLOWER CLUSTERS:** A cylindrical, long-stalked spike 1–2¼ inches high and ½ inch wide. **FLOWERS:** About ½ inch long, bright red (rarely white). **FLOWERING:** Apr–Sep. **RANGE:** Introduced from the Mediterranean region for forage, soil improvement, and erosion control and now widespread in North America. **SIMILAR TO** Red Clover (*Trifolium pratense*), which has round, dark pink or reddish-purple flower heads about 1 inch wide; it is an introduced European species found in north Georgia on roadsides and in other disturbed areas.

Buffalo Clover

Trifolium reflexum Linnaeus
Bean Family (Fabaceae)

ANNUAL OR SHORT-LIVED PERENNIAL HERB found in woodlands, stream banks, and limestone cedar glades in north Georgia and a few Coastal Plain counties. **STEMS:** 8–20 inches tall, hairy or smooth, purplish. **LEAVES:** Alternate, divided into 3 oval leaflets ½–1 inch long, very hairy; a pair of hairy, green, lance-shaped stipules, up to ¾ inch long, clasp the base of the leaf stalk. **FLOWER CLUSTERS:** Round heads about 1 inch wide, flowers at first erect, then spreading and revealing narrowly pointed calyx lobes and distinct stalks. **FLOWERS:** About ½ inch long, white tinged with green, becoming pink then brown and drooping with age. **FLOWERING:** Apr–May. **RANGE:** FL west to TX, north to NE, ON, and PA. **NOTES:** One of only two native clovers in Georgia. **SIMILAR TO** Wild White Clover (*Trifolium carolinianum*), page 208.

Carolina Vetch

Vicia caroliniana Walter
SYNONYM: *Vicia hugeri* Small
Bean Family (Fabaceae)

PERENNIAL HERB found in bottomland and upland forests, woodland borders, and disturbed areas in north Georgia and several Coastal Plain counties. **STEMS:** Up to 5 feet long, sprawling, smooth or finely hairy. **LEAVES:** Alternate, with a curling, unbranched tendril at the tip and 10–18 narrow leaflets. **FLOWER CLUSTERS:** Up to 4 inches long, with 8–20 widely spaced flowers. **FLOWERS:** About ⅓ inch long, white, sometimes tinged with lavender, the keel petal with purple spots. **FLOWERING:** Apr–Jun. **RANGE:** FL west to TX, north to ON and NY. **NOTES:** Of Georgia's 10 vetch species, only 2 are native: Carolina Vetch and Small-flower Vetch (*Vicia minutiflora*), a rare Coastal Plain species with lavender flowers. **SIMILAR TO** Large Yellow Vetch (*Vicia grandiflora*), found in disturbed areas; its flowers are yellow with purple streaks.

Winter Vetch, Hairy Vetch

Vicia villosa Roth
SYNONYM: *Vicia dasycarpa* Tenore
Bean Family (Fabaceae)

ANNUAL HERB found in disturbed areas throughout Georgia. **STEMS:** Up to 3 feet long, forming dense, tangled mats. **LEAVES:** Alternate, with a curling, branched tendril at the tip and 10–18 narrow leaflets up to 1 inch long. **FLOWER CLUSTERS:** Up to 6 inches long, with 20 or more flowers held on one side of a curved flower stalk. **FLOWERS:** About ½ inch long, purplish-pink, the calyx with a swollen base. **FLOWERING:** May–Sep. **RANGE:** Introduced from Europe, now spread throughout North America. **SIMILAR TO** Narrow-leaf Vetch (*Vicia sativa* ssp. *nigra*, synonym *Vicia angustifolia*), a widespread introduced weed found in disturbed areas throughout Georgia; its flowers have erect banner petals and are held in pairs in the upper leaf axils.

American Wisteria

Wisteria frutescens (Linnaeus) Poiret
Bean Family (Fabaceae)

PERENNIAL WOODY VINE found in swamps, floodplains, and wet thickets in the Coastal Plain and a few Piedmont counties. **STEMS:** Twining up to 50 feet high in trees. **LEAVES:** Alternate, divided into 9–15 leaflets up to 2½ inches long and 1½ inches wide, pointed, elliptic, not wavy or toothed. **FLOWER CLUSTERS:** 1½–6 inches long, drooping. **FLOWERS:** Up to ¾ inch long, lavender to bluish-purple, typical pea-flower shape with an erect banner petal; mildly fragrant. **FRUIT:** Up to 5 inches long, smooth, cylindrical, constricted between the seeds. **FLOWERING:** Apr–May. **RANGE:** FL west to TX, north to MI and MA. **SIMILAR TO** Chinese Wisteria (*Wisteria sinensis*), an extremely invasive species; its flower clusters are 4–14 inches long and crowded with very fragrant flowers; the leaflets have wavy edges, and the fruits are flat and covered with velvety hairs.

Yellow Fumewort, Yellow Harlequin

Corydalis flavula (Rafinesque) A. P. de Candolle
Fumitory Family (Fumariaceae)

ANNUAL HERB found in floodplains and other moist forests over limestone or mafic bedrock in north Georgia. **STEMS:** Up to 1 foot tall, waxy, reddish-green. **LEAVES:** Up to 2 inches long, alternate, divided into many toothed, bluish-green segments. **FLOWERS:** Less than ½ inch long, yellow, tubular with 2 flaring lips and a rounded, downcurved base (spur) extending behind the stalk. **FRUIT:** A drooping, narrow capsule up to 1 inch long. **FLOWERING:** Mar–Apr. **RANGE:** FL west to LA, north to NE, ON, and CT. **SIMILAR TO** Slender Corydalis (*Corydalis micrantha* ssp. *australis*), which occurs in dry woodlands and disturbed areas in Georgia's Coastal Plain; its flowers are more than ½ inch long; the rounded base of the flower (the spur) is straight or curved slightly upward; the fruits are smaller, less than ½ inch long, and held erect.

Dutchman's Britches

Dicentra cucullaria (Linnaeus) Bernhardi
Fumitory Family (Fumariaceae)

PERENNIAL HERB found in rich coves, boulderfields, and moist slopes in the north Georgia mountains. **LEAVES:** Up to 6 inches long and wide, divided into many toothed, waxy, silvery green segments. **FLOWERS:** About ½ inch long and wide, dangling from a stalk that arches above the leaves; each flower has 2 pointed, inflated, white petals ("britches") and 2 much smaller inner petals (white or yellow) that barely extend beyond the flower opening; the flowers are not fragrant. **FLOWERING:** Apr–May. **RANGE:** GA west to OK, north to MB and NS; ID, OR, WA. **SIMILAR TO** Squirrel Corn (*Dicentra canadensis*), a Georgia Special Concern species that often occurs with Dutchman's Britches and has nearly identical leaves. Its petals are rounded, not pointed at the tips, creating a nearly heart-shaped flower that is quite fragrant.

Screwstem Bartonia

Bartonia paniculata (Michaux) Muhlenberg
Gentian Family (Gentianaceae)

ANNUAL OR BIENNIAL HERB found infrequently in wetlands throughout Georgia, mostly in the Coastal Plain. **STEMS:** Up to 16 inches tall, erect, sprawling, or twining on other plants, twisted, purplish-green. **LEAVES:** Reduced to tiny scales that are alternate on the upper and midstem, opposite below. **FLOWER CLUSTERS:** Erect at the tip of the stem, with opposite, upright branches. **FLOWERS:** About ⅛ inch long, with 4 pointed, white, green, or pinkish petals and 4 pointed, green sepals. **FLOWERING:** Jul–Oct. **RANGE:** FL west to TX, north to WI and NL. **SIMILAR TO** Virginia Bartonia (*Bartonia virginica*), which occurs in bogs and other acidic wetlands in the Coastal Plain and a few north Georgia counties; all of its scale leaves are opposite, and it flowers Jul–Oct. Spring Bartonia (*B. verna*) occurs in wet savannas and bogs in several southeastern counties; it flowers Feb–Apr.

Columbo, Green Gentian

Frasera caroliniensis Walter
SYNONYM: *Swertia caroliniensis* (Walter) Kuntze
Gentian Family (Gentianaceae)

PERENNIAL HERB found in rich, deciduous forests in the north Georgia mountains. **FLOWERING STEMS:** Up to 10 feet or more tall, stout. **BASAL LEAVES:** Up to 14 inches long and 4 inches wide, elliptic to oblong, smooth and somewhat glossy. **STEM LEAVES:** Similar but smaller, in whorls of 3–9. **FLOWER CLUSTERS:** Up to 3 feet high and 10 inches wide, with many branches and up to 100 flowers. **FLOWERS:** About 1 inch wide, with 4 spreading, greenish-white, purple-speckled petals with a hairy, green nectar gland near the base of each petal. **FLOWERING:** Late May–Jun. **RANGE:** GA west to OK, north to MI and NY. **NOTES:** Columbo lives as a leaf rosette for up to 15 years, adding leaves each year, before flowering, fruiting, and dying in a single year. **SIMILAR TO** no other wildflower in Georgia.

Soapwort Gentian

Gentiana saponaria Linnaeus
Gentian Family (Gentianaceae)

PERENNIAL HERB found in moist to wet forests in north and southwest Georgia. **STEMS:** 1–2½ feet tall, often fallen over. **LEAVES:** Up to 3½ inches long and 1½ inches wide, opposite, linear to elliptic, widest near the middle, entire, glossy, without stalks. **FLOWERS:** 1–2 inches long and 1 inch wide, tubular with 5 short, rounded lobes closed over the opening, blue or pale purple with purple stripes; a green, cup-shaped calyx surrounds the base of the flower—its 5 lobes are lance-shaped and about the same length as the cup. **FLOWERING:** Late Sep–Nov. **RANGE:** FL west to TX, north to MI and NY. **SIMILAR TO** Appalachian Gentian (*Gentiana decora*), which occurs in high, moist forests in the Blue Ridge; its flower is pale blue to whitish, often with blue or purple stripes; the calyx lobes are very narrow, pointed, and shorter than the calyx.

Striped Gentian

Gentiana villosa Linnaeus
Gentian Family (Gentianaceae)

PERENNIAL HERB found in upland forests and clearings throughout Georgia. **STEMS:** Up to 2 feet tall, smooth. **LEAVES:** Up to 4 inches long and 1½ inches wide, opposite, elliptic or lance-shaped, entire, glossy, with no stalks. **FLOWERS:** 1–2 inches long, tubular with 5 triangular lobes erect or closed over the opening, greenish- or yellowish-white, tinged or striped with purple. **FLOWERING:** Late Aug–Nov. **RANGE:** FL west to LA, north to IN and NJ. **SIMILAR TO** Coastal Plain Gentian (*Gentiana catesbaei*), which occurs in moist woodlands, savannas, and wet clearings in the Coastal Plain; it has blue, pale purple, or maroon flowers, and its leaves are oval, widest near the base.

Stiff Gentian, Agueweed

Gentianella quinquefolia (Linnaeus) Small

SYNONYM: *Gentiana quinquefolia* Linnaeus

Gentian Family (Gentianaceae)

ANNUAL HERB found in moist, upland forests and along roadsides in Georgia's Blue Ridge. **STEMS:** 6–32 inches tall, with many stiff, upright branches. **LEAVES:** Up to 3 inches long and 1½ inches wide, opposite, oval to lance-shaped with pointed tips and no stalks. **FLOWERS:** About ¾ inch long, blue-purple (rarely white), tubular with 5 pointed lobes that are usually closed and forming a pointed tip; flowers are held in stiffly erect clusters at the tips of branches. **FLOWERING:** Late Aug–Oct. **RANGE:** GA west to KS, north to MN, ON, and ME. **NOTES:** In sunlight, the sharply angled flowers look like purple crystals. **SIMILAR TO** no other wildflower in Georgia.

Pennywort

Obolaria virginica Linnaeus

Gentian Family (Gentianaceae)

PERENNIAL HERB found in rich, upland forests in north Georgia. **STEMS:** 1–6 inches tall, smooth, fleshy. **LEAVES:** Up to ½ inch long and wide, opposite, fleshy, oval with broadly rounded tips, entire, greenish-purple, crowded near the top of the stem, with no stalks. **FLOWERS:** About ½ inch long, whitish tinged with purple or green, with 4 pointed, spreading lobes; the lower half of the flower is enclosed by 2 leafy bracts. **FLOWERING:** Mar–Apr. **RANGE:** FL west to TX, north to MO and NJ. **NOTES:** Pennywort has little chlorophyll in its leaves and is partially dependent on other plants for nutrients (see "A Forest Menage à Trois," p. 168). **SIMILAR TO** no other wildflower in Georgia.

Rose-Pink, Bitterbloom

Sabatia angularis (Linnaeus) Pursh
Gentian Family (Gentianaceae)

ANNUAL OR BIENNIAL HERB found in forests, woodlands, marshes, and clearings in north Georgia and a few Coastal Plain counties. **STEMS:** 8–32 inches tall, 4-angled, winged, with opposite branches. **LEAVES:** Up to 2 inches long and 1½ inches wide, opposite, broadly oval or lance-shaped with rounded bases, entire, smooth, with no stalk. **FLOWERS:** About 1½ inches wide with 5 spreading pink petals and a red-rimmed, yellow-green star at the center; a green calyx cup surrounds the base of the flower—its 5 narrow lobes are shorter than the petals. **FLOWERING:** Jul–Aug. **RANGE:** FL west to TX, north to KS, ON, and MA. **SIMILAR TO** Narrow-leaf Rose-pink (*Sabatia brachiata*), which occurs throughout Georgia in pine woodlands; its branches are opposite but not winged, and its leaves are elliptic, tapering at both tip and base.

Bartram's Rose-Gentian

Sabatia bartramii Wilbur
Gentian Family (Gentianaceae)

PERENNIAL HERB found in cypress ponds and wet pine flatwoods in the Coastal Plain. **STEMS:** 2–3 feet tall, with alternate branches. **BASAL LEAVES:** Up to 3 inches long and ¾ inch wide, with tapering bases and wider, bluntly rounded tips. **STEM LEAVES:** Smaller, opposite, linear, entire, succulent, pressed upward against the stem. **FLOWERS:** Held singly at the tips of stalks, up to 2½ inches wide with 10–12 spreading, dark pink petals and a red-rimmed, yellow-green "eye" at the center. **FLOWERING:** Jun–Aug. **RANGE:** FL west to MS, north to SC. **SIMILAR TO** Pineywoods Rose-gentian (*Sabatia gentianoides*), which occurs in pine flatwoods and savannas in the Coastal Plain; it is less than 20 inches tall, and its flowers are in compact clusters, each flower with 2 long, narrow bracts curving down beneath it.

Slender Rose-Pink

Sabatia campanulata (Linnaeus) Torrey
Gentian Family (Gentianaceae)

PERENNIAL HERB found in wet pine savannas and flatwoods, seepage slopes, and openings in wet woodlands in the Coastal Plain and a few Georgia Piedmont counties. **STEMS:** 8–28 inches tall, strongly ribbed and angled, with alternate branches. **LEAVES:** Up to 1 inch long and ¼ inch wide, needle-like on the upper branches. **FLOWERS:** Up to 1½ inches wide with 5 spreading, dark pink petals and a red-rimmed, yellow-green star at the center; the base of the flower is surrounded by a green calyx with 5 needle-like lobes that are about the same length as the petals. **FLOWERING:** Jun–Aug. **RANGE:** FL west to TX, north to IL and MA. **SIMILAR TO** Sea-pink (*Sabatia stellaris*), which occurs on the edges of salt marshes in Georgia's coastal counties; its calyx lobes are noticeably shorter than the petals.

Lance-Leaf Rose-Gentian

Sabatia difformis (Linnaeus) Druce
Gentian Family (Gentianaceae)

PERENNIAL HERB found in wet pine savannas and flatwoods, seepage slopes, bogs, and wet ditches in the Coastal Plain. **STEMS:** 18–32 inches tall, with several stiffly erect stems, the lower stems round, the upper angular (but not winged), and the branches mostly opposite. **LEAVES:** Up to 1½ inches long and ½ inch wide, smaller in the upper branches, opposite, pointing upward, narrowly lance-shaped to oval, with no stalk. **FLOWERS:** About 1 inch wide with 5 or 6 spreading, completely white petals. **FLOWERING:** May–Sep. **RANGE:** FL west to MS, north to NJ. **SIMILAR TO** Short-leaf Rose-gentian (*Sabatia brevifolia*), which occurs in pine savannas in Georgia's Coastal Plain; it has alternate branches and a yellow star at the center of the otherwise white flower.

Salt Marsh Rose-Pink

Sabatia dodecandra (Linnaeus) B.S.P.
Gentian Family (Gentianaceae)

PERENNIAL HERB found in salt and brackish marshes and wet interdune swales on Georgia's barrier islands. **STEMS:** 1–2⅓ feet tall, branches usually alternate; the stem length between the pairs of leaves is usually much longer than the leaves; aboveground runners between plants are absent or poorly developed. **LEAVES:** Up to 1½ inches long and ½ inch wide, opposite, lance-shaped or elliptic, entire, with no stalk. **FLOWERS:** Up to 2 inches wide with 7–14 spreading, dark pink petals and a red-rimmed, yellow-green "eye" at the center; the base of the flower is surrounded by a green calyx with 7–14 lobes about ½ inch long. **FLOWERING:** Jun–Aug. **RANGE:** FL west to TX, north to CT. **SIMILAR TO** Marsh Rose-pink (*Sabatia foliosa*), which occurs in forested freshwater wetlands in the Coastal Plain; it has conspicuous aboveground runners.

Large-Leaf Rose-Gentian

Sabatia macrophylla Hooker
Gentian Family (Gentianaceae)

PERENNIAL HERB found in wet pine savannas and flatwoods, seepage slopes, bogs, and wet ditches in the Coastal Plain and Fall Line. **STEMS:** 3–4 feet tall, with several stiffly erect stems, round in cross-section (not ribbed or angled), waxy white, with forking branches. **LEAVES:** Up to 2½ inches long and 1 inch wide, opposite, lance-shaped to oval, clasping the stem, slightly succulent, pointing upward, with a waxy-white coating. **FLOWERS:** About ½ inch wide with 5 spreading, completely white petals. **FLOWERING:** Jun–Jul. **RANGE:** GA, FL, AL, MS, and LA. **SIMILAR TO** Four-angled Rose Gentian (*Sabatia quadrangula*), which occurs in the Piedmont in granite outcrop pools and seepages, and in the Coastal Plain in sandhills, flatwoods, and seepages; it has a single stem arising from a leaf rosette, the lower half of which is 4-angled and narrowly winged.

Carolina Crane's-Bill

Geranium carolinianum Linnaeus
Geranium Family (Geraniaceae)

ANNUAL HERB found in fields, lawns, and other disturbed areas throughout Georgia. **STEMS:** Up to 1 foot tall, branches often sprawling along the ground, reddish, and very hairy. **LEAVES:** Up to 2 inches long and wide, opposite, deeply divided into 5–7 coarsely toothed lobes, hairy, grayish-green. **FLOWERS:** Up to ¾ inch wide with 5 pale pink, notched petals and 5 green sepals with stiffly pointed tips. **FRUIT:** A narrow, cylindrical capsule with a pointed, 1-inch-long beak; when the fruit is ripe, the beak splits and curls rapidly upward into 5 segments, flinging seeds for several feet. **FLOWERING:** Mar–Jul. **RANGE:** Most of North America. **SIMILAR TO** two weedy European species that have dark pink petals with purple lines. Cut-leaf Crane's-bill (*Geranium dissectum*) has deeply divided leaves. Dove's-foot Crane's-bill (*G. molle*) has nearly round, shallowly lobed leaves.

Wild Geranium, Spotted Geranium

Geranium maculatum Linnaeus
Geranium Family (Geraniaceae)

PERENNIAL HERB found in moist hardwood forests in north and central Georgia. **STEMS:** 8–24 inches tall, hairy. **LEAVES:** Up to 4 inches long and 6 inches wide, alternate, deeply lobed and toothed, upper surface hairy and spotted with pale green, lower leaf stalks long and hairy. **FLOWERS:** 1–1½ inches wide with 5 pink petals and 10 stamens. **FRUIT:** About ¼ inch long with a 1-inch-long beak; when mature, it dries and separates rapidly into 5 curled segments, flinging seeds for several feet. **FLOWERING:** Apr–Jun. **RANGE:** GA, west to LA, north to MB and QC. **NOTES:** The petals have transparent lines that act as nectar guides for pollinators. **SIMILAR TO** no other wildflower in Georgia.

Red Buckeye

Aesculus pavia Linnaeus
Horse Chestnut (Hippocastanaceae) or
Soapberry (Sapindaceae) Family

DECIDUOUS SHRUB OR SMALL TREE found in moist hardwood forests on slopes and stream banks in the Coastal Plain and lower Piedmont. **STEMS:** 5–40 feet tall, with forking branches. **LEAVES:** Opposite, with a long stalk and 5 toothed, lance-shaped leaflets up to 6 inches long and 2½ inches wide. **FLOWER CLUSTERS:** 6–10 inches long, held erect at the tips of stout twigs. **FLOWERS:** About 1½ inches long, with a red, fleshy calyx and 4 red, dissimilar petals. **FRUIT:** A round, brown, thick-husked capsule up to 2½ inches wide, splitting to release 1–3 large, glossy brown, highly toxic seeds. **FLOWERING:** Apr–May. **RANGE:** FL west to TX, north to MO, IL, and NC. **NOTES:** Pollinated mostly by Ruby-throated Hummingbirds, whose northward migration coincides with Red Buckeye flowering. **SIMILAR TO** Painted Buckeye (*Aesculus sylvatica*), below.

Painted Buckeye

Aesculus sylvatica Bartram
Horse Chestnut (Hippocastanaceae) or
Soapberry (Sapindaceae) Family

DECIDUOUS SHRUB OR SMALL TREE found in moist hardwood forests, mostly in the Piedmont. **STEMS:** 3–25 feet tall, with forking branches. **LEAVES:** Opposite, with a long stalk and 5 toothed leaflets, each 3–8 inches long and less than 3 inches wide. **FLOWER CLUSTERS:** 4–10 inches long, held at the tips of stout twigs. **FLOWERS:** 1–1½ inches long, with a fleshy, greenish-yellow calyx and 4 yellow, differently sized petals. **FRUIT:** A rounded, thick-husked, brown capsule 1–1½ inches wide, splitting to release 1–3 large, glossy brown, highly toxic seeds. **FLOWERING:** Apr–May. **RANGE:** GA, AL, north to TN and VA. **NOTES:** Migrating hummingbirds carry Red Buckeye pollen north to Painted Buckeye flowers, sometimes resulting in hybrid plants with red-tinged flowers. **SIMILAR TO** Red Buckeye (*Aesculus pavia*), above.

Climbing Hydrangea

Decumaria barbara Linnaeus
Hydrangea Family (Hydrangeaceae)

PERENNIAL WOODY VINE found in floodplain forests and on moist lower slopes above streams throughout Georgia. **STEMS:** Climbing many feet into trees by way of aerial rootlets, with rough, brown bark and drooping side branches; sometimes the stems form a nonflowering ground cover around the base of trees. **LEAVES:** Up to 4¾ inches long and 3 inches wide, oval, toothed or nearly entire, dark, glossy green on top, pale green beneath. **FLOWER CLUSTERS:** A flat- or round-topped cluster, 2–4 inches wide, held at the tips of new side branches. **FLOWERS:** Small, white, with 7–10 petals and many stamens. **FLOWERING:** May–Jun. **RANGE:** FL west to LA, north to AR and NY. **SIMILAR TO** no other woody vine in Georgia.

Wild Hydrangea

Hydrangea arborescens Linnaeus
Hydrangea Family (Hydrangeaceae)

DECIDUOUS SHRUB found in moist forests, along streams, and on seepy rock ledges in north Georgia and a few Coastal Plain counties. **STEMS:** 3–6 feet tall, brown, with peeling, papery bark on older stems. **LEAVES:** Up to 7 inches long and 5 inches wide, opposite, with toothed, oval blades and long stalks, upper surface dark green, lower leaf surface pale green, not hairy. **FLOWER CLUSTERS:** Up to 6 inches wide, with many small, white, 5-petaled flowers and sometimes several large, white, 3- or 4-lobed, sterile flowers around the edges. **FLOWERING:** May–Jul. **RANGE:** FL west to LA, north to KS and MA. **SIMILAR TO** Ashy Hydrangea (*Hydrangea cinerea*), which occurs in northwest Georgia and Elbert County over limestone or mafic bedrock; the lower leaf surface is green and thinly covered with gray (not white) hairs.

Oak-Leaf Hydrangea

Hydrangea quercifolia Bartram
Hydrangea Family (Hydrangeaceae)

DECIDUOUS SHRUB found in moist forests and along streams in west Georgia, especially over limestone bedrock. **STEMS:** Up to 10 feet tall, with brown, peeling outer bark and smooth, tan or orange inner bark; young stems covered with rusty, cottony hairs. **LEAVES:** Up to 12 inches long and 8 inches wide, opposite, with 3–7 toothed lobes, upper surface dark green, lower surface pale and hairy. **FLOWER CLUSTERS:** Up to 12 inches high and 5 inches wide, with many showy sterile flowers. **FLOWERS:** Sterile flowers have 4 large, white sepals that turn pink with age; fertile flowers have 5 tiny, white petals held in small clusters below the sterile flower. **FLOWERING:** May–Jul. **RANGE:** FL west to LA, north to TN and GA; escaped from gardens throughout the South. **SIMILAR TO** no other shrub in Georgia.

Mock-Orange

Philadelphus inodorus Linnaeus
Hydrangea Family (Hydrangeaceae)

DECIDUOUS SHRUB found in moist forests, rocky ravines, and on cliffs in north Georgia, especially over limestone or mafic bedrock. **STEMS:** Up to 13 feet tall, older branches with shredding bark. **LEAVES:** Up to 3 inches long and 1½ inches wide, opposite, oval, with a few widely separated teeth, mostly smooth. **FLOWERS:** Held at the tips of new growth, 1½–2 inches across with 4 white, spreading, rounded petals and 20 or more yellow stamens, odorless. **FLOWERING:** Apr–May. **RANGE:** FL west to AL, north to TN and VA, but planted and escaped throughout eastern North America. **SIMILAR TO** Hairy Mock-orange (*Philadelphus hirsutus*), which occurs in rocky woodlands and on bluffs over limestone or mafic bedrock in north Georgia; it is less than 6 feet tall, its leaves are hairy on both surfaces, and its fragrant flower is less than 1½ inches wide.

Skyflower

Hydrolea corymbosa J. F. Macbride ex Elliott
Waterleaf Family (Hydrophyllaceae)

PERENNIAL HERB found in wet savannas, depression ponds, forested wetlands, and wet ditches in the Coastal Plain. **STEMS:** Up to 2½ feet tall. **LEAVES:** Up to 1¼ inches long and ½ inch wide, alternate, elliptic to lance-shaped, very finely toothed, smooth, with no or very small spines in the axils. **FLOWERS:** In branched clusters at the top of the stem, each flower about 1 inch wide, bell-shaped, with 5 blue petals and 5 purple, orange-tipped stamens. **FLOWERING:** Jul–Sep. **RANGE:** GA, FL, AL, and SC. **SIMILAR TO** Blue Waterleaf (*Hydrolea ovata*), which occurs in wetlands in southwest Georgia; it has spines up to ½ inch long in the leaf axils. Waterpod (*Hydrolea quadrivalvis*), also a wetland species, occurs throughout Georgia's Coastal Plain; its stems trail across the ground and root at the nodes, with flowers and spines in the leaf axils.

Maple-Leaf Waterleaf

Hydrophyllum canadense Linnaeus
Waterleaf Family (Hydrophyllaceae)

PERENNIAL HERB found in rich, moist, hardwood forests and on seepy limestone cliffs in the north Georgia mountains. **STEMS:** 8–28 inches tall, smooth or slightly hairy. **STEM LEAVES:** 4–8 inches long and wide, alternate, shaped like a maple leaf with 5–9 lobes, long-stalked, sparsely hairy. **BASAL LEAVES:** Overwintering and dying back in the spring, elongated, deeply divided into 7–9 silver-spotted lobes. **FLOWERS:** Up to ½ inch long, bell-shaped, pale purple or pink, with 5 stamens extending well beyond the flower. **FLOWERING:** May–Aug. **RANGE:** GA and AL, north to IA, ON, and VT. **SIMILAR TO** Hairy Waterleaf (*Hydrophyllum macrophyllum*), which occurs in moist, rich forests in two northwest Georgia counties; its flower is white, and its stem leaves are elongated, deeply divided into 9–13 silver-spotted lobes, not shaped like maple leaves.

Fern-Leaf Phacelia

Phacelia bipinnatifida Michaux
Waterleaf Family (Hydrophyllaceae)

BIENNIAL HERB found in rich, moist, hardwood forests and on seepy limestone cliffs in northwest Georgia and the upper Piedmont. **STEMS:** Up to 2 feet tall, branching, covered with gland-tipped hairs (use 10× lens). **LEAVES:** Up to 3 inches long, hairy on both surfaces, divided into 3–5 toothed and segmented leaflets, upper surface green and often mottled with small, silvery patches, lower surface silvery green. **FLOWERS:** In clusters at the tips of hairy stems, each flower about ½ inch wide, 5-lobed, not fringed; lavender, blue, or violet with a white throat. **FLOWERING:** Apr–May. **RANGE:** GA west to MS, north to IA and PA. **SIMILAR TO** two Georgia Special Concern species that occur in moist forests in northwest Georgia. Fringed Phacelia (*Phacelia fimbriata*) has white flowers with fringed edges. Miami-mist (*P. purshii*) has lavender or blue flowers with fringed edges.

Flatrock Phacelia

Phacelia maculata Wood
Waterleaf Family (Hydrophyllaceae)

ANNUAL HERB found on and around granite outcrops and in nearby woods in the Piedmont. **STEMS:** Up to 14 inches long, erect or sprawling, hairy. **LEAVES:** Up to 2½ inches long, divided into many toothed and segmented leaflets, hairy on both surfaces. **FLOWERS:** In clusters at the tips of stems, each flower about ⅓ inch wide, 5-lobed, not fringed, bluish-purple with 5 blue stamens and 10 purple spots in a ring inside the throat. **FLOWERING:** Apr. **RANGE:** GA west to AL, north to NC. **SIMILAR TO** Appalachian Phacelia (*Phacelia dubia*), which occurs in floodplains, woodlands, and granite outcrops in the Piedmont and a few Coastal Plain and Blue Ridge counties; its stem is usually erect, and its flowers are pale blue or white and lack the ring of purple spots in the throat.

St. Andrew's Cross, St. Peter's-Wort

Hypericum crux-andreae (Linnaeus) Crantz

SYNONYMS: *Hypericum stans* (Michaux ex Willdenow) W. P. Adams & Robson, *Ascyrum stans* Michaux ex Willdenow

St. John's-Wort Family (Hypericaceae)

SHRUB found in upland forests, sandhills, and dry pinelands in the Coastal Plain and Piedmont. **STEMS:** Up to 3 feet tall, older stems with peeling, reddish-brown bark. **LEAVES:** Up to 1½ inches long and ¾ inch wide, opposite, oval or lance-shaped, the bases rounded and slightly clasping the stem, entire, slightly leathery. **FLOWERS:** ¾–1½ inches wide, with 4 bright yellow petals, many stamens, 2 narrow, pointed sepals, and 2 broader, oval sepals; in small clusters on twigs and solitary at the tips of stems. **FLOWERING:** Jun–Oct. **RANGE:** FL west to TX, north to OK and NY. **SIMILAR TO** Four-petal St. John's-wort (*Hypericum tetrapetalum*), which occurs in wet pine flatwoods in the Coastal Plain; its leaves are heart-shaped and strongly clasp the stem.

Peel-Bark St. John's-Wort

Hypericum fasciculatum Lamarck

St. John's-Wort Family (Hypericaceae)

SHRUB found in wet pine flatwoods, ponds, and wet ditches in the Coastal Plain. **STEMS:** Up to 6 feet tall, much-branched, the youngest twigs with narrow wings, older stems with corky or spongy bark peeling off in sheets to expose tan or cinnamon-colored inner bark. **LEAVES:** Up to 1 inch long, opposite, very narrow and needle-like, with smaller leaves clustered in the axils, the lower surface with 2 whitish grooves. **FLOWERS:** About ½ inch wide, with 5 yellow petals, 5 needle-like sepals, and many yellow stamens. **FLOWERING:** May–Sep. **RANGE:** FL west to MS, north to NC. **SIMILAR TO** Coastal Plain St. John's-wort (*Hypericum nitidum*), which occurs in wet flatwoods and along blackwater streams; it reaches 10 feet in height, and its bark is tight (not corky or spongy) and peels off in thin strips or small flakes.

Pineweed, Orange-Grass

Hypericum gentianoides (Linnaeus) B.S.P.
St. John's-Wort Family (Hypericaceae)

ANNUAL HERB found in woodland borders, granite outcrops, and disturbed areas throughout Georgia. **STEMS:** Up to 20 inches tall, wiry, much-branched, with narrow wings, the main stem reddish-brown, upper branches green. **LEAVES:** Tiny and scalelike, opposite, pressed closely against the stem. **FLOWERS:** About ⅓ inch wide, with 5 yellow petals, held at the tips of branches. **FLOWERING:** Jul–Oct. **RANGE:** FL west to TX, north to MN and NS. **NOTES:** The crushed stems smell a little like orange peel. The tough, dead stems persist through the winter and spring, looking like brown whiskbrooms. **SIMILAR TO** Nits-and-lice (*Hypericum drummondii*), which occurs infrequently throughout Georgia in similar habitats; its leaves are larger, up to ¾ inch long and ¼ inch wide, and angle slightly away from the stem. Pineweed also resembles Texas Yellow Flax, page 250.

Small-Flowered St. Andrew's Cross

Hypericum hypericoides (Linnaeus) Crantz
SYNONYM: *Ascyrum hypericoides* Linnaeus
St. John's-Wort Family (Hypericaceae)

SHRUB found in dry upland forests, sandhills, and pine flatwoods throughout Georgia. **STEMS:** 1–5 feet tall, older stems with peeling, reddish-brown bark. **LEAVES:** Up to 1 inch long and ¼ inch wide, opposite, oblong with tapering bases and rounded tips, entire, covered with tiny, clear or black dots. **FLOWERS:** Less than ¾ inch wide, with 4 pale yellow petals, 2 green, broadly oval, gland-dotted sepals up to ½ inch long, and 2 tiny sepals (not always present). **FLOWERING:** May–Aug. **RANGE:** FL west to TX, north to MO and NJ; West Indies, Central America. **SIMILAR TO** Low St. John's-wort (*Hypericum stragulum*), which occurs in dry upland forests in north Georgia, and to Pineland St. John's-wort (*H. suffruticosum*), which occurs in pine flatwoods in the Coastal Plain; both are sprawling shrubs less than 2 feet tall.

Myrtle-Leaf St. John's-Wort

Hyperricum myrtifolium Lamarck
St. John's-Wort Family (Hypericaceae)

EVERGREEN SHRUB found in wet pine flatwoods, ponds, and bogs in the Coastal Plain. **STEMS:** Up to 3 feet tall, loosely branched, young twigs waxy, pale green or reddish, bark on older stems peeling off in thin sheets. **LEAVES:** Up to 1 inch long, opposite, oval to triangular, almost clasping the stem, entire, smooth, the upper surface pale blue-green, the lower surface waxy white and gland-dotted. **FLOWERS:** Up to 1 inch wide, with 5 yellow petals, 5 oval sepals, and a large cluster of yellow stamens. **FLOWERING:** May–Sep. **RANGE:** FL west to MS, north to SC. **SIMILAR TO** Four-petal St. John's-wort (*Hypericum tetrapetalum*), which occurs in similar habitats; its flowers have 4 petals and 4 sepals. Round-pod St. John's-wort (*H. cistifolium*) has narrowly oblong leaves in pairs set at right angles to the next pair, with clusters of small leaves in the leaf axils.

Spotted St. John's-Wort

Hypericum punctatum Lamarck
St. John's-Wort Family (Hypericaceae)

PERENNIAL HERB found in woodland borders, floodplains, and disturbed areas throughout Georgia. **STEMS:** Up to 2½ feet tall, smooth, branched just below the flower cluster, covered with tiny, black dots. **LEAVES:** Up to 2½ inches long and 1 inch wide, opposite, smooth, entire, oblong with a rounded or notched tip, the lower surface and margins covered with tiny, black and clear dots (best seen in backlight with 10× lens). **FLOWERS:** About ½ inch wide, with 5 bright yellow petals, both petals and sepals with black dots and lines. **FLOWERING:** Jun–Sep. **RANGE:** FL west to TX, north to ON and NS. **SIMILAR TO** Dwarf St. John's-wort (*Hypericum mutilum*), which occurs throughout Georgia in marshes, lake edges, and wet clearings; its stem has many spreading branches, and its leaves have clear dots but no black ones.

Hairy St. John's-Wort

Hypericum setosum Linnaeus
St. John's-Wort Family (Hypericaceae)

ANNUAL OR BIENNIAL HERB found in pine savannas and flatwoods, seepage slopes, and bogs in the Coastal Plain. **STEMS:** Up to 2½ feet tall, very hairy, usually branched only in the flower cluster. **LEAVES:** Up to ½ inch long and ¼ inch wide, opposite, oval, entire, pressed upward against the stem, very hairy, gland-dotted (best seen when backlit), with a single main vein. **FLOWERS:** Pinwheel-like, about ½ inch wide, with 5 orange-yellow petals and 5 pointed sepals with hairy margins and lines of gland dots. **FLOWERING:** May–Sep. **RANGE:** FL west to TX, north to VA. **SIMILAR TO** Strict St. John's-wort (*Hypericum virgatum*), which occurs throughout Georgia in dry woodlands and on rock outcrops; its stems and leaves are smooth, and the leaves spread at right angles to the stem.

Marsh St. John's-Wort

Hypericum walteri J. G. Gmelin
SYNONYM: *Triadenum walteri* (J. G. Gmelin) Gleason
St. John's-Wort Family (Hypericaceae)

PERENNIAL HERB found in marshes, swamps, and floodplains throughout Georgia, often growing on tree trunks and logs just above the water line. **STEMS:** Up to 3 feet tall, reddish, with a branched, bushy appearance. **LEAVES:** Up to 6 inches long and 1½ inches wide, opposite, smooth, entire, with tapering bases and rounded tips, the lower surfaces pale and dotted with both black and clear glands (best seen when backlit), midvein often reddish. **FLOWERS:** About ½ inch across, pale pink or salmon-colored, with 5 pointed petals and 9 stamens. **FLOWERING:** Jul–Sep. **RANGE:** FL west to TX, north to MO and NJ. **SIMILAR TO** Virginia Marsh St. John's-Wort (*Triadenum virginicum*, *Hypericum virginicum*), which occurs in similar habitats throughout Georgia, mostly in the Coastal Plain; its leaves are rounded at the base and clasp the stem.

Virginia-Willow, Virginia Sweet-Spire

Itea virginica Linnaeus
Sweet-Spire Family (Iteaceae)

DECIDUOUS SHRUB found in swamps and moist forests and on stream banks throughout Georgia. **STEMS:** 3–10 feet tall, erect or arching, usually in clumps. **LEAVES:** Up to 4 inches long and 1½ inches wide, alternate, elliptic or oval, smooth, with small, pointed teeth. **FLOWER CLUSTERS:** A drooping or arching spike 1½–6 inches long with many closely packed flowers. **FLOWERS:** About ⅓ inch wide, with 5 white, separate, erect petals. **FRUIT:** A brown, cylindrical capsule about ⅓ inch long with a pointed tip. **FLOWERING:** May–Jun. **RANGE:** FL west to TX, north to OK, IL, and NJ. **SIMILAR TO** Sweet-pepperbush (*Clethra alnifolia*), page 157.

Trailing Ratany, Sand-Bur

Krameria lanceolata Torrey
Krameria Family (Krameriaceae)

PERENNIAL HERB found in sandhills in Georgia's Coastal Plain. **STEMS:** Sprawling and much-branched, up to 40 inches long, covered with silky, white hairs. **LEAVES:** Up to ¾ inch long, very narrow, alternate, linear to elliptic, entire, silky-hairy, with a brown, spiny tip. **FLOWERS:** Up to 1 inch across, with 5 showy, spreading, pink or maroon sepals, and 5 tiny, green-and-red petals at the center of the flower. **FRUIT:** About ¼ inch wide, rounded, covered with white, fuzzy hairs and reddish spines. **FLOWERING:** May–Jul. **RANGE:** GA, FL, AR, and TX, west to AZ and CO; Mexico. **NOTES:** Most species of *Krameria* are western desert plants; all are partially parasitic on the roots of other plants. Trailing Ratany is the only *Krameria* in the southeastern United States and is **SIMILAR TO** no other Georgia wildflower.

Mint Family (Lamiaceae or Labiatae)

The mints, with their 2-lipped flowers, opposite leaves, and square stems, are among the easiest plants to recognize. Mint flowers have a more or less consistent shape—a tube with 2 spreading lips at the end. The upper lip has 2 lobes; the lower, downcurved lip has 3 lobes. The lower lip is often marked with white patches and dark spots that guide pollinators to the nectar deep inside the flower's tube. The stems are 4-angled or square in cross-section. The leaves are usually opposite, and each pair of leaves is set at right angles to the next pair. They often have a strong scent—sometimes minty and sweet, sometimes rank and unpleasant. These odors derive from chemical compounds that discourage animals from eating the leaves and suppress the growth of neighboring plants. Humans have put these compounds to use for cooking (basil, sage, oregano, rosemary, marjoram, thyme), drinking (peppermint, spearmint), healing (hyssop, catnip, self-heal), and beautifying our homes and gardens (coleus, lavender)—just to name a few of the economically and culturally important mints. Georgia has about 80 native mint species and another 20–25 that are exotic; the exact numbers depend on which authority you ask.

Downy Wood Mint

Blephilia ciliata (Linnaeus) Bentham
Mint Family (Lamiaceae)

PERENNIAL HERB found in woodlands, rocky glades, and prairies in the Piedmont and Ridge & Valley, usually over mafic or limestone bedrock. **STEMS:** Up to 3 feet tall, 4-angled, covered with short hairs pressed against the stem. **LEAVES:** Up to 3½ inches long and 1½ inches wide, opposite, oval, shallowly toothed, hairy, very fragrant, stalked. **FLOWER CLUSTERS:** 1–3 inches wide, with many tightly packed flowers and small, leafy bracts; 2 or more clusters encircle the upper stem, separated by a short length of stem. **FLOWERS:** About ⅓ inch long, pale pink to lavender, hairy, tubular with 2 flaring lips, the upper lip narrow, the lower lip 3-lobed and speckled with purple. **FLOWERING:** May–Jul. **RANGE:** GA west to OK, north to WI, ON, and VT. **SIMILAR TO** Self-heal (*Prunella vulgaris*), page 239.

Scarlet Wild Basil

Clinopodium coccineum (Nuttall ex Hooker) Kuntze

SYNONYMS: *Calamintha coccinea* (Nuttall ex Hooker) Bentham, *Satureja coccinea* (Nuttall ex Hooker) Bertolini

Mint Family (Lamiaceae)

EVERGREEN SHRUB found in sandhills and dry, sandy scrub in Georgia's Coastal Plain. **STEMS:** Up to 4 feet tall, round or slightly 4-angled in cross-section, loosely branched, hairy, older stems with shredding bark. **LEAVES:** Up to ¾ inch long and ⅓ inch wide, opposite, elliptic, entire, leathery, with gland dots, rolled margins, and no stalks. **FLOWERS:** 1–2 inches long, red (rarely yellow), tubular with 2 flaring lips, the upper lip notched at the tip, the lower lip 3-lobed and speckled with dark red. **FLOWERING:** Spring and fall. **RANGE:** GA, FL, AL, and MS. **SIMILAR TO** Scarlet Sage (*Salvia coccinea*), a perennial herb that occurs in the Coastal Plain. Its leaves are more than 1 inch wide, triangular, toothed, and long-stalked; its scarlet flowers are about 1 inch long.

Georgia Basil

Clinopodium georgianum R. M. Harper

SYNONYMS: *Satureja georgiana* (R. M. Harper) H. E. Ahles, *Calamintha georgiana* (R. M. Harper) Shinners

Mint Family (Lamiaceae)

DECIDUOUS SHRUB found in dry woodlands, sandhills, and clearings in the Piedmont and Coastal Plain. **STEMS:** Up to 2 feet tall, woody, round or slightly 4-angled; older stems have tan, shredding bark. **LEAVES:** Up to 1½ inches long and ¾ inch wide, opposite, oval, toothed, with a short stalk and scattered gland dots. **FLOWERS:** About ½ inch long, pale pink, tubular with 2 flaring lips, the upper lip notched at the tip, the lower lip 3-lobed and speckled with purple. **FLOWERING:** Jul–Sep. **RANGE:** FL west to LA, north to NC. **NOTES:** Leaves have a sweet, minty odor when crushed. **SIMILAR TO** Ashe's Savory (*Clinopodium ashei*), a Georgia Special Concern species that occurs in sandhills in southeast Georgia; its leaves are very hairy with tightly rolled margins.

Southern Horsebalm, Anise Horsebalm

Collinsonia anisata Sims
SYNONYM: *Collinsonia serotina* Walter
Mint Family (Lamiaceae)

PERENNIAL HERB found in rich, upland forests in the Piedmont and a few southwest Georgia counties. **STEMS:** 1–2 feet tall, with 4 rounded angles, covered with gland-tipped hairs. **LEAVES:** 3–8 inches long and 2½–5 inches wide, opposite, in 3 or more pairs, elliptic to oval, coarsely toothed. **FLOWERS:** Paired in a large cluster at the top of the stem, each flower about ½ inch long, cream to pale yellow, glandular-hairy, tubular with 2 spreading lips, the lower lip much larger and fringed; 4 pink stamens and the style extend well beyond the lips. **FLOWERING:** Late Aug–Oct. **RANGE:** GA, FL, AL, and MS. **NOTES:** Leaves smell strongly of anise when crushed. **SIMILAR TO** Northern Horsebalm (*Collinsonia canadensis*), below, and Whorled Horsebalm (*C. verticillata*), page 232.

Northern Horsebalm

Collinsonia canadensis Linnaeus
Mint Family (Lamiaceae)

PERENNIAL HERB found in rich, upland forests in north Georgia and a few southwest Georgia counties. **STEMS:** Up to 3 feet tall, 4-angled, nearly hairless. **LEAVES:** 6–10 inches long and 2¾–6¾ inches wide, opposite, oval, sharply toothed, hairy or smooth, with a lemony smell when crushed. **FLOWERS:** In a large, open cluster at the top of the stem, each flower up to ½ inch long, yellow, tubular with 2 flaring lips, the lower lip fringed; 2 stamens and the style extend well beyond the lips. **FLOWERING:** Jul–Sep. **RANGE:** FL west to LA, north to WI and NH. **SIMILAR TO** Stoneroot (*Collinsonia tuberosa*), which occurs in rich forests in north Georgia; its flowers are white to cream-colored, speckled with purple, and the leaves are less than 4 inches long. Also see Whorled Horsebalm (*Collinsonia verticillata*), page 232.

Whorled Horsebalm

Collinsonia verticillata Baldwin
Mint Family (Lamiaceae)

PERENNIAL HERB found in rich upland forests over mafic or limestone bedrock in north Georgia. **STEMS:** 8–32 inches tall, with 4 rounded angles, smooth or hairy. **LEAVES:** 3½–8½ inches long and 2½–5 inches wide, opposite, oval, toothed, in 2 pairs so closely spaced that the leaves appear whorled. **FLOWER CLUSTERS:** Whorls of flowers in an unbranched spike at the top of the stem. **FLOWERS:** About ½ inch long, pale pink to lavender, short-tubular with 4 short, rounded lobes and a fringed lower lip; 4 long, pink stamens extend well beyond the lips. **FLOWERING:** Late Apr–early Jun. **RANGE:** FL west to MS, north to OH and VA. **NOTES:** All other horsebalms in Georgia have branched flower clusters. **SIMILAR TO** Northern Horsebalm (*Collinsonia canadensis*) and Southern Horsebalm (*Collinsonia anisata*), page 231.

Stone Mint, American Dittany

Cunila origanoides (Linnaeus) Britton
Mint Family (Lamiaceae)

PERENNIAL HERB found in dry, rocky woodlands and prairies in the north Georgia mountains. **STEMS:** Up to 20 inches tall, 4-angled, hairy in the upper half, smooth below, reddish-brown, wiry or almost woody. **LEAVES:** Up to 2 inches long and 1 inch wide, opposite, oval to lance-shaped, toothed, with very short or no stalk, smelling strongly of oregano when crushed. **FLOWERS:** In clusters in the leaf axils and at the tips of stems, each flower ¼ inch long, pink or lavender with purple spots, tubular with 5 short, spreading lobes and 2 pink stamens extending beyond the lobes. **FLOWERING:** Aug–Oct. **RANGE:** GA west to TX, north to KS and NY. **NOTES:** Stone Mint produces "frost flowers" in late fall when the sap freezes and ruptures the stem, extruding ribbons of ice. **SIMILAR TO** American Pennyroyal (*Hedeoma pulegioides*), page 234.

Coastal Plain Balm

Dicerandra linearifolia (Elliott) Bentham
Mint Family (Lamiaceae)

ANNUAL HERB found in sandhills, scrub, and pine flatwoods in the Coastal Plain. **STEMS:** Up to 20 inches tall, much-branched, 4-angled. **LEAVES:** Up to 1¾ inches long, needle-shaped, opposite, rough-hairy, gland-dotted, smelling of mint when crushed. **FLOWERS:** About ½ inch long; white, pale pink, or rose; funnel-shaped with 2 flaring lips, the lower lip 3-lobed, the upper lip spotted and streaked with magenta; the style and 2 of the 4 stamens extend well beyond the petals. **FLOWERING:** Sep–Nov. **RANGE:** GA, FL, and AL. **SIMILAR TO** Rose-balm (*Dicerandra odoratissima*), which occurs in sandhills in southeast Georgia; its stamens do not extend beyond the flower opening, and the leaves smell of cinnamon when crushed.

Gill-over-the-Ground, Ground-Ivy

Glecoma hederacea Linnaeus
Mint Family (Lamiaceae)

PERENNIAL HERB found in lawns, gardens, roadsides, and other disturbed areas in north Georgia. **STEMS:** Up to 2 feet long, 4-angled, creeping along the ground, rooting at the leaf nodes, and forming mats. **LEAVES:** 1–2 inches long and wide, opposite, kidney- or heart-shaped, evergreen, with scalloped margins and a short stalk. **FLOWERS:** In a small clusters at the top of short, erect stems, each flower about ½ inch long, pink or purple, funnel-shaped with 2 flaring lips, the upper lip small and 2-lobed, the lower lip larger, 3-lobed, hairy, and spotted with purple. **FLOWERING:** Mar–Jun. **RANGE:** Native to Eurasia, now spread throughout most of North America. **NOTES:** Leaves and stems have a strong, pungent smell. **SIMILAR TO** Henbit (*Lamium amplexicaule*), page 235, which has only erect stems.

American Pennyroyal

Hedeoma pulegioides (Linnaeus) Persoon
Mint Family (Lamiaceae)

ANNUAL HERB found in dry woodlands and disturbed areas in north Georgia. **STEMS:** Up to 18 inches tall, branched, round or 4-angled, covered with glandular hairs. **LEAVES:** Up to 1 inch long and ½ inch wide, opposite, oval or elliptic, with a few blunt teeth near the tip, covered with gland dots. **FLOWERS:** Held singly in the leaf axils, less than ¼ inch long; white, pink, or lavender, with a 2-lobed upper lip and a purple-spotted, 3-lobed lower lip. **FLOWERING:** Late Jul–Oct. **RANGE:** GA west to OK, north to ND and NS. **SIMILAR TO** Rough Pennyroyal (*Hedeoma hispida*), which occurs infrequently in disturbed areas and around granite outcrops throughout Georgia; its tiny flowers are arranged in whorls around the stem, and its leaves are less than ⅛ inch wide.

Musky Mint, Clustered Bush-Mint

Hyptis alata (Rafinesque) Shinners
Mint Family (Lamiaceae)

PERENNIAL HERB found in swamp forests, wet pine savannas, and disturbed wetlands in the Coastal Plain. **STEMS:** 1–10 feet tall, 4-angled, hairy on the angles. **LEAVES:** 2–6 inches long and ½–2½ inches wide, lance-shaped, tapering at both ends, toothed, with no or very short stalks. **FLOWER CLUSTERS:** Ball- or dome-shaped heads held at the tips of long stalks that arise from the leaf axils, ½–1 inch across, with small, leafy bracts surrounding the base of the head. In fruit, the head appears honeycombed with the remains of the calyxes. **FLOWERS:** About ¼ inch long, white speckled with purple, tubular with 2 flaring lips, the upper lip 2-lobed, the lower lip 3-lobed. **FLOWERING:** Late Jun–Sep. **RANGE:** FL west to TX, north to NC; West Indies. **SIMILAR TO** no other wildflower in Georgia.

Henbit

Lamium amplexicaule Linnaeus
Mint Family (Lamiaceae)

ANNUAL HERB found in pastures, lawns, and other disturbed areas throughout Georgia. **STEMS:** Up to 2 feet long, sprawling across the ground except for erect new growth, 4-angled, rough-hairy. **LEAVES:** Up to 1 inch long and wide, opposite, nearly round, hairy, with lobed and scalloped margins, lower leaves with reddish stalk, upper leaves with no stalk. **FLOWERS:** Held in whorls in the axils of the upper leaves, each about ½ inch long, pale to dark pink, tubular with 2 spreading lips, the upper lip hooded and hairy, the lower lip lobed, drooping, and white with purple speckles. **FLOWERING:** Feb–May, sometimes all year. **RANGE:** Native to Eurasia and Africa, now spread through most of North America. **SIMILAR TO** Purple Dead-nettle (*Lamium purpureum*), below.

Purple Dead-Nettle

Lamium purpureum Linnaeus
Mint Family (Lamiaceae)

ANNUAL HERB found in pastures, lawns, and other disturbed areas in north Georgia. **STEMS:** Up to 2½ feet long, sprawling across the ground except for erect new growth, hollow, 4-angled, hairy only on the angles. **LEAVES:** Up to 2 inches long and wide, opposite, scalloped, hairy; older, lower leaves round with long stalk; younger, upper leaves purplish, triangular, with short stalk. **FLOWERS:** Held in whorls in the axils of the upper leaves, each about ½ inch long, pink to purple, tubular with 2 spreading lips, the upper lip hooded and hairy, the lower lip paler and speckled with purple. **FLOWERING:** Mar–Oct. **RANGE:** Native to Eurasia, now spread through most of North America. **NOTES:** Purple Dead-nettle turns fields and pastures into seas of purple in early spring. **SIMILAR TO** Henbit (*Lamium amplexicaule*), above.

Virginia Bugle-Weed, Water Horehound

Lycopus virginicus Linnaeus
Mint Family (Lamiaceae)

PERENNIAL HERB found in wetlands in north Georgia and several Coastal Plain counties. **STEMS:** 1½–4 feet tall, with 4 rounded angles, sparsely hairy, often forming colonies. **LEAVES:** Up to 3½ inches long and 2 inches wide, more or less the same size throughout, opposite, oval to lance-shaped with a wedge-shaped base tapering to the stem, toothed above the middle, smooth or slightly hairy, often tinged with purple, without a minty smell. **FLOWER CLUSTERS:** Held in the leaf axils, densely flowered, often encircling the stem. **FLOWERS:** About ⅛ inch wide, white, tubular, with 4 erect lobes. **FLOWERING:** Jul–Nov. **RANGE:** FL west to TX, north to ON and QC. **SIMILAR TO** Stalked Bugle-weed (*Lycopus rubellus*), which occurs in wetlands in the Coastal Plain and several north Georgia counties; its flowers have 5 spreading lobes.

Basil Bee-Balm, White Bergamot

Monarda clinopodia Linnaeus
Mint Family (Lamiaceae)

PERENNIAL HERB found in moist forests in north Georgia, mostly in the mountains. **STEMS:** 1½–3 feet tall, 4-angled, smooth or hairy on the angles. **LEAVES:** 2½–5 inches long and 1–3 inches wide, opposite, oval or lance-shaped, sharply pointed, toothed, with a rounded base and stalks ½–1⅓ inches long. **FLOWERS:** In a single head (rarely 2) at the top of the stem, each flower ¾–1 inch long; white, cream, or pale pink, tubular with 2 spreading lips, the upper lip narrow and straight, without a tuft of hairs at the tip, the lower lip 3-lobed and spotted with purple; 2 stamens and the style extend beyond the end of the flower; fragrant. **FLOWERING:** May–Sep. **RANGE:** GA west to AL, north to MO, MI, and VT. **SIMILAR TO** Wild Bergamot (*Monarda fistulosa*), page 237.

Scarlet Bee-Balm, Oswego Tea

Monarda didyma Linnaeus
Mint Family (Lamiaceae)

PERENNIAL HERB found in seepy areas and on stream banks in the Blue Ridge. **STEMS:** 2–6 feet tall, 4-angled, smooth or hairy on the angles. **LEAVES:** 3–6 inches long and 1–2¾ inches wide, opposite, oval or lance-shaped, with a strong, pungent smell when crushed. **FLOWER CLUSTERS:** 1–2 heads at the top of the stem, with a whorl of reddish bracts at the base. **FLOWERS:** 1–1½ inches long, red, tubular with 2 spreading lips, the upper lip narrow and straight, without a tuft of hairs at the tip, the lower lip notched at the tip, with the style and 2 stamens extending beyond the tip; not fragrant. **FLOWERING:** Jul–Sep. **RANGE:** GA west to MO, north to ON and QC. **NOTES:** Bee-balm is pollinated mostly by hummingbirds. **SIMILAR TO** Wild Bergamot (*Monarda fistulosa*), below.

Wild Bergamot

Monarda fistulosa Linnaeus
Mint Family (Lamiaceae)

PERENNIAL HERB found in cove forests, prairies, woodlands, and clearings in north Georgia. **STEMS:** Up to 4 feet tall, 4-angled, smooth or hairy, green or reddish-brown with a whitish, waxy coating. **LEAVES:** Up to 4 inches long and 2 inches wide, opposite, lance-shaped, toothed, often red- or purple-tinged, long-stalked, with a strong, oregano-like smell. **FLOWER CLUSTERS:** A single round head about 1–3 inches wide, surrounded by leafy bracts; flowers open from the center of the head outward. **FLOWERS:** About 1 inch long, pink, lavender, purple, or white, tubular with 2 spreading lips, the upper lip narrow and straight with a tuft of hairs at the tip, the lower lip 3-lobed; the style and 2 stamens extend beyond the end of the flower. **FLOWERING:** Jun–Sep. **RANGE:** GA west to TX, north to MN and ME. **SIMILAR TO** Basil Bee-balm (*Monarda clinopodia*), page 236.

Spotted Bee-Balm, Spotted Horse-Mint

Monarda punctata Linnaeus
Mint Family (Lamiaceae)

PERENNIAL HERB found in dry woodland edges, maritime forests, clearings in Longleaf Pine forests, and clearings throughout Georgia except for the mountains. **STEMS:** Up to 3 feet tall, 4-angled, branched, softly hairy. **LEAVES:** Up to 3½ inches long and 1 inch wide, opposite, elliptic to lance-shaped, toothed, hairy. **FLOWER CLUSTERS:** Dome-shaped heads held at the tips of branches and also wrapped around the stem, each head with a whorl of pink, leafy bracts surrounding its base. **FLOWERS:** Up to 1 inch long, pale yellow with purple spots, with 2 spreading lips, the upper lip arched, the lower lip downcurved and 3-lobed. **FLOWERING:** Late Jul–Sep. **RANGE:** FL west to NM, north to MN and QC. **SIMILAR TO** no other wildflower in Georgia.

Beefsteak Plant

Perilla frutescens (Linnaeus) Britton
Mint Family (Lamiaceae)

ANNUAL HERB found in floodplains, moist forests, and disturbed areas throughout Georgia. **STEMS:** Up to 3 feet tall, 4-angled, grooved, hairy. **LEAVES:** Up to 5 inches long and 4 inches wide, opposite, oval, smooth, long-stalked, coarsely toothed, often crimped around the margins. **FLOWERS:** In elongated clusters, each flower ⅛–¼ inch long, pink, with 5 spreading lobes and 4 pink-tipped stamens. **FLOWERING:** Aug–Oct. **RANGE:** Native to India, now spread throughout most of eastern North America. **NOTES:** Stems and leaves are usually green in sunny areas and purple in shady sites. This nonnative, invasive annual spreads easily and should be pulled up or sprayed with herbicide before it sets seed. **SIMILAR TO** Mosla, or Miniature Beefsteak Plant (*Mosla dianthera*), also an invasive Asian plant. Its leaves and stems are always green, and its flowers have only 2 stamens.

Obedient Plant

Physostegia virginiana (Linnaeus) Bentham
SYNONYM: *Dracocephalum virginianum* Linnaeus
Mint Family (Lamiaceae)

PERENNIAL HERB found in seepages, glades, and open, disturbed areas in north Georgia and a few Coastal Plain counties, usually over mafic or limestone bedrock. **STEMS:** 1–5 feet tall, strongly 4-angled, unbranched, smooth. **LEAVES:** Up to 6 inches long and 1¾ inches wide, opposite, lance-shaped, toothed, smooth, with or without a stalk (but not clasping the stem). **FLOWER CLUSTERS:** Up to 10 inches high, with 4 vertical rows of flowers. **FLOWERS:** About 1 inch long and ½ inch wide, pink, tubular with a hooded upper lip and a 3-lobed, purple-spotted lower lip. Flowers "obediently" remain where pushed. **FLOWERING:** Jul–Oct. **RANGE:** FL west to NM, north to MB and NS. **SIMILAR TO** Purple Obedient Plant (*Physostegia purpurea*), which occurs in Coastal Plain wetlands; it flowers in early summer, and its leaves clasp the stem.

Self-Heal, Heal-All

Prunella vulgaris Linnaeus
Mint Family (Lamiaceae)

PERENNIAL HERB found in moist forests, woodlands, roadsides, and other disturbed areas in north Georgia and a few Coastal Plain counties. **STEMS:** 4–32 inches tall, unbranched, 4-angled, smooth or hairy, especially on the angles. **LEAVES:** Up to 3½ inches long and 1½ inches wide, opposite, oval or lance-shaped, shallowly toothed or entire. **FLOWERS:** In a short spike at the top of the stem, each flower about ½ inch long, pinkish-purple, with a hooded upper lip and a 3-lobed, fringed lower lip. **FLOWERING:** Apr–Dec. **RANGE:** Throughout North America. **NOTES:** American Self-heal (variety *lanceolata*) has rounded leaf bases and occurs in more natural settings; Eurasian Self-heal (variety *vulgaris*) has wedge-shaped leaf bases and occurs in disturbed areas. **SIMILAR TO** Florida Betony (*Stachys floridana*), page 244.

Savanna Mountain-Mint

Pycnanthemum flexuosum (Walter) B.S.P.

Mint Family (Lamiaceae)

PERENNIAL HERB found in moist to wet pinelands and seepages in the Coastal Plain. **STEMS:** 1–3½ feet tall, 4-angled, covered with white hairs. **LEAVES:** Up to 2 inches long and ½ inch wide, opposite, narrowly oblong to lance-shaped, toothed, hairy. **FLOWER CLUSTERS:** Round, slightly flattened heads with a bristly look due to the elongated white (but not hairy) tips of the calyx lobes. **FLOWERS:** Up to ¼ inch long, with 2 spreading lips, the lower lip 3-lobed, white or pale pink, speckled with purple. **FLOWERING:** Jun–Sep. **RANGE:** FL west to MS, north to TN and VA. **NOTES:** The abundance of white hairs gives the entire plant a whitish cast. **SIMILAR TO** Coastal Plain Mountain-mint (*Pycnanthemum nudum*), which occurs in wet flatwoods in the Coastal Plain; its leaves are oval, smooth, entire, and less than ¾ inch long.

Hoary Mountain-Mint

Pycnanthemum incanum (Linnaeus) Michaux

Mint Family (Lamiaceae)

PERENNIAL HERB found in woodlands and clearings in north Georgia and a few Coastal Plain counties. **STEMS:** Up to 6 feet tall, 4-angled, hairy. **LEAVES:** Up to 4 inches long and 1½ inches wide, opposite, oval or lance-shaped, toothed, hairy, upper leaves white on both surfaces, lower leaves dark green above with pale green lower surface. **FLOWER CLUSTERS:** Dome-shaped, crowded with tiny, white, hairy bracts and surrounded by white, leaflike bracts. **FLOWERS:** About ⅓ inch long, white or pale pink, speckled with purple; the calyx lobes are broadly triangular with a tuft of bristly hairs at the tip—use a 10× lens to distinguish the calyx lobes from the hairy bracts. **FLOWERING:** Jul–Aug. **RANGE:** FL west to MS, north to IL, ON, and NH. **SIMILAR TO** Southern Mountain-mint (*Pycnanthemum pycnanthemoides*), which occurs in similar habitats; its calyx lobes are elongated, narrowly triangular, and also tipped with bristles.

Narrow-Leaf Mountain-Mint

Pycnanthemum tenuifolium Schrader
Mint Family (Lamiaceae)

PERENNIAL HERB found in moist to wet forests and clearings, savannas, flatwoods, and bogs in north and southwest Georgia. **STEMS:** Up to 2½ feet tall, 4-angled, smooth. **LEAVES:** Up to 2 inches long and ⅛ inch wide, opposite, linear, entire, gland-dotted, with a minty fragrance. **FLOWER CLUSTERS:** Compact, crowded with flowers and small, white, hairy bracts. **FLOWERS:** About ¼ inch long with 2 spreading lips, the lower lip 3-lobed, white or pale pink, speckled with purple. **FLOWERING:** Jun–Aug. **RANGE:** FL west to TX, north to MN and ME. **SIMILAR TO** Virginia Mountain-mint (*Pycnanthemum virginianum*), a Georgia Special Concern species that occurs in sunny wetlands over mafic or limestone bedrock in a few north Georgia counties; its stems are hairy on the angles, and it flowers late Jul–Sep.

Blue Sage, Azure Sage

Salvia azurea Michaux ex Lamarck
Mint Family (Lamiaceae)

PERENNIAL HERB found in sandhills and dry woodlands throughout Georgia. **STEMS:** 2–5 feet tall, 4-angled, hairy. **LEAVES:** Up to 3½ inches long and 1¼ inches wide, opposite, elliptic to lance-shaped, tapering at both ends, entire or toothed, slightly hairy, grayish-green, smell like sage when crushed. **FLOWERS:** In loose, elongated spikes, each flower ½–1 inch long, pale to dark blue, white, or both; tubular with 2 spreading lips, the upper lip small and hoodlike, the lower lip much larger, 3-lobed, notched, usually with white stripes. **FLOWERING:** Aug–Oct. **RANGE:** FL west to TX, north to CO and NC. **SIMILAR TO** no other wildflower in Georgia.

Lyre-Leaved Sage

Salvia lyrata Linnaeus
Mint Family (Lamiaceae)

PERENNIAL HERB found in woodlands, roadsides, lawns, and meadows throughout Georgia. **STEMS:** 1–2 feet tall, solitary, usually unbranched, 4-angled, hairy. **BASAL LEAVES:** Up to 8 inches long and 2 inches wide, deeply lobed and divided, very hairy, with no minty odor, often marked with purple along the conspicuous veins. **STEM LEAVES:** In 1 or 2 pairs, opposite, toothed, hairy, usually absent. **FLOWERS:** In widely spaced whorls around the upper stem, each flower ½–1¼ inches long, pale blue to lavender, tubular with 2 spreading lips, the lower lip much larger and curved downward; the calyx has 5 spine-tipped lobes and is often purple. **FLOWERING:** Apr–Jun. **RANGE:** FL west to TX, north to MO and CT. With its large, lobed basal leaves and nearly leafless stems, Lyre-leaved Sage is **SIMILAR TO** no other member of the Mint Family.

Nettle-Leaf Sage

Salvia urticifolia Linnaeus
Mint Family (Lamiaceae)

PERENNIAL HERB found in woodlands and glades, usually over mafic or limestone bedrock, in north Georgia and a few upper Coastal Plain counties. **STEMS:** 8–32 inches tall, 4-angled, finely hairy. **LEAVES:** Up to 4¾ inches long and 2½ inches wide, opposite, oval to triangular with pointed tips, toothed or scalloped margins, and leaf tissue forming narrow wings on the stalks. **FLOWERS:** In whorls around the upper stem and tips of branches, each flower ½ inch long, blue or violet, with a small, hoodlike upper lip and a larger lower lip marked with 2 white bands. **FLOWERING:** Apr–Jun. **RANGE:** FL west to MS, north to KY and PA. Flower color and markings are **SIMILAR TO** some lobelias (pp. 144–146), which have alternate leaves and round stems.

Hairy Skullcap

Scutellaria elliptica Muhlenberg ex Sprengel
Mint Family (Lamiaceae)

PERENNIAL HERB found in upland hardwood forests throughout Georgia. **STEMS:** Up to 3 feet tall, 4-angled, hairy. **LEAVES:** Up to 3 inches long and 1½ inches wide, opposite, oval to elliptic, smooth and dark green on the upper surface, hairy and silvery green on the lower surface, with scalloped margins. **FLOWERS:** Up to ¾ inch long, bluish-purple, tubular with 2 spreading lips, the lower lip marked with a central white patch; the tube is bent at the base and held nearly erect; the calyx has a tiny, caplike crest on the upper side. **FLOWERING:** May–Jul. **RANGE:** FL west to TX, north to OK and NY. **NOTES:** Gland-tipped hairs cover the leaf stalks, flowers, and calyxes (use a 10× lens). **SIMILAR TO** Hoary Skullcap (*Scutellaria incana*), which occurs in similar habitats in north and southwest Georgia; it is covered with short, nonglandular hairs.

Helmet Skullcap

Scutellaria integrifolia Linnaeus
Mint Family (Lamiaceae)

PERENNIAL HERB found in dry to moist forests, pine savannas, and prairies throughout Georgia. **STEMS:** 6–32 inches tall, slender, 4-angled, hairy. **LEAVES:** Up to 2½ inches long and ¾ inch wide, opposite, oval to lance-shaped, entire or with scalloped margins, covered with gland *dots* (not gland-tipped hairs). **FLOWERS:** Up to 1 inch long, blue or purple, hairy, tubular with a hoodlike upper lip and a lower lip marked with 2 white bands; the tube is bent upward at the base; the calyx is covered with gland-tipped hairs and has a tiny, caplike crest on the upper side. **FLOWERING:** May–Jul. **RANGE:** FL west to TX, north to MO and MA. **SIMILAR TO** Small's Skullcap (*Scutellaria multiglandulosa*), which occurs in dry woodlands in the Coastal Plain and a few Piedmont counties; its leaves are covered with gland-tipped hairs (use a 10× lens).

Heart-Leaf Skullcap

Scutellaria ovata Hill
Mint Family (Lamiaceae)

PERENNIAL HERB found in dry to moist forests, limestone cedar glades, and boulderfields in north Georgia. **STEMS:** 1–5 feet tall, 4-angled, very hairy. **LEAVES:** Up to 6 inches long and 3 inches wide, opposite, heart-shaped, toothed, hairy. **FLOWERS:** Up to ¾ inch long, blue or purple, tubular with 2 spreading lips, the lower lip marked with a white, purple-speckled patch; the tube is bent at the base and nearly erect; the calyx has a tiny, caplike crest on the upper side and is covered with gland-tipped hairs. **FLOWERING:** May–Jul. **RANGE:** FL west to TX, north to MN and PA. **SIMILAR TO** Southern Showy Skullcap (*Scutellaria pseudoserrata*), a Georgia Special Concern species of moist, hardwood forests found in north Georgia; its flowers are 1 – 1½ inches long, and the upper surfaces of its leaves are gland-dotted.

Florida Betony

Stachys floridana Shuttleworth ex Bentham
Mint Family (Lamiaceae)

PERENNIAL HERB found in disturbed areas in the Coastal Plain. **STEMS:** Up to 20 inches tall, slender, 4-angled, the angles hairy. **LEAVES:** Up to 2¼ inches long and 1 inch wide, opposite, oval or lance-shaped, toothed, lower leaves with long stalks, upper leaves with short or no stalks. **FLOWERS:** In whorls near the top of the stem, each flower less than ½ inch long, pale pink with dark pink markings, tubular with 2 spreading lips, the upper lip small and curved over the stamens, the lower lip large and 3-lobed. **FLOWERING:** Apr–Jul. **RANGE:** Native to FL, weedy and spreading west to TX and north to VA. **NOTES:** The root tips bear small, white tubers. **SIMILAR TO** Smooth Hedge-nettle (*Stachys tenuifolia*), which occurs in floodplains and swamps in the mountains; its stems are usually smooth.

American Germander

Teucrium canadense Linnaeus
Mint Family (Lamiaceae)

PERENNIAL HERB found in bottomland forests, wet thickets, and wet disturbed areas, mostly in north and coastal Georgia. **STEMS:** Up to 3 feet tall, 4-angled, stout, usually unbranched, very hairy. **LEAVES:** Up to 5 inches long and 2½ inches wide, opposite, elliptic to lance-shaped, coarsely toothed, hairy, with deeply etched veins and a foul odor when crushed. **FLOWER CLUSTERS:** Densely flowered, conical, held at the top of the stem. **FLOWERS:** About ⅓ inch long, pale pink to lavender with dark pink spots, covered with tiny, gland-tipped hairs; each flower is divided into 2 erect, hornlike lobes, 2 small, forward-pointing lobes, and a large, downcurved lobe; 4 orange-tipped stamens arch from the top of the flower. **FLOWERING:** Jun–Aug. **RANGE:** Most of North America. The oddly shaped flower is **SIMILAR TO** no other species in the Mint Family.

Blue Curls

Trichostema dichotomum Linnaeus
Mint Family (Lamiaceae)

PERENNIAL HERB found in dry woodlands, sandhills, rock outcrops, and disturbed uplands throughout Georgia. **STEMS:** Up to 2 feet tall, branched, very hairy, with 4 rounded angles. **LEAVES:** Up to 2¾ inches long and 1 inch wide, opposite, elliptic, entire or toothed, 3-veined, hairy, gland-dotted, fragrant when crushed. **FLOWERS:** About ½ inch long, dark blue or purple, with 4 oval, spreading lobes and a larger, downcurved lip marked with white; 4 showy, purple stamens arch up and outward from the center of the flower. **FLOWERING:** Aug–Nov. **RANGE:** FL west to TX, north to IA, ON, and ME. **SIMILAR TO** Narrow-leaf Blue Curls (*Trichostema setaceum*), which occurs in similar habitats; its leaves are less than ¼ inch wide and have only 1 obvious vein.

Butterworts

Butterworts are in the genus *Pinguicula*, a Latin name that means "fatty," referring to the oily or buttery feel of the leaves. The "oil" is actually a musty-smelling mucilage secreted by glands on the upper surface of the leaves that attracts and traps small insects. About a minute after the victim is ensnared, another set of glands releases a digestive fluid that begins to dissolve its body. This is a speedy process—the nutrients released from the insect's body begin to enter the leaf within two hours or so. After the insect meal is digested and absorbed, the leaves dry out and their margins uncurl somewhat to prepare the trap anew. Insects provide butterworts with the nitrogen and phosphorus that are at low levels in wetland soils. Butterwort flowers also lure insects, though not to their deaths. Long-tongued insects are drawn to the brightly colored flowers and the nectar that collects in the narrow spur at the rear of the flower tube. Insects enter the mouth of the flower and push their way over the palate, a large, hairy protrusion that almost blocks the flower tube. As insects probe for nectar, pollen deposited on their bodies by previously visited flowers rubs off onto the stigma. As they leave, the insects pick up a fresh load of pollen that they will carry to the next butterwort flower they visit. There are nine species of *Pinguicula* in North America; six of these occur in the Southeast, with four in Georgia.

Blue Butterwort, Violet Butterwort

Pinguicula caerulea Walter
Bladderwort Family (Lentibulariaceae)

PERENNIAL HERB found in moist to wet pine savannas and flatwoods and wet ditches in the Coastal Plain in southeast Georgia. **LEAVES:** In a basal rosette up to 5 inches across, yellowish-green, fleshy, oval, entire, their sides curled up and forming a V-shaped tip. **FLOWERS:** Held singly at the tip of a hairy stalk up to 8 inches tall, each flower about 1 inch across, pale blue to violet, usually heavily veined with dark violet, tubular with 5 notched, spreading lobes, a backward-pointing spur, and a large, hairy palate projecting out of the tube. **FLOWERING:** Apr–May. **RANGE:** FL, GA, SC, and NC. **SIMILAR TO** Small Butterwort (*Pinguicula pumila*), which occurs in the same habitats; its rosettes are less than 1½ inches across, and the flowers are less than ½ inch across.

Yellow Butterwort

Pinguicula lutea Walter
Bladderwort Family (Lentibulariaceae)

PERENNIAL HERB found in moist to wet pine savannas and flatwoods, seepage bogs, and wet ditches in the Coastal Plain. **LEAVES:** In a basal rosette up to 6 inches across, pale yellowish-green, somewhat fleshy, oval or oblong, entire, their sides curled upward and forming a V-shaped tip. **FLOWERS:** Held singly at the tip of a hairy stalk up to 10 inches tall, each flower about 1 inch across, yellow, tubular with 5 notched, spreading lobes, a backward-pointing spur, and a large, hairy palate projecting out of the tube. **FLOWERING:** Late Mar–May. **RANGE:** FL west to LA, north to NC. **SIMILAR TO** other butterworts, although this is the only yellow-flowered butterwort in Georgia.

Long-Spurred Bladderwort

Utricularia biflora Lamarck
Bladderwort Family (Lentibulariaceae)

PERENNIAL HERB found in ponds, lakes, and ditches throughout Georgia, usually in floating mats of many intertwined plants. **STEMS:** Held erect 1–4 inches above the water, without spongy, floating branches. **LEAVES:** Underwater, forked and dissected into many segments, bearing tiny bladders and traps. **FLOWERS:** About ½ inch long, yellow, with 2 lips, a backward-pointing spur, and a hairy, red-streaked palate between the lips. **FLOWERING:** Jun–Oct. **RANGE:** FL west to TX, north to OK and MA. **SIMILAR TO** several other aquatic, yellow-flowered bladderworts that also lack floating, inflated branches. Only two are common in Georgia. Short-spurred Bladderwort (*Utricularia gibba*) has smaller flowers, less than ¼ inch long. Fibrous Bladderwort (*U. striata*, synonym *U. fibrosa*) has flower stalks held 4–10 inches above the water surface and is not usually entangled in a mat.

Bladderworts

Bladderworts, in the genus *Utricularia*, have the most sophisticated of all the carnivorous plant traps. Most bladderworts live in ponds, lakes, seepages, and slow-moving streams, but a few live in damp soil. Their highly dissected leaves are an elegant adaptation to life in the water because they vastly increase the surface area of the leaves, and thus the plant's ability to extract nutrients and carbon dioxide (necessary for photosynthesis) from the water. The branches and leaves spreading across the water surface or into the mud are adorned with small, bladder-like traps, each bearing a trigger hair. The bladders are hollow and maintain a partial vacuum. When a tiny aquatic animal trips the trigger hair, a trap-door opens and the vacuum sucks water and animal into the bladder. The plant digests the insect and absorbs the water, re-creating the vacuum, and the trap is reset for the next victim. Thirteen species of Bladderwort are found in Georgia's ponds, lakes, and wetlands.

Horned Bladderwort

Utricularia cornuta Michaux
Bladderwort Family (Lentibulariaceae)

PERENNIAL HERB found in wet pine flatwoods, seepage bogs, granite outcrop seepages, and wet ditches in the Piedmont and Coastal Plain, often in shallow water but rooted in soil, not floating. **STEMS:** 4–16 inches tall, wiry, green to yellowish. **LEAVES:** Buried in the soil. **FLOWERS:** ½–1 inch long, yellow, with 2 lips and a narrow, backward-pointing spur. **FLOWERING:** Late May–Sep. **RANGE:** FL west to TX, north to QC and NL; Bahamas, Cuba. **SIMILAR TO** 2 other nonfloating, yellow-flowered bladderworts, both with flowers less than ½ inch long: Southern Bladderwort (*Utricularia juncea*) flowers are widely spaced along the stalk and are smaller with shorter spurs; Slender Bladderwort (*U. subulata*) has reddish, wiry, slightly zigzag stalks 1½–7 inches tall.

Swollen Bladderwort

Utricularia inflata Walter
Bladderwort Family (Lentibulariaceae)

PERENNIAL HERB found in ponds, lakes, drainage ditches, and canals in the Coastal Plain of Georgia. **STEMS:** Up to 6 inches tall, held above the water by 5–10 spongy, floating branches that radiate like spokes from the base of the stem. **LEAVES:** Underwater, divided into many narrow segments bearing tiny bladders and traps. **FLOWERS:** 9–14 per stem, each about ¾ inch long, yellow, with 2 large lips and a lumpy, red-streaked palate; the spur is deeply notched at its tip. **FLOWERING:** May–Nov. **RANGE:** FL west to TX, north to OK, TN, and MA. **SIMILAR TO** Small Swollen Bladderwort (*Utricularia radiata*), the only other bladderwort in North America with spokelike floats; it has smaller flowers, only 3 or 4 per stem, and the floats have only 4–7 spokes.

Purple Bladderwort

Utricularia purpurea Walter
Bladderwort Family (Lentibulariaceae)

PERENNIAL HERB found in ponds, slow-moving streams, and ditches in the Coastal Plain of Georgia. **STEMS:** 1–8 inches tall, floating erect above the water without spokelike branches. **LEAVES:** Underwater, divided into many narrow segments tipped with tiny bladders and traps. **FLOWERS:** About ½ inch long, dark pink to purple, with a short, backward-pointing spur and 2 lips, the lower lip divided into 3 lobes. **FLOWERING:** May–Sep. **RANGE:** FL west to TX, north to MN, QC, and NS; Mexico, West Indies, Central America. **SIMILAR TO** other bladderworts, but Purple Bladderwort is the only common purple-flowered species in Georgia.

Texas Yellow Flax

Linum medium (Planchon) Britton var. *texanum* (Planchon) Fernald
SYNONYM: *Linum virginianum* Linnaeus var. *medium* Planchon
Flax Family (Linaceae)

PERENNIAL HERB found in bogs, savannas, wet ditches, and moist to dry woods throughout Georgia. **STEMS:** ½–2½ feet tall, smooth, with narrowly winged angles and stiffly erect flowering branches. **LEAVES:** Up to 1 inch long and ⅛ inch wide, alternate (except for the lowest, which are opposite), narrowly lance-shaped, sharply pointed, entire, smooth, without stalks. **FLOWERS:** About ½ inch wide, with 5 spreading, oval, pale yellow petals. **FRUIT:** A tiny, round, slightly flattened capsule. **FLOWERING:** Jun–Oct. **RANGE:** FL west to TX, north to IA, ON, and ME. **SIMILAR TO** Virginia Yellow Flax (*Linum virginianum*), which occurs in thickets, clearings, and woods in north Georgia; its leaves are blunt-tipped and its flowering branches are spreading.

Ridged Yellow Flax

Linum striatum Walter
Flax Family (Linaceae)

PERENNIAL HERB found in seepages, bogs, stream banks, and wet clearings throughout Georgia. **STEMS:** 1–4 feet tall, narrowly ribbed, much-branched, several stems arising from a single crown. **LEAVES:** Up to 1½ inches long and ½ inch wide, opposite on the lower half of the stem, alternate above, sometimes all opposite, narrowly elliptic, tapering at both ends. **FLOWERS:** About ⅓ inch wide, with 5 spreading, oval, yellow petals. **FRUIT:** A tiny, round, slightly flattened capsule. **FLOWERING:** Jun–Oct. **RANGE:** FL west to TX, north to MI, ON, and MA. **SIMILAR TO** Florida Yellow Flax (*Linum floridanum*), which occurs in similar habitats in the Coastal Plain; it has oval fruits and alternate leaves except for a few opposite leaves at the base of the stem.

Yellow Jessamine

Gelsemium sempervirens (Linnaeus) St. Hilaire
Strychnine (Loganiaceae) or Jessamine (Gelsemiaceae) Family

WOODY VINE found in upland forests, thickets, fencerows, and pine flatwoods throughout Georgia. **STEMS:** High-climbing, sprawling over other plants, or trailing on the ground; young stems reddish, older stems with gray bark. **LEAVES:** Up to 3½ inches long and ½ inch wide, opposite, oval to lance-shaped, entire, smooth, evergreen. **FLOWERS:** Up to 1½ inches long, yellow, funnel-shaped with 5 spreading lobes, fragrant; the 5 green sepals have blunt or rounded tips. **FLOWERING:** Feb–Apr. **RANGE:** FL west to TX, north to AR and VA; Mexico, Guatemala. **NOTES:** All parts of the plant contain strychnine and are toxic to livestock and humans. **SIMILAR TO** Swamp Jessamine (*Gelsemium rankinii*), which occurs in Coastal Plain swamps; its flowers are not fragrant, and its sepals have long-pointed tips. Also see Cross-vine (*Bignonia capreolata*), page 129.

Caribbean Miterwort

Mitreola petiolata (J. F. Gmelin) Torrey & A. Gray
SYNONYM: *Cynoctonum mitreola* (Linnaeus) Britton
Strychnine Family (Loganiaceae)

ANNUAL HERB found in swamps, marshes, lakeshores, and wet ditches throughout Georgia (except the Blue Ridge). **STEMS:** 4–32 inches tall, branched only in the flower cluster, slightly 4-angled or winged, smooth. **LEAVES:** Up to 3 inches long and 1⅓ inches wide, opposite, oval, tapering at both ends, entire, with stalks. **FLOWER CLUSTERS:** With curled branches bearing flowers only along the top side. **FLOWERS:** Tiny, white to pink, with 5 spreading, pointed, white petals. **FLOWERING:** Jul–frost. **RANGE:** FL west to TX, north to MO and VA; Mexico, West Indies, South America. **SIMILAR TO** Small-leaved Miterwort (*Mitreola sessilifolia*), which occurs in wetlands in Georgia's Coastal Plain; its leaves are less than ¾ inch long and have rounded bases with no or very short stalks.

Rustweed, Juniper-Leaf

Polypremum procumbens Linnaeus
Strychnine (Loganiaceae) or Tetrachondra (Tetrachondraceae) Family

ANNUAL OR PERENNIAL HERB found on exposed lakeshores, sandbars, dune swales, and both wet and dry disturbed areas throughout Georgia. **STEMS:** Up to 12 inches long, spreading in a circular mat across the ground from a central crown, much-branched, the branches usually orange near the crown. **LEAVES:** Up to ¾ inch long, very narrow and pointed, opposite, smooth, leaves near the crown usually orange. **FLOWERS:** Held in the axils of leaves and the forks of branches, less than ¼ inch wide, white, with 4 rounded, white lobes and a tuft of white hairs in the center. **FLOWERING:** Late May–Sep. **RANGE:** FL west to TX, north to MO and NY; West Indies, Central and South America. **SIMILAR TO** Bosc's Bluet (*Oldenlandia boscii*), which occurs in wetlands in southwest Georgia; it has stalked flowers with pointed petals.

Indian Pink, Pink Root

Spigelia marilandica Linnaeus
Strychnine Family (Loganiaceae)

PERENNIAL HERB found in rich, moist forests, usually over mafic or limestone bedrock, throughout Georgia. **STEMS:** 6–28 inches tall, unbranched, smooth. **LEAVES:** 2–4¾ inches long and ½–2½ inches wide, opposite, oval to lance-shaped with rounded bases and pointed tips, entire, mostly smooth. **FLOWERS:** Up to 2 inches long, tubular, red on the outer surface, yellow on the inner, with 5 short, pointed, spreading lobes. **FLOWERING:** May–Jun. **RANGE:** FL west to TX, north to OK, IL, and MD. **NOTES:** All parts of the plant are toxic. **SIMILAR TO** no other wildflower in Georgia.

Columbian Waxweed

Cuphea carthagenensis (Jacquin) J. F. Macbride
Loosestrife Family (Lythraceae)

ANNUAL HERB found in wet forests, marshes, clearings, and ditches in Georgia's Coastal Plain. **STEMS:** 4–24 inches tall, branched, sticky with glandular hairs. **LEAVES:** Up to 2½ inches long and 1 inch wide, opposite, elliptic or oval, tapering at both ends, rough-hairy, with very short or no stalk. **FLOWERS:** About ¼ inch wide, with 6 dark pink petals and a green or purple, ribbed, sticky-hairy calyx tube that is swollen on one side at the base. **FLOWERING:** Jun–Sep. **RANGE:** FL west to TX, north to AR and NC; native to South America. **SIMILAR TO** Blue Waxweed (*Cuphea viscosissima*), which occurs in north Georgia in wet or dry openings over limestone or mafic bedrock; it has leaf stalks up to ¾ inch long, deep purple petals, and a purple calyx tube ⅓–½ inch long.

Swamp Loosestrife

Decodon verticillatus (Linnaeus) Elliott
Loosestrife Family (Lythraceae)

DECIDUOUS SHRUB found in shallow waters of swamps, lakes, limesink ponds, and marshes in Georgia's Coastal Plain and a few northwest Georgia counties. **STEMS:** Up to 10 feet long, spreading and arching no higher than 4 feet, often rooting at the tips where they meet the soil; submerged stems have spongy bark, emergent stems have peeling, rust-colored bark. **LEAVES:** Up to 8 inches long and 2 inches wide, opposite or whorled, entire, lance-shaped or elliptic, tapering at both ends with ¼-inch leaf stalks. **FLOWERS:** In clusters in leaf axils, about 1 inch wide, with 4–7 (usually 5) dark pink petals and 8–10 pink, white-tipped stamens. **FLOWERING:** Jul–Sep. **RANGE:** FL west to TX, north to MN and NS. **SIMILAR TO** no other shrub or wildflower in Georgia.

Southern Winged Loosestrife

Lythrum lanceolatum Elliott
SYNONYM: *Lythrum alatum* Pursh var. *lanceolatum* (Elliott) A. Haines
Loosestrife Family (Lythraceae)

PERENNIAL HERB found in flatwoods, marshes, shorelines, and moist to wet clearings in the Piedmont and Coastal Plain. **STEMS:** Up to 4 feet tall, round, 4-angled, or slightly winged, much-branched in the upper half. **LEAVES:** Up to 2¾ inches long and ½ inch wide, alternate on the upper stem and branches, opposite on the lower stem, narrowly lance-shaped, tapering at both ends, smooth. **FLOWERS:** About ½ inch wide, with 6 dark pink (rarely white) petals and a green, ribbed calyx tube up to ¼ inch long. **FLOWERING:** May–Sep. **RANGE:** FL west to TX, north to OK and VA. **SIMILAR TO** Northern Winged Loosestrife (*Lythrum alatum*), which occurs in northwest Georgia in areas with limestone bedrock; its leaves are oval to lance-shaped with rounded or heart-shaped bases.

Toothcup

Rotala ramosior (Linnaeus) Koehne
Loosestrife Family (Lythraceae)

ANNUAL HERB found in marshes, pond shores, stream banks, and wet ditches throughout Georgia (except for the Blue Ridge and adjacent counties). **STEMS:** 8–16 inches tall, erect or sprawling, round or 4-angled, simple or much-branched, smooth. **LEAVES:** Up to 2 inches long and ½ inch wide, opposite (rarely whorled), lance-shaped, tapering at both ends, entire, smooth, with no or very short stalks. **FLOWERS:** 1 per leaf axil, about 1/16 inch long and wide, with a reddish-green, 4-lobed, cup-shaped calyx tube that encloses the ovary and fruit; the 4 white or pink petals fall soon after opening and are rarely seen. **FLOWERING:** Jun–Oct. **RANGE:** North and South America, West Indies. **NOTES:** Stems, leaves, and fruits turn red in the fall. **SIMILAR TO** Marsh Seedbox (*Ludwigia palustris*), page 266.

Comfort-Root, Pineland Hibiscus

Hibiscus aculeatus Linnaeus
Mallow Family (Malvaceae)

PERENNIAL SHRUBLIKE HERB found in pine savannas and flatwoods, seepage slopes, and wet ditches in the Coastal Plain. **STEMS:** Up to 3 feet tall, arising from a semiwoody crown, very rough-hairy. **LEAVES:** Up to 3½ inches long and wide, alternate, very rough-hairy, with 3 or 5 deeply cut lobes and coarsely toothed margins; leaf stalks are about as long as the blades. **FLOWERS:** About 4½ inches wide, cream-colored, turning yellow then pink with age, with 5 scallop-edged petals, a maroon center, and an erect column of reddish stamens surrounding the pistil; a whorl of 10–15 very narrow, green, upwardly curved bracts surrounds the base of the 5-lobed calyx. **FLOWERING:** Jun–Aug. **RANGE:** FL west to TX, north to NC. Older, pink flowers are **SIMILAR TO** those of Rose Mallow (*Hibiscus moscheutos*), page 256.

Smooth Rose Mallow, Halberd-Leaf Marsh-Mallow

Hibiscus laevis Allioni
SYNONYM: *Hibiscus militaris* Cavanilles
Mallow Family (Malvaceae)

PERENNIAL SHRUBLIKE HERB found in freshwater marshes, river banks, and floodplains throughout Georgia (except mountains). **STEMS:** 2–11 feet tall, smooth. **LEAVES:** Up to 6 inches long and 4 inches wide, alternate, arrowhead- or lance-shaped, toothed, both surfaces green and smooth; leaf stalks 6 inches or less long. **FLOWERS:** Up to 5 inches wide, pink or white, with 5 petals, a maroon center, and a tube of fused, reddish stamens surrounding the pistil; a whorl of 7–15 very narrow, green, upwardly curved bracts surrounds the base of the 5-lobed calyx. **FLOWERING:** Jun–Aug. **RANGE:** FL west to TX, north to MN, ON, and NY. **SIMILAR TO** Rose Mallow (*Hibiscus moscheutos*), page 256.

Rose Mallow

Hibiscus moscheutos Linnaeus
Mallow Family (Malvaceae)

PERENNIAL SHRUBLIKE HERB found in fresh and brackish marshes and swamps throughout Georgia. **STEMS:** 2–8 feet tall, hairy or smooth. **LEAVES:** Up to 8 inches long and 2½ inches wide, alternate, oval or lance-shaped, toothed; the upper surfaces are green and smooth or gray and very hairy, the lower surfaces are densely coated with white hairs; stalks are 1–4 inches long. **FLOWERS:** 5–8 inches wide, white, cream, or pink, with 5 petals, a maroon center, and a tube of fused reddish stamens surrounding the pistil; a whorl of 10–15 very narrow, green, upwardly curved bracts surrounds the base of the 5-lobed calyx. **FLOWERING:** Jun–Sep. **RANGE:** FL west to TX, north to MI, ON, and MA. **SIMILAR TO** Comfort-root (*Hibiscus aculeatus*), page 255.

Seashore Mallow, Saltmarsh Mallow

Kosteletzkya virginica (Linnaeus) K. Presl
SYNONYM: *Kosteletzkya pentacarpos* (Linnaeus) Ledebour
Mallow Family (Malvaceae)

PERENNIAL SHRUBLIKE HERB found in salt, brackish, and freshwater tidal marshes and dune swales on Georgia's coast. **STEMS:** 1–8 feet tall, arising from a semiwoody crown, covered with branched hairs. **LEAVES:** Blades up to 5½ inches long, stalks up to 4 inches long, alternate, heart-shaped, triangular or lance-shaped, toothed, sometimes lobed, covered with branched hairs (use a 10× lens). **FLOWERS:** Up to 3 inches wide, with 5 pink, spreading petals and an erect tube of fused yellow stamens surrounding the pistil; a whorl of 7–10 very narrow, green, upwardly curved bracts surrounds the 5-lobed calyx. **FLOWERING:** Jul–Oct. **RANGE:** FL west to TX, north to NY. **SIMILAR TO** Rose Mallow (*Hibiscus moscheutos*), above.

Bristle Mallow

Modiola caroliniana (Linnaeus) G. Don
Mallow Family (Malvaceae)

ANNUAL OR BIENNIAL HERB found in fields, lawns, and other disturbed areas throughout Georgia. **STEMS:** Up to 2½ feet long, hairy, usually sprawling across the ground and rooting at the nodes. **LEAVES:** Up to 2¾ inches long and 1 inch wide, with stalks as long as the leaves, alternate, divided into 3–7 toothed lobes. **FLOWERS:** About ⅓ inch wide, with 5 oval, salmon-colored petals. **FRUIT:** About ¼ inch wide, flat, wheel-like, with 15–25 bristly, kidney-shaped segments. **FLOWERING:** Late Mar–Jun. **RANGE:** FL west to CA and OR, north to OK and MA; probably native to the southeastern United States but now globally widespread. **SIMILAR TO** Scarlet Pimpernel (*Anagallis arvensis*, synonym *Lysimachia arvensis*), which occurs in similar habitats; it has opposite, entire leaves without stalks and a nearly square stem.

Arrow-Leaf Sida

Sida rhombifolia Linnaeus
Mallow Family (Malvaceae)

ANNUAL HERB found in fields, roadsides, and other disturbed areas throughout Georgia. **STEMS:** Up to 3 feet tall, hairy, with upright branches and tough, stringy bark. **LEAVES:** Up to 3 inches long and 1¾ inches wide, alternate, roughly diamond-shaped, toothed except near the base of the leaf, densely hairy on the lower surface. **FLOWERS:** About ¾ inch wide, cream-colored, yellow, or pale orange, with 5 overlapping and lop-sided petals and a short tube of fused yellow stamens surrounding the pistil. **FLOWERING:** Apr–Oct. **RANGE:** FL west to CA, north to KS and PA; native to the tropics, now globally widespread. **SIMILAR TO** Prickly Sida (*Sida spinosa*), another tropical weed found in disturbed areas throughout Georgia; it has a spine beneath each leaf. Elliott's Fan-petals (*Sida elliottii*), a Georgia Special Concern species, has narrowly elliptic or linear leaves.

Meadow-Beauties (Melastomataceae Family)

Meadow-beauties grow in open or thinly wooded wetlands such as pine flatwoods and savannas, mostly in the Coastal Plain. Meadow-beauty flowers always have 4 petals and, except for Yellow Meadow-beauty, are pink or pale pinkish-white. The asymmetrical petals are attached to the rim of a vase-shaped floral tube topped with 4 triangular sepals. The 8 stamens are showy, with yellow, usually curved, anthers about ¼ inch long that are sometimes described as boat- or banana-shaped. Bumblebees collect the pollen from a pore at the tip of the anther by grasping the anther and vibrating their thoracic muscles, a process called buzz pollination (see "The BUZZZZZZZZ on Pollination," p. 175).

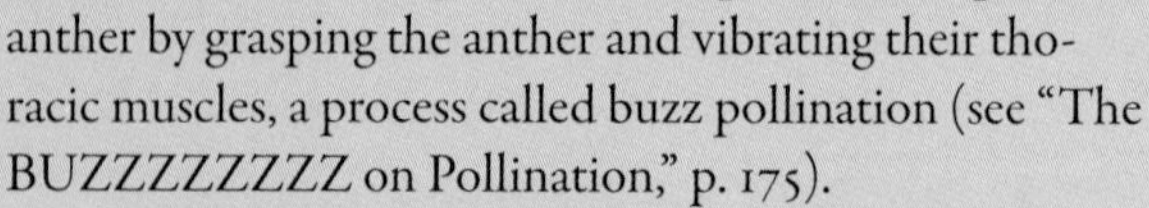

Meadow-beauty stems are erect, 4-angled, and often narrowly winged. The leaves are opposite and arranged so that each pair is oriented 90 degrees from the next pair. The leaves usually have 3 conspicuous veins. The floral tube completely encases the fruit, and seeds are shed in late autumn from an opening at the top of the tube. The stems persist through the winter, and the vase-shaped floral tubes stand out in bare, brown wetlands. There are 11 species of Meadow-beauty in Georgia.

Smooth Meadow-Beauty

Rhexia alifanus Walter
Meadow-Beauty Family
(Melastomataceae)

PERENNIAL HERB found in wet savannas and flatwoods, bogs, and seepage areas in the Coastal Plain. **STEMS:** 1½–3 feet tall, usually unbranched, smooth, bluish-green. **LEAVES:** Up to 2¾ inches long and ½ inch wide, opposite, tapered at both ends, upwardly pointing, mostly entire, bluish-green, with 3 conspicuous veins. **FLOWERS:** Up to 2 inches wide, held at the top of a glandular-hairy, vase-shaped floral tube about ⅓ inch long with 4 dark pink petals and 8 stamens with curved, yellow anthers. **FLOWERING:** May–Sep. **RANGE:** FL west to TX, north to NC. **SIMILAR TO** Virginia Meadow-beauty (*Rhexia virginica*), page 260.

Yellow Meadow-Beauty

Rhexia lutea Walter
Meadow-Beauty Family
(Melastomataceae)

PERENNIAL HERB found in wet savannas and flatwoods, bogs, and seepage slopes in the Coastal Plain. **STEMS:** 4–16 inches tall, branched, bushy, and hairy, forming large colonies. **LEAVES:** Up to 1¼ inches long, opposite, 3-veined, narrowly elliptic to lance-shaped, toothed or entire, covered with stiff, yellow hairs. **FLOWERS:** Up to 1 inch wide, held at the top of a hairy, vase-shaped floral tube about ¼ inch long, with 4 golden yellow petals and 8 stamens with straight, yellow anthers. **FLOWERING:** Apr–Jul, later in response to fire. **RANGE:** FL west to TX, north to NC. **SIMILAR TO** other meadow-beauties, but with yellow flowers. Also compare with the seedboxes, pages 264–266, which do not have vase-shaped floral tubes.

Maryland Meadow-Beauty

Rhexia mariana Linnaeus
Meadow-Beauty Family
(Melastomataceae)

PERENNIAL HERB found in wetlands and wet disturbed areas throughout Georgia. **STEMS:** Up to 3 feet tall, branched, sometimes winged, hairy, often forming large colonies. **LEAVES:** Up to 1½ inches long, opposite, linear, narrowly oval, or lance-shaped, 3-veined, hairy, with finely toothed margins. **FLOWERS:** Up to 1 inch wide, held at the top of a vase-shaped floral tube about ⅓ inch long, with 4 white or pale pink petals and 8 yellow stamens with curved, yellow anthers. **FLOWERING:** May–Oct. **RANGE:** FL west to TX, north to KS, MI, and MA. **SIMILAR TO** Hairy Meadow-beauty (*Rhexia nashii*), which occurs in similar habitats in the Coastal Plain; it is up to 5 feet tall, with bristle-edged leaves up to 2¾ inches long; its dark pink flowers are 1½–2 inches wide.

Bog Meadow-Beauty

Rhexia petiolata Walter
Meadow-Beauty Family
(Melastomataceae)

PERENNIAL HERB found in wet savannas and flatwoods, bogs, and borders of ponds and depressions in the Coastal Plain. **STEMS:** 4–20 inches tall, usually unbranched, smooth. **LEAVES:** Up to ½ inch long and wide, opposite, oval to nearly round, finely toothed, with 3 conspicuous veins and bristly margins. **FLOWERS:** Up to 1½ inches wide, held at the top of a vase-shaped floral tube about ¼ inch long with 4 pink petals and 8 stamens with straight, yellow anthers. **FLOWERING:** Jun–Sep. **RANGE:** FL west to TX, north to VA. **SIMILAR TO** other meadow-beauties except the flowers face upward, not sideways, and the anthers are not curved.

Virginia Meadow-Beauty

Rhexia virginica Linnaeus
Meadow-Beauty Family
(Melastomataceae)

PERENNIAL HERB found in wet savannas and flatwoods, bogs, marshes, pond shores, and wet ditches throughout Georgia. **STEMS:** Up to 3 feet tall, branched at the top, hairy at least at the leaf nodes, with narrowly winged angles. **LEAVES:** Up to 2¾ inches long and 1 inch wide, gradually reduced in size up the stem, opposite, conspicuously 3-veined, margins finely toothed and hairy. **FLOWERS:** About 1½ inches wide, held at the top of a vase-shaped floral tube about ⅓ inch long with 4 pinkish-purple petals and 8 stamens with curved, yellow anthers. **FLOWERING:** May–Oct. **RANGE:** FL west to TX, north to MI and NS. **SIMILAR TO** Smooth Meadow-beauty (*Rhexia alifanus*), page 258.

Coral Beads, Carolina Moonseed

Cocculus carolinus (Linnaeus) A. P. de Candolle
Moonseed Family (Menispermaceae)

PERENNIAL HERBACEOUS VINE found in woodlands, thickets, fencerows, and disturbed areas throughout Georgia. **STEMS:** Up to 16 feet long, sprawling over other plants. **LEAVES:** Up to 7 inches long and 5½ inches wide, alternate, oval to heart-shaped, often lobed; stalks are attached at the edge of the leaf. **FEMALE AND MALE FLOWERS** are in drooping clusters on separate plants; both have 6 tiny, yellow petals. **FRUIT:** Round, red, translucent, about ¼ inch wide with circular, snail-shaped seeds. **FLOWERING:** Jun–Aug; fruits ripen Aug–Nov. **RANGE:** FL west to TX and KS, north to IN and VA. **SIMILAR TO** Canada Moonseed (*Menispermum canadensis*), which has similar leaves except the stalks are attached about ¼ inch in from the edge of the leaf; it has white flowers and bluish-black berries. Also compare with the leaves of Yellow Passionflower (*Passiflora lutea*), page 278.

Little Floating Hearts

Nymphoides cordata (Elliott) Fernald
Buckbean Family (Menyanthaceae)

PERENNIAL AQUATIC HERB found in sandhill ponds, beaver ponds, and slow-moving streams in the Coastal Plain and a few lower Piedmont counties. **STEMS:** Underwater, slender and unbranched, arising from rhizomes buried in the soil below the water, usually solid green, bearing 1 or 2 floating leaves at the tip. **LEAVES:** Floating, up to 2¾ inches long and wide, heart-shaped, the upper surface green and purple, the lower surface smooth and maroon. **FLOWERS:** Floating, about ⅓ inch across, with 5 white, delicate petals. **FLOWERING:** Apr–Aug. **RANGE:** FL west to LA, north to NL. **NOTES:** Small, fleshy, green, banana-shaped tubers develop beneath the flowers. **SIMILAR TO** Big Floating Hearts (*Nymphoides aquatica*), which occurs in similar habitats; its leaves are 2–6 inches long, with a green upper surface and a purple, pebbly lower surface; the stems are mottled with purple.

Spatterdock, Yellow Pond Lily

Nuphar advena (Aiton) R. Brown ex Aiton

SYNONYM: *Nuphar luteum* (Linnaeus) Sibthorp & J. E. Smith ssp. *macrophyllum* (Small) E. O. Beal

Water Lily Family (Nymphaeaceae)

PERENNIAL HERB found in lakes, marshes, and slow-moving streams mostly in the Coastal Plain. **LEAVES:** Floating or held above the water surface on stout stalks, up to 16 inches long and 12 inches wide, heart-shaped, green on both surfaces, not hairy. **FLOWERS:** Floating or held above the water surface, 1–3 inches wide, cup-shaped with 3 thick, yellow or green sepals enclosing many small, yellow petals and stamens and a yellow, column-like pistil. **FLOWERING:** Apr–Oct. **RANGE:** FL west to TX, north to WI and ME; Cuba, northern Mexico. **SIMILAR TO** Round-leaf Pond Lily (*Nuphar orbiculata*), which occurs in similar habitats; its leaves are round, and the lower leaf surface is densely hairy.

White Waterlily

Nymphaea odorata W. T. Aiton

Water Lily Family (Nymphaeaceae)

PERENNIAL AQUATIC HERB found in lakes, canals, and slow-moving streams in the Coastal Plain and a few north Georgia counties. **LEAVES:** Floating, 4–16 inches long and wide, round except for a deep cleft at the top, maroon on the lower surface; stalks arise from thick rhizomes buried in the soil. **FLOWERS:** Floating, up to 7½ inches wide, with many white (rarely pink) petals, very fragrant, opening each morning and closing at night. **FLOWERING:** Apr–Oct. **RANGE:** FL west to TX, north to WI and ME; Cuba, northern Mexico. **NOTES:** When a flower opens, it secretes fluid at the base of the petals. Beetles searching for pollen fall into the fluid, which washes pollen picked up from other flowers off their bodies and onto the stigma. **SIMILAR TO** Yellow Waterlily (*Nymphaea mexicana*), which has yellow flowers.

Enchanter's Nightshade

Circaea canadensis (Linnaeus) Hill
SYNONYM: *Circaea lutetiana* Linnaeus ssp. *canadensis* (Linnaeus) Ascherson & Magnus
Evening-Primrose Family (Onagraceae)

PERENNIAL HERB found in moist, rich forests in the north Georgia mountains. **STEMS:** 1–3 feet tall, with few or no branches, smooth or hairy. **LEAVES:** Up to 4¾ inches long and 2¾ inches wide, opposite, each pair at a 90-degree angle to the pair below, oval, toothed, mostly smooth; the stalks are not winged. **FLOWERS:** Less than ¼ inch wide, on ½-inch stalks, with 2 white petals and 2 green sepals. **FRUIT:** About ¼ inch long, bristly, deeply grooved, with corky ribs. **FLOWERING:** Jun–Aug. **RANGE:** GA west to LA and OK, north to MB and NS. **SIMILAR TO** Small Enchanter's Nightshade (*Circaea alpina*), which occurs on spray cliffs and in seepages in high-elevation northern hardwood and boulderfield forests; it is less than 1 foot tall and has winged leaf stalks.

Eastern Willow-Herb

Epilobium coloratum Biehler
Evening-Primrose Family (Onagraceae)

PERENNIAL HERB found in swamps, beaver ponds, marshes, and bogs in the north Georgia mountains. **STEMS:** Up to 3 feet tall, much-branched in the upper half, usually with vertical lines of hairs on the lower half. **LEAVES:** Up to 6 inches long and ¾ inch wide, opposite, lance-shaped, short-stalked, toothed, smooth, often with purplish markings. **FLOWERS:** About ⅓ inch wide, with 4 pink or white, notched petals alternating with 4 green sepals atop a slender, reddish-green tube up to 2½ inches long. **FRUIT:** Slender pods up to 2½ inches long; mature fruits split lengthwise and release tiny seeds, each tipped with a reddish-brown tuft of hair. **FLOWERING:** Jun–Sep. **RANGE:** GA west to TX, north to SD and NL. **SIMILAR TO** no other wildflower in Georgia.

Slender Gaura, Slender Beeblossom

Gaura filipes Spach
SYNONYM: *Oenothera filipes* (Spach) W. L. Wagner & Hoch
Evening-Primrose Family (Onagraceae)

PERENNIAL HERB found in sandy oak-pine woodlands and dry, disturbed areas throughout Georgia. **STEMS:** Up to 5 feet tall, reddish, hairy, with many slender branches. **LEAVES:** Up to 2½ inches long and ¼ inch wide, alternate, tapered at both ends, hairy or smooth, with irregularly toothed or wavy margins and no stalk. **FLOWERS:** About ½ inch wide, with 4 white (aging to pink) petals, 4 green or pink, sharply downcurved sepals, and 8 showy stamens. **FRUIT:** About ⅛ inch long, shorter than its stalk, sharply 4-angled. **FLOWERING:** Apr–Jul. **RANGE:** FL west to LA, north to IL and SC. **SIMILAR TO** Southeastern Gaura (*Gaura angustifolia*, synonym *Oenothera simulans*), which occurs in similar habitats in the Coastal Plain; its flowers often have only 3 petals, and its fruits have no stalks.

Alternate-Leaf Seedbox

Ludwigia alternifolia Linnaeus
Evening-Primrose Family (Onagraceae)

PERENNIAL HERB found in marshes and other open wetlands, wet ditches, and disturbed wetlands throughout Georgia. **STEMS:** Up to 4 feet tall, widely branched, smooth or hairy. **LEAVES:** Up to 4 inches long and ¾ inch wide, alternate, lance-shaped, hairy or smooth, entire, tapering to short or no stalk. **FLOWERS:** About ½ inch wide, with 4 yellow, rounded petals, 4 stamens, 4 broadly oval sepals clearly visible between the petals, and a cubical, 4-winged ovary below the sepals. **FRUIT:** About ¼ inch long and wide, cubical, with a pore at the top. **FLOWERING:** May–Oct. **RANGE:** FL west to TX, north to NE and QC. **SIMILAR TO** Savannah Seedbox (*Ludwigia virgata*), which occurs in wet to dry pinelands in the Coastal Plain; its leaves are rounded at the base, and the sepals are bent downward and barely visible between the petals.

Wingstem Water-Primrose

Ludwigia decurrens Walter
Evening-Primrose Family (Onagraceae)

ANNUAL OR SHORT-LIVED PERENNIAL HERB found in swamps, marshes, and disturbed wetlands throughout Georgia. **STEMS:** 1–8 feet tall, smooth, with 2 conspicuous wings running between the leaves. **LEAVES:** Up to 7 inches long and 1 inch wide, alternate, lance-shaped, hairy or smooth, with rough margins and short or no stalks. **FLOWERS:** About ¾ inch wide, with 4 yellow, rounded petals, 8 stamens, 4 narrow, pointed sepals visible between the petals, and an elongated, 4-angled ovary. **FRUIT:** About ½ inch long and ⅛ inch wide, square in cross-section, narrowly winged. **FLOWERING:** Jun–Oct. **RANGE:** FL west to TX, north to MO and MD. **SIMILAR TO** Narrow-leaf Seedbox (*Ludwigia linearis*), which occurs in similar habitats in the Coastal Plain and northwest Georgia; its stems are angled but not winged, its leaves are narrowly linear, and the sepals are tiny.

Water-Willow

Ludwigia leptocarpa (Nuttall) Hara
Evening-Primrose Family (Onagraceae)

ANNUAL OR PERENNIAL HERB found in marshes and open, disturbed wetlands throughout Georgia (except the mountains). **STEMS:** 1–6½ feet tall, often bushy-branched, shaggy-hairy, lower stem woody and smooth. **LEAVES:** Up to 6 inches long and 1¼ inches wide, alternate, elliptic to lance-shaped, hairy. **FLOWERS:** About ½ inch wide, with 5–7 yellow, rounded or notched petals, 5–7 pointed sepals visible between the petals, and an elongated ovary. **FRUIT:** 1–2 inches long and ⅛ inch wide, reddish, narrowly cylindrical, ribbed, hairy, with 5 sepals around the top. **FLOWERING:** Jun–Sep. **RANGE:** FL west to TX, north to MO and PA. **SIMILAR TO** Common Water-primrose (*Ludwigia grandiflora*, synonym *L. uruguayensis*), which occurs in similar habitats; its lower stems root at the nodes, and its flowers are 1–2 inches wide.

Marsh Seedbox, Water-Purslane

Ludwigia palustris (Linnaeus) Elliott
Evening-Primrose Family (Onagraceae)

PERENNIAL HERB found in swamps, pond shores, and disturbed wetlands throughout Georgia. **STEMS:** Up to 1 foot long, sprawling or floating, mat-forming, smooth or slightly hairy, reddish, often rooting at the leaf nodes. **LEAVES:** Up to 1½ inches long and ¾ inch wide, opposite, oval to elliptic, smooth, often reddish, with winged, 1-inch stalks. **FLOWERS:** Held in the leaf axils, very small, with no petals, 4 triangular sepals, 4 white-tipped stamens, and a nearly cubical, 4-angled ovary. **FRUIT:** A capsule less than ⅛ inch high and wide, pale green with dark green bands on the angles. **FLOWERING:** May–Nov. **RANGE:** Most of North America, Eurasia, and Africa. **SIMILAR TO** Creeping Seedbox (*Ludwigia repens*), which occurs in a few Coastal Plain counties; its flowers have 4 yellow, quickly shed petals, and there are no green bands on the ovary or fruit.

Common Evening-Primrose

Oenothera biennis Linnaeus
Evening-Primrose Family (Onagraceae)

BIENNIAL HERB found on roadsides and in other disturbed areas throughout Georgia. **STEMS:** Up to 5 feet tall, hairy, reddish or light green. **LEAVES:** Up to 6 inches long and 1½ inches wide, alternate, lance-shaped, entire or toothed, with short or no stalks. **FLOWERS:** 1–2½ inches wide, with 4 yellow, heart-shaped petals, 8 stamens, a cross-shaped stigma, and an elongated calyx tube topped with 4 sharply downturned sepals; the buds are yellow and pointed. **FLOWERING:** Jun–Oct. **RANGE:** Most of North America. **NOTES:** The flowers, which open at dusk and close during the heat of the day, are pollinated mostly by sphinx moths. **SIMILAR TO** Large-flowered Evening Primrose (*Oenothera grandiflora*), which occurs in similar habitats throughout Georgia; its flowers are 3–4¾ inches wide.

Southern Sundrops

Oenothera fruticosa Linnaeus
Evening-Primrose Family (Onagraceae)

PERENNIAL HERB found in dry woodlands, limestone and granite outcrops, and dry, disturbed areas throughout Georgia. **STEMS:** Up to 2½ feet tall, hairy, often dark red. **LEAVES:** Up to 4⅓ inches long and 1 inch wide, alternate, elliptic to lance-shaped, entire or toothed, with short or no stalks. **FLOWERS:** Up to 1½ inches wide, with 4 yellow, notched petals, 8 stamens, a cross-shaped stigma, and an elongated calyx tube topped with 4 spreading or downturned sepals; buds are pale to reddish-orange and pointed. **FRUIT:** A narrow, hairy capsule about ⅓ inch long, widest near the top, the hairs not gland-tipped. **FLOWERING:** Apr–Aug. **RANGE:** FL west to OK, north to ON and NS. **SIMILAR TO** Appalachian Sundrops (*Oenothera tetragona*), which occurs in north Georgia woodlands and on roadsides; the fruits are covered with a mix of glandular and nonglandular hairs.

Cutleaf Evening-Primrose

Oenothera laciniata Hill
Evening-Primrose Family (Onagraceae)

ANNUAL OR BIENNIAL HERB found in fields, roadsides, and other dry, disturbed areas throughout Georgia. **STEMS:** Up to 2½ feet tall, branching near the base with the lowest branches sprawling over the ground, hairy. **LEAVES:** Up to 4 inches long and 1 inch wide, alternate, elliptic to lance-shaped, irregularly toothed and lobed, hairy, with short or no stalks. **FLOWERS:** Up to 2 inches wide, with 4 yellow petals, 8 stamens, a cross-shaped stigma, and a 1-inch-long calyx tube topped with 4 downturned sepals. **FLOWERING:** Feb–Nov. **RANGE:** FL west to NM, north to WY and ME; CA. **NOTES:** The flowers bloom at night and are pollinated primarily by sphinx moths. **SIMILAR TO** Sea-beach Evening-primrose (*Oenothera humifusa*), which occurs on beach dunes; the entire plant is sprawling and covered with long, white hairs, and the leaves are wavy-margined.

Showy Evening-Primrose, Pink Ladies

Oenothera speciosa Nuttall
Evening-Primrose Family (Onagraceae)

PERENNIAL HERB found on roadsides and in other disturbed areas throughout Georgia. **STEMS:** Up to 2 feet tall, hairy or smooth, often forming large colonies. **LEAVES:** Up to 3 inches long and 1 inch wide, alternate, irregularly toothed and lobed, hairy or smooth. **FLOWERS:** 2–3½ inches wide, with 4 pink-and-white petals, 8 stamens, a cross-shaped stigma, and a yellow throat; flower buds are spindle-shaped and nodding. **FLOWERING:** May–Aug. **RANGE:** FL west to CA, north to UT and CT. **NOTES:** The flowers open before dawn and wither soon after sunlight hits them. The size of the flowers suggests that they are pollinated by hummingbirds, large butterflies, or day-flying sphinx moths, but this has not been confirmed. **SIMILAR TO** no other wildflower in Georgia.

Broomrape Family (Orobanchaceae)

Broomrape Family plants are parasites that attach themselves to the roots of other plants and extract some or all of their nutrients from their "host." Some broomrapes are hemiparasites—they have green leaves and supply some of their energy needs through photosynthesis; others have no chlorophyll at all and are totally parasitic (holoparasites). All broomrapes, whether fully or partially parasitic, attach to their hosts by way of specialized root tips called haustoria that penetrate the roots of the host plant. Native broomrapes and their native hosts coexist—the broomrapes would die if they killed their hosts—but some members of the Broomrape Family are introduced agricultural pests that kill their host plants. Botanists once limited the Broomrape Family to holoparasites, but in the last 15 years or so, many of the hemiparasitic species that were in the Snapdragon ("scroph") Family have been moved to the Broomrape Family. Thirty-five species in the Broomrape Family are native to Georgia; only one, Lesser Broomrape (*Orobanche minor*), is an introduced agricultural pest.

Tall False Foxglove

Agalinis fasciculata (Linnaeus) Pennell
Broomrape Family (Orobanchaceae)

ANNUAL HERB found in sandhills, pinelands, and disturbed areas in the Coastal Plain and lower Piedmont. **STEMS:** 2–4 feet tall, stout, rough-hairy, branched, very leafy. **LEAVES:** Up to 1½ inches long, very narrow, opposite, clustered, linear, rough-hairy on both surfaces. **FLOWERS:** 1–1½ inches long, pinkish-purple, bell-shaped with 5 rounded, spreading lobes, hairy on the outside, with dark pink spots and yellow stripes on the inside. **FLOWERING:** Aug–Oct. **RANGE:** FL west to TX, north to KS and NY. **SIMILAR TO** Blunt-leaved False Foxglove (*Agalinis obtusifolia*), which occurs in similar habitats in the Coastal Plain and northwest Georgia; its pale pink flowers are smooth on the outer surface, less than ½ inch long, and lack yellow stripes on the inside; its stems and leaves are light yellow-green. Also see Purple False Foxglove (*Agalinis purpurea*), page 270.

Flax-Leaf False Foxglove

Agalinis linifolia (Nuttall) Britton
Broomrape Family (Orobanchaceae)

PERENNIAL HERB found in wet pine savannas, cypress-gum ponds, and marshes in the Coastal Plain. **STEMS:** 2–4 feet tall, unbranched or with a few erect branches, smooth, green. **LEAVES:** 1–2 inches long, very narrow, opposite, without leaf clusters in the axils. **FLOWERS:** 1–1½ inches long, on delicate stalks up to 1 inch long, pink, with 5 rounded, spreading lobes, hairy in the throat and on the outside and edges of the flower, with pale dots but no yellow lines in the throat. **FLOWERING:** Aug–Sep. **RANGE:** FL west to LA, north to DE. **SIMILAR TO** Seminole False Foxglove (*Agalinis filifolia*), which occurs infrequently in pinelands and dunes in southeast Georgia; its stems and leaves are reddish-purple, the leaves are alternate, and there are clusters of leaves in the leaf axils, giving the stems a leafy appearance.

Purple False Foxglove

Agalinis purpurea (Linnaeus) Pennell
Broomrape Family (Orobanchaceae)

ANNUAL HERB found in woodlands, pine flatwoods, wetlands, old fields, and disturbed areas throughout Georgia. **STEMS:** 1–4 feet tall, smooth or rough, angled, much-branched in the upper half, often sprawling. **LEAVES:** Up to 1½ inches long, very narrow, opposite, linear, dark green or purplish, with a conspicuous midvein; there are few or no clusters of leaves in the leaf axils. **FLOWERS:** Up to 1½ inches long, pinkish-purple, bell-shaped with 5 spreading lobes; the inner surface is hairy with purple spots and 2 yellow lines. **FLOWERING:** Aug–Oct. **RANGE:** FL west to TX, north to NE, ON, and NH. **SIMILAR TO** Tall False Foxglove (*Agalinis fasciculata*), page 269, and Salt Marsh False Foxglove (*Agalinis maritima*), which occurs infrequently in salt marshes on Georgia's coast; its stems and leaves are fleshy, and the flowers are about ½ inch long.

Slender-Leaf False Foxglove

Agalinis tenuifolia (Vahl) Rafinesque
Broomrape Family (Orobanchaceae)

ANNUAL HERB found in open woodlands, prairies, and clearings throughout Georgia. **STEMS:** 1–3 feet tall, slender, angled, smooth or slightly hairy, dull green to reddish-purple. **LEAVES:** Up to 3 inches long, narrow and needle-like, opposite, smooth, with a prominent midvein. **FLOWERS:** About ½ inch long, pinkish-purple, bell-shaped with 5 rounded lobes, the upper lobe folded forward over the throat, which is white with purple spots and 2 yellow lines; the stamens (but not the throat) are very hairy. **FLOWERING:** Aug–Oct. **RANGE:** FL west to NM, north to MB and NB. **SIMILAR TO** Thread-leaf False Foxglove (*Agalinis setacea*), which occurs in sandhills and other dry forests in Georgia's eastern Coastal Plain; the upper lobes of its flowers are erect or curved backward, and the throat is very hairy.

Sticky False Foxglove, Southern Oak-Leech

Aureolaria pectinata (Nuttall) Pennell
Broomrape Family (Orobanchaceae)

ANNUAL HERB found in sandhills and other dry woodlands throughout Georgia. **STEMS:** Up to 3 feet tall, glandular-hairy, much-branched and bushy. **LEAVES:** Up to 2½ inches long, opposite, divided into 10–12 toothed segments. **FLOWERS:** About 1½ inches long, yellow (sometimes tinged with red), tubular with 5 spreading lobes. **FLOWERING:** May–Oct. **RANGE:** FL west to TX, north to MO and VA. **NOTES:** Sticky, gland-tipped hairs cover all parts of the plant. **SIMILAR TO** Oak-leech (*Aureolaria pedicularia*), which occurs in dry woodlands mostly in northeast Georgia; the upper parts of the plant are smooth, the lower parts glandular-hairy. Also see Downy False foxglove (*Aureolaria virginica*), below. All species in the genus *Aureolaria* are partially parasitic on the roots of oak trees.

Downy False Foxglove, Virginia Oak-Leech

Aureolaria virginica (Linnaeus) Pennell
Broomrape Family (Orobanchaceae)

PERENNIAL HERB found in dry oak forests and woodlands throughout Georgia. **STEMS:** Up to 5 feet long, leaning on other plants, with few or no branches. **LEAVES:** Up to 4¾ inches long and 1¾ inches wide, opposite, oval or lance-shaped with pointed tips; lower leaves with 1–3 pairs of large lobes, midstem leaves with shallow lobes, upper leaves unlobed. **FLOWERS:** About 1½ inches long, yellow, tubular with 5 spreading lobes. **FLOWERING:** May–Jul. **RANGE:** FL west to TX, north to MI and NH. **NOTES:** Fine, soft, nonglandular hairs cover most of the plant. **SIMILAR TO** Smooth Oak-leech (*Aureolaria flava*), which has a waxy, smooth stem and deeply dissected lower leaves; and Appalachian Oak-leech (*A. laevigata*), which has a smooth stem and usually entire leaves.

Florida Bluehearts

Buchnera floridana Gandoger
Broomrape Family (Orobanchaceae)

PERENNIAL HERB found in pine savannas, seepage slopes, and moist, sandy roadsides in the Coastal Plain. **STEMS:** Up to 2½ feet tall, unbranched, rough-hairy near the base, smooth above. **LEAVES:** Up to 2¾ inches long and ½ inch wide, rough-hairy, elliptic with blunt tips and few or no teeth. **FLOWERS:** In a loose spike at the top of the stem, each about ½ inch long and ⅓ inch wide, bluish-purple, tubular with 5 spreading, slightly unequal lobes and a tuft of white hairs in the center. **FLOWERING:** Apr–Oct. **RANGE:** FL west to TX, north to NC; West Indies. **NOTES:** Bluehearts are semiparasitic on the roots of grasses and woody plants. **SIMILAR TO** American Bluehearts (*Buchnera americana*), a Georgia Special Concern species that occurs in limestone cedar glades in northwest Georgia; its leaves are pointed and toothed, with 3 conspicuous veins.

Bear Corn, American Cancer-Root

Conopholis americana (Linnaeus) Wallroth
Broomrape Family (Orobanchaceae)

PERENNIAL HERB found in rich, moist forests throughout Georgia, though largely absent from the Piedmont. **STEMS:** 4–10 inches tall and up to 1 inch thick, fleshy, yellow or cream-colored, resembling an ear of corn. **LEAVES:** None. **FLOWERS:** Up to ½ inch long, tubular with 5 cream-colored petals, densely packed onto the stem. **FRUIT:** A pale, oval capsule about ½ inch long. **FLOWERING:** Mar–Jun. **RANGE:** FL west to MS, north to MB and NS; eastern Mexico. **NOTES:** Bear Corn has no roots or chlorophyll and derives all its water and nutrients from a parasitic attachment to red oak tree roots; it does little or no harm to its host. It is a popular food for black bear and deer, which disperse its seeds. **SIMILAR TO** no other wildflower in Georgia.

Beech Drops

Epifagus virginiana (Linnaeus) W. Barton
Broomrape Family (Orobanchaceae)

ANNUAL HERB found in moist forests throughout Georgia except for the southeastern counties. **STEMS:** 4–18 inches tall; tan, purplish, or brown, with stiff, upright branches that often persist, dried, through the winter. **LEAVES:** A few tiny, brownish scales scattered along the stem and branches. **FLOWERS:** Of 2 types: open flowers (few) with 4 flaring lobes, and closed flowers (numerous) with closed, pointed tips; both types are about ⅓ inch long, white with purple stripes. **FRUIT:** An oval capsule about ¼ inch long filled with dustlike seeds; only the closed flowers produce fruit, apparently by self-pollination. **FLOWERING:** Sep–Nov. **RANGE:** FL west to TX, north to ON and NS; Mexico. **NOTES:** Beech Drops have no chlorophyll or roots and derive water and nutrients from their parasitic connection to the roots of beech trees. **SIMILAR TO** no other wildflower in Georgia.

Cow-Wheat

Melampyrum lineare Desrousseaux
Broomrape Family (Orobanchaceae)

ANNUAL HERB found in acidic cove forests and dry woodlands, and around rock outcrops in Georgia's Blue Ridge. **STEMS:** 4–16 inches tall, finely hairy. **LEAVES:** Up to 3 inches long and 1 inch wide, opposite, linear to lance-shaped, entire except the uppermost leaves, which have pointed lobes near the base. **FLOWERS:** Solitary in the leaf axils, up to ½ inch long, with a white or pale yellow tube and 2 bright yellow lips. **FLOWERING:** Late Apr–Jul. **RANGE:** GA north to MN and ME; MT, ID, WA, Canada. **NOTES:** Cow-wheat occurs in poor, acidic soils and derives most of its nutrients through attachment to the roots of oaks, maples, and pines. Its seeds are dispersed by ants (see "The Skinny on Fat Bodies—Elaiosomes," p. 44). **SIMILAR TO** no other wildflower in Georgia.

One-Flowered Cancer-Root

Orobanche uniflora Linnaeus
Broomrape Family (Orobanchaceae)

PERENNIAL HERB found in moist forests and on stream banks in north Georgia. **STEMS:** Underground, sending up flower stalks 1–8 inches tall, slender, leafless, grayish-tan, very hairy. **LEAVES:** None. **FLOWERS:** One at the top of each stalk, nodding or horizontal, less than 1 inch long, hairy, white with purple stripes and 5 spreading lobes. **FLOWERING:** Apr–May. **RANGE:** Throughout North America. **NOTES:** One-flowered Cancer-Root is parasitic on the roots of a wide range of host plants, including sunflowers and goldenrods; it lacks roots and chlorophyll and derives all its water and nutrients from the roots of its host plant. **SIMILAR TO** no other wildflower in Georgia.

Lousewort, Wood Betony

Pedicularis canadensis Linnaeus
Broomrape Family (Orobanchaceae)

PERENNIAL HERB found in moist to dry forests and woodlands in north Georgia and a few Coastal Plain counties. **STEMS:** 4–16 inches tall, very hairy, green or reddish, often forming small colonies. **BASAL LEAVES:** Up to 6 inches long and 2 inches wide, deeply lobed and toothed, fernlike, often tinged with red. **STEM LEAVES:** Similar to basal leaves but smaller, alternate. **FLOWERS:** In a spikelike cluster, each about ¾ inch long, yellow, reddish-purple, or both; tubular with 2 lips, the upper lip hoodlike; the flowers are twisted sideways so they form a pinwheel when viewed from above. **FLOWERING:** Apr–May. **RANGE:** FL west to NM, north to MB and NB. **NOTES:** Lousewort is a partial parasite, drawing some of its nutrients from the roots of a wide variety of host plants. **SIMILAR TO** no other wildflower in Georgia.

Black Senna, Senna Seymeria

Seymeria cassioides (J. F. Gmelin) Blake
Broomrape Family (Orobanchaceae)

ANNUAL HERB found in sandhills, back dunes, pine savannas and flatwoods, and roadsides in the Coastal Plain and a few north Georgia counties. **STEMS:** Up to 3 feet tall, smooth or finely hairy, bushy, with many opposite branches. **LEAVES:** Less than ¾ inch long, opposite, divided into many very narrow segments, often tinged with dark red. **FLOWERS:** About ½ inch wide, yellow, smooth, with 5 spreading lobes, the 2 upper lobes with red markings; 4 stamens extend well above the flower. **FLOWERING:** Aug–Oct. **RANGE:** FL west to TX and AR, north to VA. **NOTES:** Both species of *Seymeria* are parasitic on the roots of pine trees. **SIMILAR TO** Comb-leaf Seymeria (*Seymeria pectinata*), which occurs in the Coastal Plain in similar habitats; its flower is hairy, the leaf segments are wider and lance-shaped, and the stem is very hairy.

Regular or Irregular: The Basic Patterns of Flowers

Flowers come in many shapes but generally follow one of two patterns in the arrangement of their petals and sepals.

REGULAR FLOWERS have radial symmetry, meaning that the petals are all more or less alike and radiate symmetrically from the center of the flower. The flower can be divided into equal halves (mirror images) along several lines. Another term for this flower pattern is actinomorphic. Classic examples are Bloodroot, buttercups, cacti, evening-primroses, lilies, and roses.

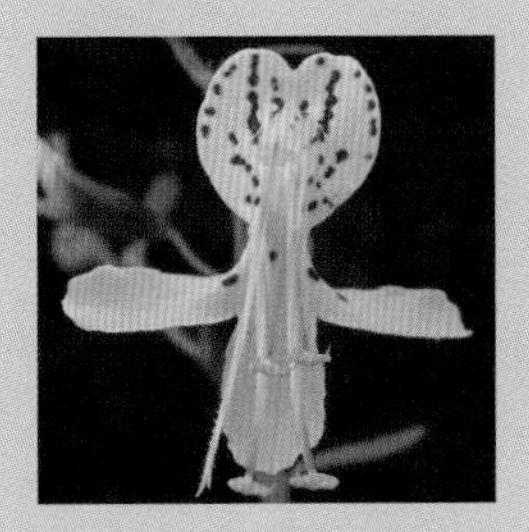

IRREGULAR FLOWERS have bilateral symmetry, meaning that the petals are of different sizes and shapes and form a flower that can be divided into equal halves (mirror images) along only one line. Another term for this flower pattern is zygomorphic. Examples include orchids, violets, beans and peas, mints, and lobelias.

Yellow Wood Sorrel

Oxalis stricta Linnaeus
Wood Sorrel Family (Oxalidaceae)

PERENNIAL HERB found in open woodlands, meadows, thickets, and disturbed areas throughout Georgia. **STEMS:** 3–20 inches tall, usually much-branched from the base, softly hairy or smooth. **LEAVES:** Alternate, clover-like with 3 heart-shaped leaflets up to ½ inch long, solid green, often folded lengthwise along the central vein. **FLOWERS:** About ½ inch wide, with 5 rounded, pure yellow petals. **FRUIT:** Hairy, 5-angled capsules up to ¾ inch long. **FLOWERING:** May–Oct. **RANGE:** Throughout North America and worldwide. **NOTES:** When mature, the capsules split along the angles and fling their seeds several feet away. All parts of the plant have a tangy taste. **SIMILAR TO** Southern Yellow Wood Sorrel (*Oxalis dillenii*), which occurs in a variety of disturbed habitats; its stems are covered with stiff, straight hairs.

Violet Wood Sorrel

Oxalis violacea Linnaeus
Wood Sorrel Family (Oxalidaceae)

PERENNIAL HERB found in dry to moist forests throughout Georgia. **STEMS:** Underground, sending up leaves and flower stalks. **LEAVES:** Clover-like with 3 heart-shaped leaflets about ⅓ inch long, grayish-green or purplish. **FLOWERS:** On long, smooth stalks, each flower about ½ inch long, bell-shaped, with 5 pink or lavender petals, a white-and-green "eye," and 5 smooth sepals. **FLOWERING:** Mar–May, Aug–Oct. **RANGE:** Most of North America. **NOTES:** Flowers and leaflets close up on overcast days and reopen when the sun shines. Fall-flowering plants lack leaves. **SIMILAR TO** Red Wood Sorrel (*Oxalis rubra*, synonym *Oxalis articulata*), a common garden plant from South America that has escaped to roadsides and other disturbed areas; it has dark pink or white flowers with hairy stalks and sepals.

Bloodroot

Sanguinaria canadensis Linnaeus
Poppy Family (Papaveraceae)

PERENNIAL HERB found in moist hardwood forests and bottomlands throughout Georgia. **STEMS:** Underground, sending up leaves and flower stalks. **LEAVES:** Small and wrapped around the flower stalk when young, expanding to 5–8 inches wide by midsummer; waxy and pale green, with 3–9 lobes and wavy, lobed margins. **FLOWERS:** 1–2 inches wide, with 6–12 bright white petals and many golden stamens, solitary at the top of a stalk 1–6 inches long. **FRUIT:** An erect, waxy, green, spindle-shaped capsule about 2 inches long that splits lengthwise along both sides to release seeds. **FLOWERING:** Feb–Apr. **RANGE:** FL west to TX, north to MB and NS. **NOTES:** The genus is named for the red sap in the rhizome, which contains the poisonous alkaloid sanguinarine. **SIMILAR TO** no other common wildflower in Georgia.

Grass-of-Parnassus

Parnassia asarifolia Ventenat
Grass-of-Parnassus Family
(Parnassiaceae)

PERENNIAL HERB found in bogs, seepages, and wet woods in north Georgia. **STEMS:** Up to 16 inches tall, with a single flower at the top and a small leaf about midway. **BASAL LEAVES:** Up to 2⅓ inches long and 4 inches wide, kidney-shaped (wider than long), smooth, with conspicuously curved veins and long stalks. **FLOWERS:** Up to 1⅓ inches wide, with 5 white petals, each with 11–15 green or tan, branching veins. **FLOWERING:** Aug–Oct. **RANGE:** GA west to TX and AR, north to KY and MD. **SIMILAR TO** Big-leaf Grass-of-Parnassus (*Parnassia grandifolia*), a Georgia Special Concern species found only in high-elevation seepages over mafic bedrock; it has oval leaves (longer than wide), and each petal has only 9–11 veins.

May-Pop, Purple Passion-Flower

Passiflora incarnata Linnaeus
Passion-Flower Family (Passifloraceae)

PERENNIAL HERBACEOUS VINE found in woodlands, stream banks, and disturbed areas throughout Georgia. **STEMS:** Up to 12 feet long, sprawling over other plants, and climbing with tightly coiled tendrils. **LEAVES:** 2½–8 inches long and 2–6 inches wide, alternate, 3- or 5-lobed, with toothed margins. **FLOWERS:** About 2½ inches across, showy and complicated, with 5 sepals (oblong, green on the outer surface, white on the inside, with a pointed beak), 5 petals (similar to sepals except purplish and without beak), a corona of purple-and-white, threadlike segments, and a stalk that supports 5 stamens and a fleshy, green, 3-parted style. **FRUIT:** An oval berry 1½–2¾ inches long. **FLOWERING:** May–Jul. **RANGE:** FL west to TX, north to KS and PA. **NOTES:** May-pop is the only larval host for Gulf Fritillary butterflies. **SIMILAR TO** no other wildflower in Georgia.

Yellow Passion-Flower

Passiflora lutea Linnaeus
Passion-Flower Family (Passifloraceae)

PERENNIAL HERBACEOUS VINE found in forests and thickets throughout Georgia. **STEMS:** Up to 15 feet long, twining and sprawling, with tightly coiled tendrils. **LEAVES:** 1–3 inches long and 1–6 inches wide, alternate, 3-lobed, entire, often mottled silvery gray. **FLOWERS:** About ½ inch across, pale greenish-yellow, with 5 green sepals, 5 greenish-yellow or white petals, and a corona of green, threadlike segments (sometimes purple at the base) surrounding a stalk that supports 5 stamens and a fleshy, green, 3-parted style. **FRUIT:** A green, oval or round berry, ½ inch long, black when mature. **FLOWERING:** Jun–Sep. **RANGE:** FL west to TX, north to KS and PA. **NOTES:** The flowers open for only one day and are pollinated by bees. **SIMILAR TO** no other wildflower in Georgia.

Ditch Stonecrop

Penthorum sedoides Linnaeus
Ditch Stonecrop Family (Penthoraceae)

PERENNIAL HERB found in floodplains, marshes, beaver ponds, and wet disturbed areas in north Georgia and a few Coastal Plain counties. **STEMS:** Up to 3 feet tall, reddish, smooth except for stalked glandular hairs on the upper half. **LEAVES:** Up to 7 inches long and 1½ inches wide, alternate, lance-shaped to elliptic, tapering at both ends, finely toothed, the midvein white or tinged red. **FLOWER CLUSTERS:** On curved, glandular-hairy branches rising from the upper leaf axils, the flowers held on the upper side of the branches. **FLOWERS:** About ¼ inch across, with 5 green or reddish sepals (no petals) and 10 showy, white stamens. **FRUIT:** Clusters of red, pointed fruits cover the branches. **FLOWERING:** Jun–Oct. **RANGE:** FL west to TX, north to MB and NB; introduced in the Pacific Northwest. **SIMILAR TO** no other wildflower in Georgia.

American Lopseed

Phryma leptostachya Linnaeus
Lopseed Family (Phrymaceae)

PERENNIAL HERB found in moist, rich, deciduous forests in north Georgia and a few Coastal Plain counties. **STEMS:** 1½–3 feet tall, 4-angled, with swollen purple areas above the leaf stalks. **LEAVES:** Up to 6 inches long and 3½ inches wide, opposite, oval to lance-shaped, with toothed margins and stalks up to 1½ inches long. **FLOWER CLUSTERS:** 2–8 inches long, narrow and spikelike, flowers opening from the bottom upward. **FLOWERS:** About ¼ inch long, in pairs, pale pink and white, 2-lipped, upper lip notched, lower lip 3-lobed. **FLOWERING:** Jun–Aug. **RANGE:** FL west to TX, north to MB and NB. **NOTES:** The flowers are at first held horizontally; after the petals drop, the calyx droops, or "lops," downward and is held close to the stem as the fruit develops. **SIMILAR TO** Mint Family species, pages 229–245.

Pokeweed, Poke Salad

Phytolacca americana Linnaeus
Pokeweed Family (Phytolaccaceae)

PERENNIAL HERB found in disturbed areas throughout Georgia. **STEMS:** 3–10 feet tall, stout, dark red, smooth. **LEAVES:** 3–12 inches long and 1–5 inches wide, alternate, oval to lance-shaped, with stalks up to 2¼ inches long. **FLOWER CLUSTERS:** 2–8 inches long, leaning or nodding. **FLOWERS:** About ¼ inch wide, with 5 white (or pinkish) sepals but no petals, a bright green ovary, and a stalk longer than ¼ inch. **FRUIT:** A round, purple berry with red juice, about ⅓ inch wide on a stalk longer than ¼ inch. **FLOWERING:** May–Sep. **RANGE:** Most of North America. **SIMILAR TO** Seaside Pokeweed (*Phytolacca rigida*), which occurs in dunes, salt marshes, and disturbed areas on barrier islands; it has erect flower clusters less than 3½ inches long, and the flower and fruit stalks are usually less than ¼ inch long.

Long-Bracted Plantain

Plantago aristata Michaux
Plantain Family (Plantaginaceae)

ANNUAL OR SHORT-LIVED PERENNIAL HERB found in disturbed areas throughout Georgia. **BASAL LEAVES:** Up to 8 inches long and ½ inch wide, linear, smooth or hairy, entire, with conspicuous parallel veins. **FLOWER CLUSTERS:** A slender, hairy spike up to 5 inches long, held at the top of a stalk up to 6 inches long, crowded with flowers and narrow, green, pointed bracts up to 1¾ inches long. **FLOWERS:** About ⅛ inch wide, with 4 papery, whitish, translucent petals and 4 tiny, green sepals. **FLOWERING:** May–Nov. **RANGE:** Nearly throughout North America. **SIMILAR TO** English Plantain (*Plantago lanceolata*), a European species that occurs in disturbed areas throughout Georgia; its leaves are narrowly lance-shaped, and the flower spikes lack the long bracts. Also see Virginia Plantain (*Plantago virginica*), page 281.

Virginia Plantain

Plantago virginica Linnaeus
Plantain Family (Plantaginaceae)

ANNUAL HERB found in disturbed areas throughout Georgia. **LEAVES:** Basal, up to 6 inches long and 1½ inches wide, spoon-shaped with winged stalks, hairy on both surfaces, slightly toothed or entire, with conspicuous parallel veins. **FLOWER CLUSTERS:** A slender, hairy spike 1–6 inches long at the top of a hairy stalk 4–8 inches tall, crowded with flowers but with no visible bracts. **FLOWERS:** About ⅛ inch wide, with 4 papery, tan, erect petals and 4 tan, papery sepals. **FLOWERING:** Late Mar–Jul. **RANGE:** Most of North America. **SIMILAR TO** two species with wider, broadly oval leaf blades and distinct leaf stalks; both occur in disturbed areas. Common Plantain (*Plantago major*) leaf stalks are green and hairy at the base. American Plantain (*Plantago rugelii*) leaf stalks are purple and smooth at the base.

Sea Lavender

Limonium carolinianum (Walter) Britton
Leadwort Family (Plumbaginaceae)

PERENNIAL HERB found in salt marshes, salt flats, and dune swales near the coast. **STEMS:** 1–2½ feet tall. **LEAVES:** Up to 12 inches long and 3 inches wide, in a basal rosette, spoon-shaped with rounded tips and tapering bases, entire, leathery, fleshy, smooth. **FLOWERS:** In a widely branched, fanlike cluster, each flower about ¼ inch long; pink, lavender, or blue, with 5 rounded petals and a reddish, 5-lobed calyx. **FLOWERING:** Aug–Oct. **RANGE:** FL west to TX and Mexico, north to Labrador. **NOTES:** Tiny glands on the leaves (use 10× lens) secrete salt that the plant's roots absorb from the soils of its coastal habitat. **SIMILAR TO** no other wildflower on the coast.

Standing Cypress

Ipomopsis rubra Linnaeus
Phlox Family (Polemoniaceae)

BIENNIAL HERB found in dry woodlands, sandhills, dunes, riverbanks, and disturbed areas in the Piedmont. **STEMS:** 2–6 feet tall, hairy. **LEAVES:** Up to 1¼ inches long, divided into many very narrow segments. **FLOWERS:** Up to 1½ inches long, bright red (rarely yellow), tubular with 5 spreading, pointed lobes; in a tall, narrow cluster up to one-third the total height of the plant. **FLOWERING:** Jun–Aug. **RANGE:** FL west to TX, north to ON and MA. **NOTES:** Standing Cypress produces a low rosette of ferny leaves the first year, then sends up the leafy flowering stem its second year, after which it dies. The odd common name may refer to the leaf segments, which vaguely resemble the slender needles of Bald Cypress. Standing Cypress flowers are magnets for hummingbirds and butterflies. **SIMILAR TO** no other wildflower in Georgia.

Hairy Phlox

Phlox amoena Sims
Phlox Family (Polemoniaceae)

PERENNIAL HERB found in dry woodlands and sandhills and on dry road banks throughout Georgia. **STEMS:** Up to 18 inches tall, hairy. **LEAVES:** Up to 2 inches long and ½ inch wide, opposite, elliptic to lance-shaped, hairy on both surfaces, with long hairs on the margins. **FLOWER CLUSTERS:** Compact, with green, hairy bracts surrounding the base of the cluster. **FLOWERS:** About ½ inch long, with 5 spreading lobes, pink or magenta with a darker star at the center and a hairless flower tube. **FLOWERING:** Apr–Jun. **RANGE:** FL west to MS, north to KY and NC. **SIMILAR TO** Downy or Prairie Phlox (*Phlox pilosa*), which occurs in similar habitats, especially where there is mafic or limestone bedrock; its leaves are narrowly lance-shaped, the flowers are loosely clustered, and the flower tube is hairy; it flowers Apr–May.

Carolina Phlox, Smooth Phlox

Phlox carolina Linnaeus
Phlox Family (Polemoniaceae)

PERENNIAL HERB found in openings and edges of deciduous forests, clearings, and prairies throughout Georgia. **STEMS:** Up to 3 feet tall, smooth. **LEAVES:** Up to 5 inches long and 1 inch wide, opposite, linear, elliptic, or lance-shaped, smooth or slightly hairy, entire with smooth margins. **FLOWER CLUSTERS:** Broadly round-topped, not compact, the lower branches elongated. **FLOWERS:** ½–1 inch long and wide, with 5 spreading lobes, pink or magenta (rarely white) with a darker star at the center and a hairless flower tube. **FLOWERING:** May–Jul. **RANGE:** FL west to TX, north to MO and MD. **SIMILAR TO** Broadleaf Phlox (*Phlox amplifolia*), which flowers Jun–Sep; it occurs in moist forests, especially over mafic or limestone bedrock, in northwest Georgia; its leaves are broadly oval, 1½–3 inches wide, hairy, and slightly toothed.

Blue Phlox

Phlox divaricata Linnaeus
Phlox Family (Polemoniaceae)

PERENNIAL HERB found in moist, deciduous forests, especially over limestone or mafic bedrock, in western and central Georgia. **FLOWERING STEMS:** Up to 20 inches tall, smooth to very hairy, with 4 pairs of widely spaced leaves; nonflowering stems sprawl along the ground. **LEAVES:** Up to 1½ inches long and ½ inch wide, opposite, elliptic or lance-shaped, hairy, entire. **FLOWERS:** Up to ¾ inch long and 1¼ inches wide, with 5 spreading lobes, lavender or bluish-pink with a darker star at the center and a hairless flower tube. **FLOWERING:** Apr–May. **RANGE:** FL west to TX, north to MN, QC, and VT. **SIMILAR TO** Smooth Phlox (*Phlox glaberrima*), which occurs in bottomlands and other wet forests, mostly in the mountains and a few central Georgia counties. It has 7–15 pairs of leaves, pinkish-purple flowers, and smooth or only slightly hairy stems and leaves. It flowers May–Jul.

Meadow Phlox

Phlox maculata Linnaeus var. *pyramidalis* (J. E. Smith) Wherry
Phlox Family (Polemoniaceae)

PERENNIAL HERB found in moist meadows and forests and on stream banks and moist roadsides in the north Georgia mountains. **STEMS:** 1–5 feet tall, smooth, red-spotted, with 16–35 pairs of closely spaced leaves. **LEAVES:** Up to 6 inches long and 2 inches wide, opposite, lance-shaped with a long, tapering tip, smooth. **FLOWER CLUSTERS:** 3–12 inches tall, pyramid-shaped, densely flowered. **FLOWERS:** Up to 1 inch long; pink, lavender, or white, paler at the center; with 5 spreading lobes and a hairless flower tube. **FLOWERING:** Jul–Sep. **RANGE (THIS VARIETY):** GA and TN, north to MO and PA. **SIMILAR TO** Summer Phlox (*Phlox paniculata*), which occurs in the Blue Ridge in moist forests and on stream banks; its leaf margins have low, hairy teeth, and the flower tube is hairy; it flowers Jul–Sep.

Trailing Phlox

Phlox nivalis Loddiges ex Sweet
Phlox Family (Polemoniaceae)

PERENNIAL found in sandhills, dry woodlands, rock outcrops, road banks, and old fields in the Piedmont and Coastal Plain. **STEMS:** Semiwoody, trailing, forming mounds, and rooting at the nodes; erect flowering stems are up to 12 inches tall. **LEAVES:** Up to 1 inch long, very narrow, evergreen, opposite or in clusters in the axils of opposite leaves, stiff, linear or narrowly lance-shaped, with bristly margins. **FLOWERS:** Up to ¾ inch long and 1 inch wide, with 5 spreading lobes, pink, lavender, or white with a darker star in the center and a hairless flower tube. **FLOWERING:** Mar–May. **RANGE:** FL west to TX, north to VA. **SIMILAR TO** Creeping Phlox (*Phlox stolonifera*), which occurs in moist forests in northeast Georgia; its stems are not woody, and its leaves are spoon-shaped. Also see Carolina Sandwort (*Minuartia caroliniana*), page 150.

Boykin's Milkwort

Polygala boykinii Nuttall
Milkwort Family (Polygalaceae)

PERENNIAL HERB found in woodlands, prairies, and Longleaf Pine flatwoods and savannas in southwest Georgia and the Coosa Valley prairies in northwest Georgia. **STEMS:** 8–24 inches tall, slender, unbranched, often several stems arising from a single crown. **LEAVES:** Mostly whorled, some alternate, linear, oval, lance- or spoon-shaped, smooth. **FLOWER CLUSTERS:** Spikelike, 2–6 inches long, with closely spaced flowers opening from the bottom upward. **FLOWERS:** About ¼ inch wide, white, with 2 spreading, nearly round or oval wings (sepals) and 3 petals united into a tiny tube with a fringed tip. **FLOWERING:** Mar–Jul. **RANGE:** FL west to LA, north to TN and GA. **SIMILAR TO** Whorled Milkwort (*Polygala verticillata*), page 289.

Drumheads

Polygala cruciata Linnaeus
Milkwort Family (Polygalaceae)

ANNUAL HERB found in marshes, wet savannas, and bogs in the Coastal Plain and a few north Georgia counties. **STEMS:** Up to 12 inches tall, 4-angled, smooth. **LEAVES:** Up to 2 inches long and ¼ inch wide, in whorls of 4, linear, pointed. **FLOWER CLUSTERS:** Up to 2 inches long, cylinder-shaped, densely flowered; after the flowers wither and fall (from the bottom up), tiny green bracts remain on the stem. **FLOWERS:** About ⅓ inch wide, with 2 showy, spreading, pink, green, and white wings (sepals) and 3 petals united into a tiny, green, pink, or yellow tube with a fringed tip. **FLOWERING:** Jun–Oct. **RANGE:** FL west to TX, north to ON and ME. **SIMILAR TO** Blood Milkwort (*Polygala sanguinea*), a rare plant in Georgia that occurs in woodland openings; it has very narrow, alternate leaves and flowers that range from dark red to greenish-white.

Curtiss' Milkwort

Polygala curtissii A. Gray
Milkwort Family (Polygalaceae)

ANNUAL HERB found on rock outcrops and in woodland borders, old fields, and thickets in north Georgia and a few Coastal Plain counties. **STEMS:** Up to 16 inches tall. **LEAVES:** Up to 1½ inches long, very narrow, alternate, linear, pointed. **FLOWER CLUSTERS:** Up to ¾ inch high, loosely flowered, with unopened buds forming a greenish-yellow, pointed tip; after the flowers wither and fall (from the bottom up), tiny green bracts remain on the stem. **FLOWERS:** About ¼ inch wide, pink, white, or green, with 2 wings (sepals) and 3 petals united into a tiny tube with a knobby tip. **FLOWERING:** Jun–Oct. **RANGE:** GA west to MS, north to OH and PA. **SIMILAR TO** Maryland Milkwort (*Polygala mariana*), which occurs in bogs and pine savannas in the Coastal Plain; there are no tiny bracts remaining on the stem after the flowers fall.

Tall Pine Barren Milkwort

Polygala cymosa Walter
Milkwort Family (Polygalaceae)

BIENNIAL HERB found in wet pine savannas and flatwoods, bogs, and wet ditches in the Coastal Plain. **STEMS:** 1½–4 feet tall, smooth. **LEAVES:** Mostly in a basal rosette, each leaf up to 5½ inches long and ¼ inch wide, linear; stem leaves few, alternate, and much smaller. **FLOWER CLUSTERS:** Flat-topped, many-branched, with flowers densely crowded at the tips of branches. **FLOWERS:** Very small, bright yellow, with 2 wings (sepals) and 3 petals united into a tiny tube with a fringed tip. **FLOWERING:** Apr–Jul. **RANGE:** FL west to LA, north to DE. **SIMILAR TO** Short Pine Barren Milkwort (*Polygala ramosa*), an annual species that occurs in similar habitats in the Coastal Plain; it is always less than 1 foot tall and has few basal leaves. Also see Pineland Rayless Goldenrod (*Bigelowia nudata*), page 62.

Pink Milkwort, Procession Flower

Polygala incarnata Linnaeus
Milkwort Family (Polygalaceae)

ANNUAL HERB found in savannas, wet prairies, woodlands, old fields, and clearings throughout Georgia. **STEMS:** Up to 1½ feet tall, slender, ribbed, waxy blue-green, rarely branched. **LEAVES:** Up to ½ inch long, very narrow, alternate, widely spaced, fleshy, waxy blue-green. **FLOWER CLUSTERS:** A short, densely flowered spike held at the top of the stem, flowers opening from the bottom to the top. **FLOWERS:** About ¼ inch long, dark pink (rarely white), with 2 very narrow sepals, 2 very narrow petals, and 1 tubular, deeply fringed petal. **FLOWERING:** Jun–Jul. **RANGE:** FL west to TX, north to WI, ON, and NY. **SIMILAR TO** Nuttall's Milkwort (*Polygala nuttallii*), which occurs infrequently in savannas and depression ponds in the Coastal Plain; it is widely branched, and the tips of the flowers are not deeply fringed.

Orange Milkwort, Bog Cheetos

Polygala lutea Linnaeus
Milkwort Family (Polygalaceae)

ANNUAL OR BIENNIAL HERB found in wet savannas, flatwoods, and bogs in the Coastal Plain. **STEMS:** 2–16 inches tall. **LEAVES:** Up to 2½ inches long and ¾ inch wide, alternate, somewhat succulent, spoon-shaped. **FLOWER CLUSTERS:** Up to 1½ inches long and ½ inch wide, cylindrical, held at the top of the stem and tips of branches, crowded with many flowers. **FLOWERS:** About ½ inch wide, orange, with 2 spreading wings (sepals) and 3 petals united into a tiny tube with a fringed tip. **FLOWERING:** Apr–Oct. **RANGE:** FL west to LA, north to NY. **NOTES:** A third common name, Orange Candy-root, refers to the sweet, wintergreen smell of the roots. **SIMILAR TO** Candy-root (*Polygala nana*), page 288.

Candy-Root

Polygala nana (Michaux) A. P. de Candolle
Milkwort Family (Polygalaceae)

ANNUAL OR BIENNIAL HERB found in wet savannas and prairies, pine flatwoods, bogs, and wet ditches in the Coastal Plain and northwest Georgia. **STEMS:** 2–6 inches tall, usually branched from the base. **LEAVES:** Up to 2 inches long and ¾ inch wide, most in a basal rosette, spoon-shaped, somewhat succulent. **FLOWER CLUSTERS:** Up to 1½ inches long and ½ inch wide, cylindrical, held at the top of the stem, crowded with many flowers. **FLOWERS:** About ½ inch wide, yellow, with 2 spreading wings (sepals) and 3 petals united into a tiny tube with a fringed tip. **FLOWERING:** Mar–Oct. **RANGE:** FL west to TX, north to AR and NC. **NOTES:** The roots smell like wintergreen. **SIMILAR TO** Orange Milkwort (*Polygala lutea*), page 287.

Bitter Milkwort

Polygala polygama Walter
Milkwort Family (Polygalaceae)

BIENNIAL OR SHORT-LIVED PERENNIAL HERB found in sandhills, woodland openings, and disturbed areas throughout Georgia. **STEMS:** Up to 20 inches tall, angled, smooth, several arising from a crown. **LEAVES:** Up to 1½ inches long, narrow, alternate, linear or oblong, smooth, with a sharply pointed tip. **FLOWER CLUSTERS:** Up to 3 inches tall, spike-like, with loosely spaced flowers. **FLOWERS:** About ½ inch wide, on short stalks, with 2 spreading, pink wings (sepals) and 3 petals united into a tiny tube with a fringed tip; the wings are twice as long as wide. **FLOWERING:** May–Jul. **RANGE:** FL west to TX, north to MN, ON, and NS. **SIMILAR TO** Showy Milkwort (*Polygala grandiflora*, synonym *Asemeia grandiflora*), which occurs mostly in the Coastal Plain in similar habitats; its stems and leaves are hairy, the wings are about as long as they are wide, and the tube is not fringed.

Whorled Milkwort

Polygala verticillata Linnaeus
Milkwort Family (Polygalaceae)

PERENNIAL HERB found in dry woodlands, prairies, and clearings in north and southwest Georgia. **STEMS:** Less than 8 inches tall, slender, branched, 4-angled, smooth. **LEAVES:** Mostly whorled, upper leaves sometimes alternate, linear to elliptic, smooth. **FLOWER CLUSTERS:** Spikelike, up to ½ inch long, with closely spaced flowers opening from the bottom upward. **FLOWERS:** About ⅛ inch wide, white or pink, shaded with green, with 2 oval wings (sepals) and 3 petals united into a tiny tube with a fringed tip. **FLOWERING:** Jun–Sep. **RANGE:** FL west to TX, north to MB and VT. **SIMILAR TO** Loose Milkwort (*Polygala ambigua*, synonym *P. verticillata* variety *ambigua*), which occurs in similar habitats in north Georgia; its upper leaves are whorled, lower leaves alternate; the flower spike is ¾–2 inches long. Seneca Snakeroot (*P. senega*), which occurs in the mountains, has only alternate leaves.

Gay-Wings

Polygaloides paucifolia (Willdenow) J. R. Abbott
SYNONYM: *Polygala paucifolia* Willdenow
Milkwort Family (Polygalaceae)

PERENNIAL HERB found in moist forests in the Blue Ridge mountains and foothills of northeast Georgia. **STEMS:** 2–6 inches tall, smooth, spreading by rhizomes and forming large colonies. **LEAVES:** Up to 1½ inches long, oval, smooth except for hairy margins and midveins, clustered at the top of the stem. **FLOWERS:** ¾–1½ inches wide, dark pink, with 2 spreading wings (sepals) and 3 petals united into a tube with a fringed tip. **FLOWERING:** Apr–May. **RANGE:** GA north to AB and NL. **SIMILAR TO** no other wildflower in northeast Georgia.

Smartweed Family (Polygonaceae)

The Smartweed Family is one of the easiest plant families to recognize. In nearly all of its species, a sheath called an ocrea wraps around the leaf node and the adjacent portion of the stem. In Georgia's species, the ocrea is papery and pale green, tan, reddish, or white. There may or may not be bristles along its upper edge. The flowers are in spikes or compact clusters and have 3–6 colorful tepals that may be marked with minute dots or depressions that are best seen with a 10× lens. The name "smartweed" refers to the leaves of some species that have a biting, peppery taste that makes your tongue smart. The family is sometimes called the Buckwheat Family, after its most economically important species, or the Jointweed Family because the swollen leaf nodes give the stems a jointed appearance. Though inconspicuous, smartweed flowers are an important nectar source for a surprising number of bees, wasps, butterflies, and skippers, and several moth species lay their eggs on the leaves. There are about 37 smartweed species in Georgia.

Ocrea without bristles

Ocrea with bristles

Dog-Tongue, Wild Buckwheat

Eriogonum tomentosum Michaux
Smartweed Family (Polygonaceae)

PERENNIAL HERB found in very dry, sandy soils of river dunes, sandhills, and Turkey Oak scrub in the Coastal Plain. **STEMS:** ½–3 feet tall, much-branched near the top, rusty tan, very hairy. **BASAL LEAVES:** Up to 4¾ inches long and 1½ inches wide, elliptic or oblong, entire, the upper surface dark green and smooth, the lower surface densely covered with white or tan hairs. **STEM LEAVES:** Smaller, in widely spaced whorls of 3 or 4 leaves; there is no ocrea at the base of the leaf stalk. **FLOWER CLUSTERS:** Held at the tips of branches in tight clusters of 10–20 flowers, the stalks and bracts densely covered with tan hairs. **FLOWERS:** Up to ½ inch wide, with 6 white or pinkish tepals. **FLOWERING:** Jul–Oct. **RANGE:** GA, FL, AL, SC, and NC. **SIMILAR TO** no other Georgia wildflower.

Climbing Buckwheat

Fallopia scandens (Linnaeus) Holub
SYNONYM: *Polygonum scandens* Linnaeus
Smartweed Family (Polygonaceae)

PERENNIAL HERBACEOUS VINE found in bottomlands and disturbed areas in north Georgia and a few Coastal Plain counties. **STEMS:** Up to 20 feet long, twining over other plants or sprawling across the ground, often reddish. **LEAVES:** Up to 5½ inches long and 2¾ inches wide, alternate, heart-shaped; the ocrea is shed as the leaf matures. **FLOWERS:** Held in clusters on stalks arising from the leaf axils, each about ¼ inch wide, with 6 white, green, or pink tepals; the 3 outer tepals expand into showy, crinkled wings that enclose the shiny, black fruit. **FLOWERING:** Jul–Oct. **RANGE:** FL west to TX, north to MB and NS. **SIMILAR TO** Black Bindweed (*Fallopia convolvulus*), a Eurasian species that occurs in disturbed areas in north Georgia; its tepals are not winged.

Swamp Smartweed, Mild Waterpepper

Persicaria hydropiperoides (Michaux) Small
SYNONYM: *Polygonum hydropiperoides* Michaux
Smartweed Family (Polygonaceae)

PERENNIAL HERB found in wetlands throughout Georgia. **STEMS:** Up to 3 feet tall, erect or sprawling, usually smooth. **LEAVES:** 2–10 inches long and ¼–1½ inches wide, alternate, lance-shaped, smooth or hairy, without a reddish blotch on the upper surface; the ocrea has ¼–½-inch-long bristles. **FLOWER CLUSTERS:** Spikelike, erect, up to 3 inches long, with closely spaced flowers. **FLOWERS:** About ⅛ inch wide, with 5 white, pink, or greenish-white tepals; the outer surfaces of the tepals lack dots (use a 10× lens). **FLOWERING:** May–Nov. **RANGE:** Most of North America, south into the tropics. **SIMILAR TO** Common Smartweed or Waterpepper (*Persicaria hydropiper*), a European species that occurs in disturbed wetlands in north Georgia; its nodding flower cluster has irregularly spaced flowers, and the tepals are covered with tiny, clear dots or pits.

Pennsylvania Smartweed

Persicaria pensylvanica (Linnaeus) M. Gómez
SYNONYM: *Polygonum pensylvanicum* Linnaeus
Smartweed Family (Polygonaceae)

ANNUAL HERB found in floodplains and wet, disturbed areas throughout Georgia. **STEMS:** Usually 1–2 feet tall, slightly zigzag, hairy or smooth, sometimes with gland-tipped hairs. **LEAVES:** Up to 7 inches long and 2 inches wide, alternate, lance-shaped, hairy on the margins and sometimes the surfaces, often with a reddish blotch on the upper surface; the upper edge of the ocrea has no bristles. **FLOWER CLUSTERS:** Up to 2 inches tall, erect, with a hairy or glandular stalk up to 3 inches long. **FLOWERS:** About ⅛ inch wide, with 5 pink tepals, usually closed; there are no dots or pits on the tepals. **FLOWERING:** Jul–Dec. **RANGE:** Most of North America. **SIMILAR TO** Dense-flowered Smartweed (*Persicaria glabra*, synonym *Polygonum densiflorum*), which occurs in swamps in Georgia's Coastal Plain; its leaves are not blotched.

Dotted Smartweed

Persicaria punctata (Elliott) Small
SYNONYM: *Polygonum punctatum* Elliott
Smartweed Family (Polygonaceae)

ANNUAL OR SHORT-LIVED PERENNIAL HERB found in wetlands throughout Georgia. **STEMS:** Up to 4 feet tall, usually about 1 foot, branched, smooth, gland-dotted. **LEAVES:** 1½–6 inches long and ¼–1 inch wide, alternate, lance-shaped, hairy only on the margins, without a reddish blotch; the upper edge of the ocrea has bristles ¼–½ inch long. **FLOWER CLUSTERS:** Spikelike, 2–6 inches tall, erect, with widely and irregularly spaced flowers. **FLOWERS:** About ⅛ inch long, with 5 white or greenish-white tepals, closed (rarely open), covered with tiny dots or pits. **FLOWERING:** Jul–Nov. **RANGE:** Most of North America, south into the tropics. **SIMILAR TO** other smartweeds, but the combination of dotted tepals, erect spikes with irregularly spaced flowers, and bristly ocrea is distinctive.

Arrow-Leaf Tearthumb

Persicaria sagittata (Linnaeus) Gross ex Nakai
SYNONYM: *Polygonum sagittatum* Linnaeus
Smartweed Family (Polygonaceae)

ANNUAL HERB found in wetlands in north Georgia and a few coastal counties. **STEMS:** Up to 6 feet long, sprawling over other plants, hollow, brittle, 4-angled with sharp, downward-pointing barbs on the angles. **LEAVES:** Up to 3 inches long and 1 inch wide, alternate, arrowhead-shaped, with a barbed midvein on the lower surface; the ocrea has very short bristles along the upper edge. **FLOWER CLUSTERS:** A small, rounded head with 2–10 flowers. **FLOWERS:** About ⅛ inch long with 5 white, pink, or greenish tepals. **FLOWERING:** May–Dec. **RANGE:** FL west to CO, north to MB and NL. **SIMILAR TO** Halberd-leaf Tearthumb (*Persicaria arifolia*), a Georgia Special Concern species found in wetlands in two coastal counties; its leaves are not barbed on the midvein, and its flowers have 4 tepals.

Virginia Jumpseed

Persicaria virginiana (Linnaeus) Gaertner
SYNONYMS: *Tovara virginiana* (Linnaeus) Rafinesque, *Polygonum virginianum* Linnaeus
Smartweed Family (Polygonaceae)

PERENNIAL HERB found in rich, moist forests and floodplain forests in north Georgia and a few Coastal Plain counties. **STEMS:** 1½–4 feet tall, smooth or hairy. **LEAVES:** Up to 7 inches long and 4 inches wide, alternate, oval to lance-shaped, hairy on both surfaces, with a dark blotch on the upper surface; the upper edge of the ocrea has tiny bristles. **FLOWER CLUSTERS:** A slender spike up to 14 inches tall, with widely spaced flowers. **FLOWERS:** About ¼ inch wide when open, usually closed and curved downward, with 4 white or greenish-white tepals. **FLOWERING:** Jul–Oct. **RANGE:** FL west to TX, north to MN, ON, and NH. The flowers and ocrea are **SIMILAR TO** those of other knotweed species, but Virginia Jumpseed has much longer, sparsely flowered spikes and larger leaves.

Southern Jointweed

Polygonella americana (Fischer & C. A. Meyer) Small

SYNONYM: *Polygonum americanum* (Fischer & C. A. Meyer) T. M. Schuster & Reveal

Smartweed Family (Polygonaceae)

PERENNIAL HERB found in sandhills and other dry woodlands in the Fall Line and nearby counties. **STEMS:** Up to 3 feet tall, much-branched, smooth, with a woody base. **LEAVES:** Up to ¾ inch long, alternate, needle- or scale-like, fleshy, smooth, lighter-colored at the tip (but not with a clear margin); the ocrea is green, and its rounded upper edge is whitish or yellowish with no bristles. **FLOWERS:** About ¼ inch wide, with 5 white or pink tepals and 8 stamens that extend beyond the tepals. **FLOWERING:** Jun–Oct. **RANGE:** GA west to NM, north to MO and NC. **SIMILAR TO** October Flower (*Polygonella polygama*), which grows in similar habitats; its leaves are often spoon-shaped and have a narrow, clear margin for about half their length (use a 10× lens).

Sandhill Jointweed

Polygonella fimbriata (Elliott) Horton

SYNONYM: *Polygonum fimbriatum* Elliott

Smartweed Family (Polygonaceae)

ANNUAL HERB found in sandhills, river dunes, and Turkey Oak scrub in the Coastal Plain. **STEMS:** Up to 2 feet tall, rough-hairy on the lower half, much-branched, brittle, with a woody base. **LEAVES:** Up to 2 inches long and less than ⅛ inch wide, alternate, needle-like, fleshy, scattered, not clustered; the upper edge of the ocrea has a few long bristles. **FLOWER CLUSTERS:** Up to 1¼ inches long, spikelike, crowded with flowers. **FLOWERS:** Less than ⅛ inch long, with 5 frilly, white or pink tepals and 8 pink-tipped stamens that extend beyond the tepals; flowers dry to a reddish-brown color. **FLOWERING:** Jul–Nov. **RANGE:** GA, FL, and AL. **SIMILAR TO** Wireweed (*Polygonella gracilis*), which grows in similar habitats; it is taller (up to 5 feet), rarely branched, and has loosely flowered spikes up to 2¾ inches long; its ocrea lack bristles.

Curly Dock

Rumex crispus Linnaeus
Smartweed Family (Polygonaceae)

PERENNIAL HERB found in pastures, roadsides, and other disturbed areas throughout Georgia. **STEMS:** 1–5 feet tall, ribbed, red or green, branched. **LEAVES:** Up to 12 inches long and 2½ inches wide, mostly in a basal rosette, lance-shaped, with wavy, crinkled ("curly") margins. **FLOWER CLUSTERS:** 6–24 inches tall, erect, the flowers in whorls around the stem. **FLOWERS:** About ¼ inch long, reddish- or yellowish-green, with 6 oval tepals. **FRUIT:** Tiny, brown, enclosed by 3 papery, greenish-pink, heart-shaped, ¼-inch wings, each wing with an oval swelling in the center. **FLOWERING:** Mar–May. **RANGE:** Throughout North America; native to Eurasia, now spread worldwide. **SIMILAR TO** Fiddle Dock (*Rumex pulcher*), which also occurs in disturbed areas; its leaves are fiddle-shaped with smooth margins, and the wings on its fruits are deeply fringed.

Wild Dock

Rumex hastatulus Baldwin
Smartweed Family (Polygonaceae)

ANNUAL OR SHORT-LIVED PERENNIAL HERB found in pastures, roadsides, and other disturbed areas throughout Georgia. **STEMS:** Up to 1½ feet tall, angled, often vertically striped red and green. **BASAL LEAVES:** Up to 4 inches long and ¾ inch wide, lance-shaped with 2 spreading lobes at the base, entire. **STEM LEAVES:** Smaller, alternate, mostly without lobes. **FLOWER CLUSTERS:** Up to 6 inches long, slender, erect or drooping, the flowers in whorls; female and male flowers are on separate plants. **FLOWERS:** Tiny; red, white, or green, with 6 oval tepals. **FRUITS:** Tiny, dark brown, enclosed by 3 enlarged tepals that form heart-shaped wings. **FLOWERING:** Mar–May. **RANGE:** FL west to NM, north to KS and NY. **NOTES:** Wild Dock leaves are pleasantly sour-tasting. **SIMILAR TO** Sheep Sorrel (*Rumex acetosella*), a Eurasian species that occurs in similar habitats; its fruits are not winged.

Eastern Spring-Beauty

Claytonia virginica Linnaeus
Purslane (Portulacaceae) or
Montia (Montiaceae) Family

PERENNIAL HERB found in rich, moist forests and bottomlands throughout Georgia (except the Blue Ridge). **STEMS:** 2–15 inches tall. **LEAVES:** 2–8 inches long, including the stalk, which merges into the blade without a clear demarcation; about ¼ inch wide, entire, succulent. **FLOWERS:** ½–1 inch wide, white with pink stripes, with 5 oval petals and 5 stamens tipped with pink anthers. **FLOWERING:** Mar–Apr. **RANGE:** GA west to TX, north to MN and ON. **NOTES:** Both species of Spring-beauty are pollinated by the Spring Beauty Bee (*Andrena erigeniae*), which feeds its larvae only on Spring-beauty's pink pollen. **SIMILAR TO** Carolina Spring-beauty (*Claytonia caroliniana*), which occurs in the Blue Ridge in moist, high-elevation forests; it has distinct leaf stalks and a leaf blade that is widest at the middle, with an elongated diamond shape.

Fame-Flower, Rock-Pink

Phemeranthus teretifolius (Pursh) Rafinesque
SYNONYM: *Talinum teretifolium* Pursh
Purslane (Portulacaceae) or Montia (Montiaceae) Family

PERENNIAL HERB found on granite, sandstone, and mafic outcrops in north Georgia and Altamaha Grit in the Coastal Plain. **STEMS:** 2–20 inches tall, fleshy, reddish, smooth. **LEAVES:** 1–2½ inches long, narrow, alternate, round in cross-section and tapered to a point, fleshy, smooth. **FLOWERS:** Held at the top of a reddish stalk up to 10 inches long, each about ½ inch wide, with 5 oval, dark pink petals and 12–30 stamens. **FLOWERING:** May–Sep. **RANGE:** GA west to AL, north to KY and PA. **NOTES:** Flowers open from 3 p.m. to 7 p.m. EDT and last only one afternoon. **SIMILAR TO** Large-flowered Rock-pink (*Phemeranthus mengesii*), which occurs in similar habitats; its flowers are ¾–1¼ inches across and have more than 40 stamens; they open around 1 p.m.

Common Purslane

Portulaca oleracea Linnaeus
Purslane Family (Portulacaceae)

ANNUAL HERB found in lawns, gardens, and other disturbed areas throughout Georgia. **STEMS:** Up to 2 feet long, sprawling across the ground, fleshy, smooth, reddish-green. **LEAVES:** Up to 1 inch long and ½ inch wide, alternate to nearly opposite, oval to spoon-shaped, fleshy, smooth, reddish-green, mostly clustered at the tips of branches. **FLOWERS:** ¼–½ inch wide with 5 oval, yellow petals and many yellow stamens; flowers open in the morning and wither after a few hours. **FRUIT:** A tiny, oval capsule with a cap that flips open to release seeds; there is no wing or rim on the cap (use 10× lens). **FLOWERING:** May–Nov. **RANGE:** Most of North America and worldwide. **SIMILAR TO** Flatrock Portulaca (*Portulaca coronata*), which occurs on Piedmont granite outcrops and Coastal Plain Altamaha Grit outcrops; the cap on its fruit is encircled by a narrow wing or rim.

Kiss-Me-Quick, Hairy Portulaca

Portulaca pilosa Linnaeus
Purslane Family (Portulacaceae)

ANNUAL HERB found in sandy soil on beaches, and in dunes swales, pinelands, and disturbed areas in the Coastal Plain. **STEMS:** Sprawling or erect, with fleshy branches up to 10 inches long, with long, coarse hairs at the leaf nodes and beneath the flowers. **LEAVES:** Up to 3 inches long and ⅓ inch wide, mostly alternate, linear to lance-shaped, fleshy, smooth. **FLOWERS:** ¼–½ inch wide with 5–7 oval, dark pink or magenta petals and many yellow stamens; petals wither and drop after a few hours. **FLOWERING:** May–Nov. **RANGE:** FL west to AZ, north to KS and NC; Central and South America. **SIMILAR TO** Small's Portulaca (*Portulaca smallii*), which occurs on Piedmont granite outcrops and in adjacent disturbed areas; its flowers are medium pink to nearly white.

Fringed Loosestrife

Lysimachia ciliata Linnaeus
Primrose Family (Primulaceae)

PERENNIAL HERB found in moist, deciduous forests and bottomlands in north Georgia and a few Fall Line counties. **STEMS:** 1–4 feet tall, usually branched, smooth or hairy only at the nodes. **LEAVES:** 1½–6 inches long and ½–2½ inches wide, opposite, oval or lance-shaped with a rounded base and short stalk, smooth except for a fringe of hairs on the margins and the stalk. **FLOWERS:** Up to 1 inch wide, nodding on long stalks, with 5 oval, yellow petals with red bases, sharply pointed tips, and slightly fringed margins; the sepals have reddish veins. **FLOWERING:** Jun–Aug. **RANGE:** Most of North America. **SIMILAR TO** Lance-leaf Loosestrife (*Lysimachia lanceolata*), which occurs in moist to dry forests throughout Georgia; its leaves have tapering bases and no stalks, and its sepals have green veins.

Whorled Loosestrife

Lysimachia quadrifolia Linnaeus
Primrose Family (Primulaceae)

PERENNIAL HERB found in dry to moist deciduous forests in the mountains and upper Piedmont. **STEMS:** 1–3 feet tall, hairy. **LEAVES:** 1–4¾ inches long and ⅓–1¾ inches wide, in whorls, lance-shaped, entire, with fringelike hairs on the margin; the upper surface is covered with tiny, dark pits (use 10× lens). **FLOWERS:** About ½ inch wide, nodding on long stalks that arise from the leaf axils, with 5 yellow, sharply pointed petals; the petals are streaked with black resin canals and usually have red bases and margins. **FLOWERING:** May–Aug. **RANGE:** GA west to OK, north to MN and NB. **SIMILAR TO** Fraser's Loosestrife (*Lysimachia fraseri*), a rare species that occurs in similar habitats; its leaves have a narrow, translucent red border, and its flowers are in branched clusters at the top of the stem.

Eastern Shooting-Star

Primula meadii (Linnaeus) A. R. Mast and Reveal
SYNONYM: *Dodecatheon meadia* Linnaeus
Primrose Family (Primulaceae)

PERENNIAL HERB found in moist, hardwood forests, especially over mafic or limestone bedrock, in northwest Georgia and a few Piedmont and Fall Line counties. **STEMS:** 4–20 inches tall, smooth. **LEAVES:** 3–12 inches long and up to 3 inches wide, in a basal rosette, oblong or spoon-shaped, smooth, entire, with a prominent midvein. **FLOWER CLUSTERS:** At the top of the stem, with 10 or more flowers nodding at the tips of downturned stalks. **FLOWERS:** With 5 white or pink, sharply upturned petals, each marked at the base with yellow and maroon; the stamens form a pointed, yellow beak. **FLOWERING:** Apr–Jun. **RANGE:** FL west to TX, north to MB and CT. **NOTES:** Bumblebees use buzz pollination to shake pollen from the tubular anthers (see "The BUZZZZZZZZ on Pollination," p. 175). **SIMILAR TO** no other wildflower in Georgia.

Water Pimpernel

Samolus parviflorus Rafinesque
SYNONYM: *Samolus valerandi* Linnaeus ssp. *parviflorus* (Rafinesque) Hultén
Primrose Family (Primulaceae)

PERENNIAL HERB found in floodplains, marshes, and on stream banks throughout Georgia. **STEMS:** Up to 16 inches tall, smooth, with elongated flowering branches. **LEAVES:** Up to 3 inches long and 1 inch wide, in a basal cluster and alternating along the branches, spoon-shaped or elliptic, entire, smooth, glossy. **FLOWERS:** Less than ⅛ inch wide, with 5 white, oval or notched petals; the wiry flower stalks are up to ½ inch long and have a tiny bract about halfway. **FLOWERING:** Apr–Oct. **RANGE:** Nearly throughout North America. **SIMILAR TO** no other wetland species in Georgia.

Blue Monks-Hood

Aconitum uncinatum Linnaeus
Buttercup Family (Ranunculaceae)

PERENNIAL HERB found in rich, moist, deciduous forests and seepage areas in Georgia's Blue Ridge mountains and foothills. **STEMS:** Up to 5 feet tall, usually leaning on other plants. **LEAVES:** About 4 inches long and wide, alternate, deeply divided into 3–5 coarsely toothed segments. **FLOWERS:** 1–2 inches long, with 5 showy, blue or purple sepals that enclose 2 small, blue petals; the upper sepal is shaped like a peaked hood, the 2 lateral sepals are rounded, and the 2 lowest sepals are drooping and narrow. **FLOWERING:** Aug–Oct. **RANGE:** GA and AL, north to IN and PA. **NOTES:** All parts of the plant contain aconitine, a deadly toxin that can be absorbed through the skin while picking the leaves. **SIMILAR TO** White Monks-hood (*Aconitum reclinatum*), which has yellow or white flowers and occurs north of Georgia.

Doll's Eyes, White Baneberry

Actaea pachypoda Elliott
Buttercup Family (Ranunculaceae)

PERENNIAL HERB found in rich coves and other moist, upland, hardwood forests in north Georgia. **STEMS:** Up to 2½ feet tall, smooth. **LEAVES:** 2 large, compound leaves per stem, each with 20–70 coarsely toothed leaflets, the leaflets up to 2 inches long, smooth. **FLOWER CLUSTERS:** About 3 inches long, held at the top of a long, leafless stalk, each cluster with 10–25 white flowers. **FLOWERS:** About ½ inch wide, with 4–10 narrow, white petals and many white, petal-like stamens. **FRUITS:** Round, white berries, about ⅓ inch long, with red stalks and black "pupils" (dried remains of the stigma). **FLOWERING:** Apr–May. **FRUITING:** Aug–Oct. **RANGE:** FL west to LA, north to ON and NS. Flowers are **SIMILAR TO** Black Cohosh (*Actaea racemosa*), page 301, and their leaves are nearly identical.

Black Cohosh, Bugbane

Actaea racemosa Linnaeus
SYNONYM: *Cimicifuga racemosa* (Linnaeus) Nuttall
Buttercup Family (Ranunculaceae)

PERENNIAL HERB found in cove forests and other moist, deciduous forests and bottomlands in north Georgia. **STEMS:** 3–8 feet tall, smooth. **LEAVES:** Large, alternate, each with 20–70 sharply pointed, coarsely toothed leaflets 2–4 inches long and wide. **FLOWER CLUSTERS:** Erect spikes up to 1 foot long at the top of tall, leafless stalks; often nodding at the tip, sometimes branched. **FLOWERS:** About ½ inch wide, with many white, petal-like stamens and a single white ovary at the center (there are no petals). **FLOWERING:** May–Aug. **RANGE:** GA west to MS, north to ON and QC. **SIMILAR TO** Mountain Black Cohosh (*Actaea podocarpa*, synonym *Cimicifuga americana*), which has 3–8 white ovaries in the center of the flower; it occurs in rich cove forests at higher elevations.

Round-Lobed Hepatica, Round-Lobed Liverleaf

Anemone americana (A. P. de Candolle) H. Hara
SYNONYMS: *Hepatica americana* (A. P. de Candolle) Ker-Gawler, *Hepatica nobilis* P. Miller var. *obtusa* (Pursh) Steyermark
Buttercup Family (Ranunculaceae)

PERENNIAL HERB found in moist, deciduous forests in north Georgia and a few Fall Line and southwest Coastal Plain counties. **LEAVES:** Up to 4 inches wide and 2¾ inches long, with 3 rounded lobes and hairy stalks, often mottled green and brown on top and purplish beneath, overwintering. **FLOWERS:** About 1 inch wide, held at the tips of very hairy stalks, with 5–12 (usually 6) white, pink, or lavender sepals (there are no petals) and 3 oval, green, hairy bracts beneath the flower. **FLOWERING:** Feb–Apr. **RANGE:** FL west to MS, north to MB and NS. **SIMILAR TO** Sharp-lobed Hepatica (*Anemone acutiloba*), which occurs in north Georgia in deciduous forests over limestone or mafic bedrock; its leaves have pointed lobes.

Wood Anemone

Anemone quinquefolia Linnaeus
Buttercup Family (Ranunculaceae)

PERENNIAL HERB found in rich, moist forests in north Georgia. **STEMS:** 2–12 inches tall, solitary, mostly smooth. **LEAVES:** Divided into 3–5 toothed leaflets, the largest up to 3 inches long and 1¼ inches wide; there is a single basal leaf and 2 or 3 opposite or whorled stem leaves; leaves are often tinged red. **FLOWERS:** About 1 inch wide, with 4–10 (usually 5) white, petal-like sepals, often tinged red on the lower surface, and many white stamens and green pistils in the center. **FLOWERING:** Mar–May. **RANGE:** GA west to AR, north to AB and NS. **NOTES:** Wood Anemone spreads quickly from rhizomes, forming large patches of plants. Plants may take 5 years to reach maturity and flower. **SIMILAR TO** Rue-Anemone (*Thalictrum thalictroides*), page 308.

Tall Thimbleweed

Anemone virginiana Linnaeus
Buttercup Family (Ranunculaceae)

PERENNIAL HERB found in upland forests over limestone or mafic bedrock in north and central Georgia. **STEMS:** Up to 2½ feet tall, hairy. **BASAL LEAVES:** Up to 3½ inches long and wide, divided into 3 lobed and toothed, lance-shaped, hairy segments, with long, purplish stalks. **STEM LEAVES:** In 1 or 2 whorls at the top of the stem, similar to basal leaves but with 4–9 lobes. **FLOWERS:** Solitary at the top of a long stalk, up to 1½ inches wide, with 5 white sepals (no petals) and many stamens surrounding a bristly, green central cone. **FRUITS:** In a spiky, cylinder-shaped head about 1 inch long and ½ inch wide. **FLOWERING:** May–Jul. **RANGE:** GA west to OK, north to ND and NL. **SIMILAR TO** Glade Windflower (*Anemone berlandieri*), a Georgia Special Concern species found on rock outcrops; its stem leaves have narrow, linear segments.

Eastern Columbine

Aquilegia canadensis Linnaeus
Buttercup Family (Ranunculaceae)

PERENNIAL HERB found in rich, rocky forests and glades and on limestone cliffs in north Georgia. **STEMS:** 1–3 feet tall, smooth or hairy. **LEAVES:** Alternate, divided into many rounded and lobed leaflets up to 3 inches long and 2 inches wide, often whitened on the lower surface. **FLOWERS:** About 1½ inches long, nodding at the tip of a long, downcurved stalk; 5 red, oval sepals alternate with 5 red and yellow petals with upward-pointing, spurred tips; many yellow stamens hang well below the mouth of the flower. **FLOWERING:** Mar–May. **RANGE:** FL west to TX, north to MB and ON. **NOTES:** Hummingbirds and long-tongued insects pollinate Eastern Columbine flowers while probing the spurs for nectar. **SIMILAR TO** no other wildflower when in bloom; Meadow-rues, pages 306–307, have similar leaves.

Southern Leatherflower

Clematis crispa Linnaeus
Buttercup Family (Ranunculaceae)

PERENNIAL HERBACEOUS VINE found in marshes, swamps, and floodplains in central and south Georgia. **STEMS:** Up to 10 feet long, angled, smooth or hairy, sprawling over other plants. **LEAVES:** Opposite, thin-textured, smooth, each leaf divided into 4–10 lobed, oval or lance-shaped leaflets up to 4 inches long and 2 inches wide, with a curling tendril between the outermost leaflets. **FLOWERS:** 1–2 inches long, bell-shaped and nodding at the tip of a long, downcurved stalk, with 4 thick, leathery, pink or violet sepals with upwardly curved and crinkled margins; there are no petals. **FLOWERING:** Apr–Aug. **RANGE:** FL west to TX, north to MO and VA. **SIMILAR TO** Northern Leatherflower (*Clematis viorna*), page 304.

Northern Leatherflower

Clematis viorna Linnaeus
Buttercup Family (Ranunculaceae)

PERENNIAL HERBACEOUS VINE found in moist forests, especially over mafic bedrock, in north Georgia. **STEMS:** Up to 13 feet long, smooth or hairy, sprawling over other plants. **LEAVES:** Opposite, thin-textured, each leaf divided into 4–8 lobed or unlobed leaflets up to 5 inches long and 2 inches wide, often with a curling tendril between the outermost leaflets. **FLOWERS:** About 1 inch long, nodding at the tip of a long stalk, the 4 thick, pink or violet sepals forming a bell with pale yellow, upwardly curled tips. **FLOWERING:** May–Sep. **RANGE:** GA west to AR, north to MO and PA. **SIMILAR TO** Net-leaf Leatherflower (*Clematis reticulata*), which occurs in sandhills and dry woodlands in the Coastal Plain; its leaflets are leathery with conspicuous veins. White-leaved Leatherflower (*Clematis glaucophylla*) leaflets are waxy white on the lower surface.

Virgin's Bower

Clematis virginiana Linnaeus
Buttercup Family (Ranunculaceae)

PERENNIAL HERBACEOUS VINE found in thickets, floodplains, and low, moist forests throughout most of north Georgia. **STEMS:** Up to 20 feet long, sprawling over fences and other plants, green or reddish when young, woody and brown when older. **LEAVES:** Opposite, with 3–5 coarsely toothed leaflets, each up to 4 inches long and 2 inches wide, hairy on both or only the lower surfaces. **FLOWERS:** In showy clusters, each flower about ¾ inch wide, with 4 white sepals (no petals) and many showy stamens; fragrant. **FLOWERING:** Jun–Jul. **RANGE:** FL west to TX, north to MB and NS. **SIMILAR TO** Sweet Autumn Clematis (*Clematis terniflora*), an invasive plant from east Asia found in disturbed areas and floodplains throughout Georgia; it flowers Jul–Sep, and its leaflets are oval and have smooth rather than toothed margins.

Blue Larkspur

Delphinium carolinianum Walter ssp. *carolinianum*
Buttercup Family (Ranunculaceae)

PERENNIAL HERB found in rocky woodlands in the Piedmont and in blackland prairies and Altamaha Grit outcrops in the Coastal Plain. **STEMS:** 1–5 feet tall, hairy. **LEAVES:** Up to 4 inches long and wide, alternate, divided into many narrow, hairy segments. **FLOWERS:** About 1 inch wide, blue-violet, with 5 spreading petals, a central "beard," and a 1-inch spur extending up and behind. **FLOWERING:** May–Jul. **RANGE:** GA west to OK, north to IA and SC. **SIMILAR TO** Glade Larkspur (*Delphinium carolinianum* ssp. *calciphilum*), which occurs in limestone cedar glades and prairies in northwest Georgia; its flowers are pale blue or white. Dwarf Larkspur (*Delphinium tricorne*) occurs in moist forests over mafic or limestone bedrock in northwest Georgia; its stems are less than 2 feet tall and smooth near the base; its flowers may be blue, pink, or white, and it blooms Mar–May.

Bristly Buttercup, Swamp Buttercup

Ranunculus carolinianus A. P. de Candolle
SYNONYM: *Ranunculus hispidus* Michaux var. *nitidus* (Chapman) T. Duncan
Buttercup Family (Ranunculaceae)

PERENNIAL HERB found in swamps, floodplains, and marshes in north and southwest Georgia. **STEMS:** Up to 1 foot long, hairy, sprawling across the ground and rooting at the nodes. **LEAVES:** Mostly basal, up to 3 inches long and wide, alternate, deeply divided into 3 lobes or 3 separate leaflets, coarsely toothed, with long stalks. **FLOWERS:** About 1 inch wide, with 5 shiny, oval, yellow petals, 5 down-pointing sepals, and many yellow stamens. **FLOWERING:** Apr–Aug. **RANGE:** FL west to TX, north to MN and NY. **SIMILAR TO** Hairy Buttercup (*Ranunculus hispidus*), which occurs in north Georgia in bottomlands and dry to moist upland forests; its stems are erect, and its sepals spread horizontally; it flowers Mar–Jun.

Hooked Buttercup

Ranunculus recurvatus Poiret
Buttercup Family (Ranunculaceae)

PERENNIAL HERB found in moist forests and bottomlands and on stream banks in north Georgia and a few Coastal Plain counties. **STEMS:** ½–2 feet tall, hairy. **LEAVES:** Up to 5 inches long and wide, divided into 3 toothed and lobed segments, the upper surface dull green with deeply etched veins, the lower surface shiny green, the margins with long hairs. **FLOWERS:** About ½ inch wide, with 5 tiny, glossy yellow petals, 5 green, downcurved sepals, and a central cluster of green, hooked, immature fruits. **FLOWERING:** Apr–Jun. **RANGE:** FL west to TX, north to MN and NL. **SIMILAR TO** Spearwort (*Ranunculus pusillus*), which occurs in wetlands throughout Georgia; its leaves are oval, linear, or lance-shaped with irregularly spaced teeth on the margins. Kidney-leaf Buttercup (*R. abortivus*) occurs statewide in disturbed areas; its basal leaves are kidney-shaped and its flowers are about ¼ inch wide.

Mountain Meadow-Rue, Lady-Rue

Thalictrum clavatum A. P. de Candolle
Buttercup Family (Ranunculaceae)

PERENNIAL HERB found in moist forests, seepages, rocky creek banks, and wet cliffs beside waterfalls in the Blue Ridge and Cumberland Plateau. **STEMS:** 6–24 inches tall, smooth. **LEAVES:** Divided into many oval or rounded leaflets, each ½–1½ inches wide, smooth, blue-green, lobed at the tip. **FLOWERS:** In loose, branched clusters, each flower about ⅓ inch wide, white, with many flat, petal-like stamens (there are no petals), 5 small, white sepals (sometimes absent), and 3–8 erect, stalked, flat, curved immature fruits in the center. **FLOWERING:** May–Jul. **RANGE:** GA north to KY, WV, and VA, only in the southern Appalachian Mountains. **SIMILAR TO** Wind-flower (*Thalictrum thalictroides*), page 308.

Early Meadow-Rue

Thalictrum dioicum Linnaeus
Buttercup Family (Ranunculaceae)

PERENNIAL HERB found in moist forests and seepages, and on stream banks in north Georgia. **STEMS:** Up to 2½ feet tall, smooth, waxy, pale green. **LEAVES:** Divided into many oval or rounded leaflets, each ½–2 inches long and wide, smooth, grayish- or purplish-green, with 3–9 lobes. **FLOWERS:** Female and male flowers are on separate plants in loose, branched clusters up to 1 foot high; neither have petals, but both have 5 tiny, oval, pale green or purplish sepals, though these drop off early from the female flowers; female flowers have 3–8 white, immature fruits with hooked tips; male flowers have many showy, dangling stamens with long, yellow anthers (see photo). **FLOWERING:** Mar–Apr. **RANGE:** GA west to KS, north to ND, QC, and ME. **SIMILAR TO** Skunk Meadow-rue (*Thalictrum revolutum*), below, which blooms in the summer.

Skunk Meadow-Rue

Thalictrum revolutum A. P. de Candolle
Buttercup Family (Ranunculaceae)

PERENNIAL HERB found in forests over mafic, ultramafic, or limestone bedrock throughout Georgia. **STEMS:** 2–5 feet tall, green or reddish-purple, waxy. **LEAVES:** Divided into many rounded, often 3-lobed leaflets up to 1½ inches long; the margins are tightly rolled under, and the lower surface is pale, waxy, and smooth. **FLOWERS:** Female flowers, with many white, hooked ovaries, and male flowers, with many dangling stamens, are on separate plants in loose, long-branched clusters up to 2 feet high. **FLOWERING:** May–Jul. **RANGE:** FL west to NV, north to MB and QC. **NOTES:** Stems and lower leaf surfaces are usually covered with tiny, golden glands (use 10× lens). The leaves smell like skunk when crushed. **SIMILAR TO** Tall Meadow-rue (*Thalictrum pubescens*, synonym *T. polygamum*), which occurs in wet forests and bogs in north Georgia; its leaflets are hairy on the lower surface.

Rue-Anemone, Wind-Flower

Thalictrum thalictroides (Linnaeus) Eames & Boivin

SYNONYM: *Anemonella thalictroides* (Linnaeus) Spach

Buttercup Family (Ranunculaceae)

PERENNIAL HERB found in moist, deciduous forests, often near creeks, in north Georgia and a few southwest Georgia counties. **STEMS:** 4–12 inches tall, smooth, green to purplish. **LEAVES:** Divided into 3, 6, or 9 delicate, grayish- to purplish-green leaflets, each about 1 inch long and wide, rounded, often with 3 shallow lobes at the tip. **FLOWERS:** ½–1 inch across, with 5–10 white to pale pink sepals (there are no petals), many showy, white stamens, and a cluster of tiny, green immature fruits in the center. **FLOWERING:** Mar–May. **RANGE:** FL west to OK, north to MN and ME. **SIMILAR TO** Mountain Meadow-rue (*Thalictrum clavatum*), page 306.

Tassel-Rue, False Bugbane

Trautvetteria caroliniensis (Walter) Vail

Buttercup Family (Ranunculaceae)

PERENNIAL HERB found in moist cove forests, swamp forests, seepages, and stream banks in north Georgia. **STEMS:** 1½–5 feet tall. **BASAL LEAVES:** Up to 16 inches long and wide, kidney-shaped, deeply lobed and toothed, with long stalks. **STEM LEAVES:** Smaller and alternate, without stalks. **FLOWER CLUSTERS:** Flat- or round-topped, with many short branches. **FLOWERS:** ½–1 inch wide, with many narrow, white, petal-like stamens (there are no petals) surrounding a small cluster of green or yellow immature fruits. **FLOWERING:** May–Jul. **RANGE:** FL west to AR, north to IL and PA; several western states. Flowers are **SIMILAR TO** those of Black Cohosh (*Actaea racemosa*), page 301, and Doll's Eyes (*Actaea pachypoda*), page 300.

Yellow-Root

Xanthorhiza simplicissima Marshall
Buttercup Family (Ranunculaceae)

SHRUB forming colonies along stream banks in north and southwest Georgia. **STEMS:** Woody, ½–2½ feet tall, about ¼ inch in diameter, unbranched. **LEAVES:** In a topknot at the top of the stem, each leaf up to 7 inches long, divided into 3–5 sharply toothed and lobed leaflets. **FLOWER CLUSTERS:** Up to 8 inches long, narrow, drooping from the top of the stem, with hairy stalks. **FLOWERS:** About ¼ inch wide, with 5 pointed, maroon sepals (there are no petals), each with a 2-lobed nectar gland at the base (use a 10× lens). **FLOWERING:** Apr–May. **RANGE:** FL west to TX, north to TN, OH, and ME. **NOTES:** The stems and rhizomes contain a bright yellow compound long used as a dye and for treating stomach ailments. **SIMILAR TO** no other wildflower in Georgia.

New Jersey Tea

Ceanothus americanus Linnaeus
Buckthorn Family (Rhamnaceae)

PERENNIAL SHRUB found in woodlands, glades, and prairies throughout Georgia. **STEMS:** 20–40 inches tall, forming a low, rounded shrub. **LEAVES:** Up to 3 inches long and 1½ inches wide, alternate, oval to lance-shaped, hairy, toothed, with 3 main veins. **FLOWER CLUSTERS:** 1–3 inches long, rounded or cylinder-shaped, held at the tips of shoots arising from the leaf axils. **FLOWERS:** About ⅓ inch wide, white, with a tiny tube topped by 5 short sepals alternating with 5 longer petals that are narrowed at the base and hooded at the tip. **FLOWERING:** May–Jun. **RANGE:** FL west to TX, north to MN and QC. **SIMILAR TO** Little-leaf Buckbrush (*Ceanothus microphyllus*), which occurs in sandhills and dry pine flatwoods; it has yellow stems and narrow leaves that are less than ½ inch long and pressed upward against the stem.

Rose Family (Rosaceae)

The Rose Family is large, with about 3,000 species worldwide and more than 100 native species in Georgia. Most of Georgia's Rose Family species are trees and shrubs such as hawthorns, cherries, plums, blackberries, and roses. About 20 are plants meeting the traditional definition of "wildflower"—an herbaceous plant with showy, colorful, or interesting flowers. Nearly all of our Rose Family plants have flowers with 5 white, pink, or yellow petals and a conspicuous tuft of stamens in the center. What sets flowers in the Rose Family apart from similar flowers in other families, such as the buttercups, is the presence of the hypanthium—a cup-shaped structure that surrounds the base of each flower. The petals, sepals, and stamens are perched on the rim of the cup, and nectar-producing tissue lines the inside. In roses, the fleshy hypanthium develops into a conspicuous rose hip, but in most species it is much less obvious. Almost all members of the Rose Family have stipules—small, leaflike appendages at the base of the leaf stalk—and most have compound leaves composed of 3 to many leaflets.

Southern Agrimony

Agrimonia parviflora Aiton
Rose Family (Rosaceae)

PERENNIAL HERB found in bottomland forests and other wet areas in north Georgia and a few Coastal Plain counties. **STEMS:** 2½–6 feet tall, stout, very hairy, and covered with tiny, glistening glands. **LEAVES:** Up to 18 inches long and 6 inches wide, alternate, composed of 7–19 large, lance-shaped, toothed leaflets alternating with similar but much smaller minor leaflets. **FLOWER CLUSTERS:** Spikelike, up to 2½ feet tall, with a few widely scattered flowers. **FLOWERS:** Less than ¼ inch wide, with 5 yellow petals. **FRUIT:** About ¼ inch wide, topped with hooked bristles. **FLOWERING:** Jul–Sep. **RANGE:** FL west to TX, north to SD, ON, and MA. **SIMILAR TO** Woodland Agrimony (*Agrimonia rostellata*), which occurs in moist to wet forests throughout Georgia; it has 5–7 large leaflets, and its flowering stem is smooth or slightly hairy.

Goat's Beard

Aruncus dioicus (Walter) Fernald
Rose Family (Rosaceae)

PERENNIAL HERB found in rich, moist forests and on shady roadsides in north Georgia. **STEMS:** 3–6 feet tall, several arching upward from a crown. **LEAVES:** Up to 20 inches long, alternate, with many toothed (but not lobed), somewhat glossy leaflets; the ultimate leaflet is shaped like the other leaflets. **FLOWER CLUSTERS:** Large, branched clusters with many spikes of closely spaced flowers. **FLOWERS:** Female and male flowers are on separate plants; female flowers have 5 tiny, greenish-white petals and 3 or 4 pistils; male flowers have 5 tiny, white petals and 15–20 stamens. **FLOWERING:** May–Jun. **RANGE:** GA west to OK, north to IN and NY. **SIMILAR TO** False Goat's Beard (*Astilbe biternata*), which occurs in rich coves in the north Georgia mountains; its female flowers have 2 pistils, its male flowers have 10 stamens, and the ultimate leaflet is lobed.

Wild Strawberry

Fragaria virginiana P. Miller
Rose Family (Rosaceae)

PERENNIAL HERB found in grasslands, woodlands, and roadsides in north Georgia and a few inner Coastal Plain counties. **STEMS:** Runners up to 2 feet long, spreading across the ground and forming plantlets at nodes. **LEAVES:** Divided into 3 oval leaflets up to 2 inches long and 1 inch wide, coarsely and sharply toothed. **FLOWERS:** About ¾ inch wide, with 5 white petals, 5 green, hairy, pointed sepals, and 5 narrow, green bracts. **FRUIT:** A red, ½-inch strawberry, very sweet and juicy. **FLOWERING:** Apr–May. **RANGE:** Most of North America. **NOTES:** Commercial strawberries are hybrids between this species and a European strawberry. Leaves are **SIMILAR TO** those of Mock Strawberry (*Potentilla indica,* synonym *Duchesnea indica*), an Asian species found in disturbed areas throughout Georgia; it has yellow flowers and conspicuous, scalloped bracts; its fruit is a red, tasteless "strawberry."

White Avens

Geum canadense Jacquin
Rose Family (Rosaceae)

PERENNIAL HERB found in moist forests and floodplains in north Georgia. **STEMS:** Up to 3 feet tall, hairy. **BASAL LEAVES:** Overwintering leaves held flat on the ground in a rosette up to 6 inches wide, each leaf with several pairs of lobed and toothed leaflets, dark green with pale silvery green markings and a wrinkled look. **STEM LEAVES:** Up to 4 inches long and 3 inches wide, decreasing in size up the stem, alternate, divided into lobed and toothed segments. **FLOWERS:** About ½ inch wide, with 5 white, oval petals alternating with 5 green, pointed sepals; yellow stamens surround a central dome of green, hooked, immature fruits. **FRUITS:** Round, ½-inch heads covered with hooked bristles. **FLOWERING:** May–Jul. **RANGE:** GA west to TX, north to MT and NS. **SIMILAR TO** no other wildflower in Georgia.

Southern Barren Strawberry

Geum donianum (Trattinick) Weakley & Gandhi
SYNONYM: *Waldsteinia fragarioides* (Michaux) Trattinick
Rose Family (Rosaceae)

PERENNIAL HERB found in moist forests and on stream banks in the mountains and upper Piedmont of Georgia. **STEMS:** Underground rhizomes. **LEAVES:** Arising from the rhizome on hairy stalks, with 3 wedge-shaped, lobed, and toothed leaflets up to 3 inches long; leaves turn bronze-maroon in winter. **FLOWERS:** In clusters held at the tips of hairy stalks, each flower about ¼ inch wide, with 5 oval, yellow petals, 5 triangular, green sepals, and many yellow stamens. **FRUIT:** Dry, seed-like. **FLOWERING:** Mar–May. **RANGE:** GA and AL north to TN and VA. **SIMILAR TO** Barren Strawberry (*Waldsteinia lobata*), a Georgia Special Concern species found in floodplains and moist lower slopes in the Piedmont; its leaves are oval to nearly round, with 3–5 rounded lobes.

Indian Physic, Bowman's Root

Gillenia trifoliata (Linnaeus) Moench

SYNONYM: *Porteranthus trifoliatus* (Linnaeus) Britton

Rose Family (Rosaceae)

PERENNIAL HERB found in moist forests and on road banks in the mountains and upper Piedmont. **STEMS:** Up to 2 feet tall, branched. **LEAVES:** Alternate, divided into 3 pointed, finely toothed leaflets up to 4 inches long. **FLOWERS:** In widely branched, open clusters, each flower 1 – 1½ inches wide, with 5 widely spreading, white, narrowly pointed petals of slightly unequal lengths; red stalks and sepals persist after the petals fall. **FLOWERING:** Apr–Jun. **RANGE:** GA west to AR, north to MI and MA. **SIMILAR TO** Midwestern Indian Physic (*Gillenia stipulata*), which occurs in woodlands over mafic or limestone bedrock in northwest Georgia and a few Piedmont counties; its leaves have 3 deeply toothed leaflets and 2 leaflike, toothed stipules at the base of the leaf stalk.

Dwarf Cinquefoil, Five-Fingers

Potentilla canadensis Linnaeus

Rose Family (Rosaceae)

PERENNIAL HERB found in dry woodlands and disturbed areas in north Georgia. **STEMS:** Runners up to 20 inches long, spreading across the ground, covered with long, silky hairs. **LEAVES:** With 5 leaflets arranged like fingers on a hand, each leaflet ½–2½ inches long, the lower half entire and tapering, the upper half sharply toothed. **FLOWERS:** About ¾ inch wide, with 5 yellow, rounded petals, 5 pointed, hairy sepals, and 5 similar bracts; the first flower arises from the leaf axil closest to the bottom of the runner. **FLOWERING:** Mar–May. **RANGE:** GA west to TX, north to WI and NL. **SIMILAR TO** Common Cinquefoil (*Potentilla simplex*), which occurs in north Georgia and the upper Coastal Plain; its leaflets are toothed nearly to the base, and the first flower arises from the axil of the second leaf from the bottom of the runner.

Sulfur Cinquefoil

Potentilla recta Linnaeus
Rose Family (Rosaceae)

PERENNIAL HERB found in fields, roadsides, and other disturbed areas in north Georgia and a few Coastal Plain counties. **STEMS:** Up to 2½ feet tall, several arising from a crown and branching in the upper third, very hairy. **LEAVES:** Alternate, with 5–7 leaflets arranged like fingers on a hand, leaflets up to 4 inches long and 1¼ inches wide, elliptic to spoon-shaped, coarsely and regularly toothed, very hairy. **FLOWERS:** In a flat-topped cluster at the top of the stem, each 1–1½ inches wide, with 5 pale yellow, rounded petals alternating with 5 pointed, hairy sepals. **FLOWERING:** Apr–Jul. **RANGE:** Native to Europe, now spread throughout North America. **SIMILAR TO** Rough Cinquefoil (*Potentilla norvegica*), which occurs in north Georgia in similar habitats; it has 3 leaflets per leaf and bright yellow flowers.

Chickasaw Plum

Prunus angustifolia Marshall
Rose Family (Rosaceae)

SHRUB found in sandy woodlands, pastures, and fencerows throughout Georgia. **STEMS:** Forming thickets of many stems up to 12 feet tall, twigs spine-tipped. **LEAVES:** Up to 2¾ inches long and 1½ inches wide, alternate, elliptic or lance-shaped, often folded up along the midvein, usually smooth, finely toothed, each tooth usually tipped with a tiny red or yellow gland (use 10× lens). **FLOWERS:** About ½ inch wide, with 5 white petals and up to 20 red-tipped stamens, opening before the leaves emerge. **FRUIT:** ½–1 inch wide, round or oval, red or yellow when ripe, with a waxy coating. **FLOWERING:** Mar–Apr. **RANGE:** FL west to NM, north to NE and NJ. **SIMILAR TO** Hog Plum (*Prunus umbellata*), which occurs in dry woodlands throughout Georgia; it is usually a small, single-trunked tree; there are no glands on the leaf teeth, and its fruit is blue-black when ripe.

Carolina Rose

Rosa carolina Linnaeus
Rose Family (Rosaceae)

SHRUB found in dry woodlands, glades, pastures, and roadsides throughout Georgia. **STEMS:** Up to 5 feet tall, often bristly even on new growth, with slender, straight thorns near the leaf nodes. **LEAVES:** Alternate, with 5–9 oval, coarsely toothed leaflets, each ¾–2½ inches long. **FLOWERS:** 1½–3 inches wide, with 5 pink petals, many yellow stamens, and 5 green, pointed sepals. **FRUIT:** An oval rose hip about ½ inch long, red when ripe, with dried flower parts on top. **FLOWERING:** May–Jun. **RANGE:** FL west to TX, north to MN and NS. **SIMILAR TO** Swamp Rose (*Rosa palustris*), which occurs in swamps and bogs and on stream banks throughout Georgia; its new growth lacks bristles, the thorns are curved, and its leaflets are finely toothed.

Common Dewberry

Rubus flagellaris Willdenow
Rose Family (Rosaceae)

LOW, TRAILING SHRUB OR WOODY VINE found in disturbed areas throughout Georgia. **STEMS:** Trailing up to 15 feet long or arching up to 4 feet tall, covered with stout, curved prickles but without sharp bristles. **LEAVES:** Alternate, deciduous, with 3 or 5 oval, toothed leaflets up to 3 inches long and 1 inch wide, mostly smooth, dark green on the upper surface, pale green below. **FLOWERS:** In clusters at the tips of branches, each about 1 inch across, with 5 white, oval petals and many stamens. **FRUIT:** About 1 inch long, purple-black, juicy, sweet-tart. **FLOWERING:** Apr–May; fruiting May–Jul. **RANGE:** FL west to TX, north to ON and QC. **SIMILAR TO** Southern Dewberry (*Rubus trivialis*), which occurs in disturbed areas throughout Georgia; it has bristly stems, leathery, evergreen leaves, and solitary flowers. Swamp Dewberry (*Rubus hispidus*) occurs in wet areas; it has bristly stems and clustered flowers.

Purple-Flowering Raspberry

Rubus odoratus Linnaeus
Rose Family (Rosaceae)

SHRUB found in moist forests and woodland borders, rocky slopes, and stream banks in the north Georgia mountains. **STEMS:** 3–6 feet tall, without thorns, older stems with peeling bark. **LEAVES:** 3½–12 inches long and wide, alternate, maple leaf–shaped, hairy or smooth, with 3–5 pointed, finely toothed lobes. **FLOWERS:** Up to 2½ inches wide, with 5 dark pink to magenta petals, many yellow stamens, and 5 green, pointed sepals. **FRUIT:** A somewhat flattened, fuzzy, dry, and tasteless raspberry. **FLOWERING:** Jun–Aug. **RANGE:** GA and AL, north to MI and NS. **NOTES:** Bristly red hairs cover the young twigs, leaf stalks, main veins on lower leaf surfaces, flower stalks, and sepals. **SIMILAR TO** no other native plant in Georgia.

Highbush Blackberry, Sawtooth Blackberry

Rubus pensilvanicus Poiret
SYNONYM: *Rubus argutus* Link
Rose Family (Rosaceae)

SHRUB found in wetlands and open disturbed areas throughout Georgia. **STEMS:** Up to 10 feet tall, covered with curved prickles; first-year stems (primocanes) erect and unbranched, second-year stems (floricanes) arching with many flowering branches. **LEAVES:** Alternate, with 3 or 5 oval, prickly, toothed leaflets 2½–4 inches long, green on both surfaces. **FLOWERS:** 1½–2 inches across, with 5 white, oval petals (pink in bud) and many white stamens; in clusters at the tips of branchlets. **FRUIT:** About ½ inch long, black, juicy, sweet or tart. **FLOWERING:** Apr–May. **RANGE:** FL west to TX, north to MO and ME. **SIMILAR TO** Sand Blackberry (*Rubus cuneifolius*), which occurs in dry woodlands and disturbed areas; its leaflets are densely white-hairy on the lower surface.

Rough Buttonweed, Poor Joe

Diodia teres Walter
Bedstraw Family (Rubiaceae)

ANNUAL HERB found in dry to moist disturbed areas throughout Georgia. **STEMS:** 4–24 inches tall, hairy. **LEAVES:** Up to 1½ inches long and ⅓ inch wide, opposite, narrowly lance-shaped, with several stiff, reddish bristles at the base. **FLOWERS:** Solitary in the leaf axils, about ¼ inch long, pink, with a short tube, 4 pointed lobes, and 4 green, triangular sepals at the base of the tube. **FRUIT:** About ⅛ inch long, round, 2-seeded, surrounded by bristles, with the 4 sepals attached to the top. **FLOWERING:** Jun–Dec. **RANGE:** FL west to CA, north to WI and MA. **SIMILAR TO** Smooth Buttonweed (*Diodia virginiana*), which occurs in wetlands throughout Georgia; it has sprawling stems and white flowers with only 2 sepals. Two white-flowered species of *Oldenlandia* occur in wetlands; both have many-seeded fruits. *Oldenlandia uniflora* has hairy stems; *O. boscii* has smooth stems.

Cleavers, Bedstraw

Galium aparine Linnaeus
Bedstraw Family (Rubiaceae)

ANNUAL HERB found in forests, meadows, and disturbed areas throughout Georgia. **STEMS:** Up to 5 feet long, sprawling, hollow, 4-angled. **LEAVES:** Up to 3 inches long and ⅓ inch wide, linear, in whorls of 6–8 leaves (usually 8). **FLOWERS:** In clusters held at the tips of short, forking branches, each flower about 1/10 inch wide, with 4 white, pointed petals. **FRUIT:** Less than ¼ inch wide, rounded, 2-lobed, covered with hooked hairs. **FLOWERING:** Apr–Jun. **RANGE:** Throughout North America, nearly worldwide. **NOTES:** The stem angles, leaf margins, and leaf midveins have stiff, sharp, downcurved prickles that help the plant grow up and over other plants. **SIMILAR TO** Sweet-scented Bedstraw (*Galium triflorum*), which occurs in north Georgia in similar habitats; its stems may be slightly rough-textured but are not lined with prickles; it has 6 leaves per whorl.

Hairy Bedstraw

Galium pilosum Aiton
Bedstraw Family (Rubiaceae)

PERENNIAL HERB found in forest and woodland borders and openings throughout Georgia. **STEMS:** Up to 30 inches long, weakly erect, 4-angled, softly hairy. **LEAVES:** In whorls of 4, each up to 1 inch long and ½ inch wide, elliptic, hairy on both surfaces, with 3 distinct veins and gland pits on the lower surface (use 10× lens). **FLOWERS:** Held on stalks at the tips of short, hairy, forking branches, each flower about ⅛ inch wide, with 4 white, green, or maroon petals. **FRUIT:** About ¼ inch wide, rounded, 2-lobed, brown or black, bristly. **FLOWERING:** May–Aug. **RANGE:** FL west to TX, north to KS and NH. **SIMILAR TO** Wide-leaf Bedstraw (*Galium latifolium*), a north Georgia mountain species; it has maroon flowers and smooth fruits, and its leaves are 1½–3 inches long and lack gland pits.

Field Madder

Galium sherardia E. H. L. Krause
SYNONYM: *Sherardia arvensis* Linnaeus
Bedstraw Family (Rubiaceae)

ANNUAL HERB found in lawns, roadsides, and other disturbed areas in north Georgia and a few Coastal Plain counties. **STEMS:** 4–10 inches tall, 4-angled, erect or sprawling, rough-hairy. **LEAVES:** About ½ inch long and less than ¼ inch wide, in whorls of 6 leaves (rarely 4), elliptic to lance-shaped, with a sharp, stiff point and finely toothed margins. **FLOWERS:** In clusters at the tips of branches, each flower about ⅛ inch long, pink or lavender, funnel-shaped with 4 spreading, pointed lobes. **FLOWERING:** Feb–Aug. **RANGE:** Most of North America; native to South America. **SIMILAR TO** Coastal Bedstraw (*Galium hispidulum*), which occurs in dry woodlands in Georgia's coastal counties and a few north Georgia counties; it has 4 leaves per whorl, very rough-hairy stems, and 2 white flowers held at the tips of short branchlets.

Long-Leaf Bluet

Houstonia longifolia Gaertner
SYNONYM: *Hedyotis longifolia* (Gaertner) Hooker
Bedstraw Family (Rubiaceae)

PERENNIAL HERB found in dry woodlands, prairies, and glades in north Georgia and a few Coastal Plain counties. **STEMS:** Up to 12 inches tall, 4-angled, narrowly winged, hairy or smooth. **LEAVES:** Up to 1½ inches long and ¼ inch wide, opposite, smooth, linear to narrowly lance-shaped, tapering at both ends. **FLOWERS:** About ¼ inch across, lavender, pale pink, or white, with a short tube and 4 spreading, pointed lobes; in compact clusters with ¼-inch stalks. **FLOWERING:** Jun–Aug. **RANGE:** FL west to OK, north to AB and ME. **SIMILAR TO** Summer Bluet (*Houstonia purpurea*), which occurs in moist and dry woodlands, around rock outcrops, and in disturbed areas in north Georgia and a few Coastal Plain counties; it has hairy, oval leaves up to 1 inch wide with rounded bases.

Innocence

Houstonia procumbens (Walter ex J. F. Gmelin) Standley
SYNONYM: *Hedyotis procumbens* (Walter ex J. F. Gmelin) Fosberg
Bedstraw Family (Rubiaceae)

PERENNIAL HERB found in moist to dry pinelands and on beach dunes in the Coastal Plain. **STEMS:** Up to 12 inches long, creeping across the ground and rooting at the nodes. **LEAVES:** About ⅓ inch long and wide, opposite, nearly round, somewhat succulent, held flat against the ground, more or less evergreen. **FLOWERS:** About ⅓ inch across, bright white, with a short tube and 4 spreading, oval, pointed lobes; held at the tips of short, erect stalks. **FLOWERING:** Nearly all year. **RANGE:** FL west to LA, north to SC. **SIMILAR TO** Partridge-berry (*Mitchella repens*), page 321.

Tiny Bluet

Houstonia pusilla Schoepf
SYNONYM: *Hedyotis crassifolia* Rafinesque
Bedstraw Family (Rubiaceae)

PERENNIAL HERB found in woodlands, rock outcrops, and dry, disturbed areas throughout Georgia. **STEMS:** 1–4 inches tall, slender. **LEAVES:** About ¼ inch long and ⅛ inch wide, in a basal rosette and opposite on the stem, oval to lance-shaped, with tiny, transparent teeth along the margins (use a 10× lens). **FLOWERS:** About ¼ inch wide, with a short tube and 4 spreading, pointed lobes, blue with a red or purple eye. **FLOWERING:** Feb–Apr. **RANGE:** FL west to TX, north to NE and NY. **SIMILAR TO** Quaker Ladies/Common Bluet (*Houstonia caerulea*), which occurs in similar habitats in north Georgia and a few Coastal Plain counties; its pale blue flowers have a yellow eye.

Appalachian Bluet, Thyme-Leaf Bluet

Houstonia serpyllifolia Michaux
SYNONYM: *Hedyotis michauxii* Fosberg
Bedstraw Family (Rubiaceae)

PERENNIAL HERB found on moist to wet stream banks, seepy rock outcrops, and cliffs around waterfalls in the Blue Ridge and adjacent Piedmont of Georgia. **STEMS:** Creeping along the ground, rooting at nodes, and forming mats with semierect flowering branches. **LEAVES:** Up to ¼ inch long and wide, opposite, oval to nearly round. **FLOWERS:** On long, threadlike stalks, each flower up to ½ inch across, blue, with 4 spreading, pointed lobes and a white or yellow eye. **FLOWERING:** Apr–May. **RANGE:** GA north to KY and PA. **SIMILAR TO** Tiny Bluet (*Houstonia pusilla*), above, and Quaker Ladies (*H. caerulea*).

Partridge-Berry

Mitchella repens Linnaeus
Bedstraw Family (Rubiaceae)

PERENNIAL HERB found in forests with moist or dry, acidic soils throughout Georgia. **STEMS:** Trailing along the ground and rooting at the nodes, sometimes forming mats. **LEAVES:** Up to 1 inch long and wide, opposite, oval to heart-shaped, leathery, evergreen, with a dark green, veiny upper surface and short stalk. **FLOWERS:** Held in pairs in the leaf axils, white, sometimes tinged with pink, hairy inside with a ½-inch tube and 4 pointed, downcurved lobes. **FRUIT:** About ¼ inch wide, round, red. **FLOWERING:** May–Jun. **RANGE:** FL west to TX, north to MN and NS; Guatemala. **NOTES:** The paired flowers and their ovaries are fused at the base, producing a single fruit with 2 calyx scars. **SIMILAR TO** Innocence (*Houstonia procumbens*), page 319.

Rough Mexican Clover

Richardia scabra Linnaeus
Bedstraw Family (Rubiaceae)

ANNUAL HERB found in fields, roadsides, and other disturbed areas in the Coastal Plain and a few lower Piedmont counties. **STEMS:** Up to 28 inches tall, very hairy near the top, less so near the base. **LEAVES:** 1–2¾ inches long and up to ¾ inch wide, opposite, elliptic to lance-shaped, hairy only on the midvein and margins. **FLOWERS:** In domed clusters at the tips of branches, each flower about ¼ inch long, white, funnel-shaped with 6 (rarely 4 or 5) spreading, pointed lobes. **FLOWERING:** Jun–Dec. **RANGE:** FL west to TX, north to IN and NJ; native to South America. **SIMILAR TO** Brazilian Clover (*Richardia brasiliense*), a native of South America found in similar habitats; its leaves are hairy on both surfaces and its flowers are less than ¼ inch long. Neither species is related to the true clovers (*Trifolium* spp.).

Pitcherplants

Pitcherplants, in the genus *Sarracenia*, are nearly endemic to the southeastern United States, with only Purple Pitcherplant making its way into northern wetlands. There are 11 species of pitcherplants in North America, and Georgia's bogs, seepage slopes, savannas, and flatwoods support 8 of them. They have a number of features in common.

- Instead of flat leaves, pitcherplants have hollow pitchers or tubes that lure, capture, and digest insects and the occasional small animal. Nectar and bright colors near the opening of the pitcher lure the animals, and downward-pointing hairs and slick surfaces inside the pitcher prevent their escape. Enzymes dissolved in water at the base of the pitcher digest the insects, releasing nutrients—especially nitrogen and phosphorus, which are typically at low levels in wetland soils.
- A single plant produces a clump of pitchers from an underground stem. The pitcher has a hood that partially covers the opening at the top and prevents dilution of the digestive juices inside the pitcher by rainwater.
- Each plant sends up one stalk with a single flower nodding at the top; flowering occurs before the pitchers are fully functional in order to protect pollinators. The flower consists of 5 drooping petals, 5 stiffly curved sepals, and an umbrella-like disk that has stigmas at its 5 points. The fruit is a round, warty capsule, ½–1 inch wide.
- The Georgia Department of Natural Resources lists all of Georgia's pitcherplant species as Endangered, Threatened, Unusual, or Special Concern. Some are threatened by poaching; all suffer from fire suppression and habitat destruction.

Yellow Pitcherplant

Sarracenia flava Linnaeus
Pitcherplant Family (Sarraceniaceae)

PERENNIAL HERB found in wet pine savannas and flatwoods, seepage slopes, and bogs in the Coastal Plain. **LEAVES:** Pitchers up to 3 feet tall, yellow, often streaked with red, with a red band around the base of the hood; a narrow wing runs the length of the pitcher. **FLOWERS:** Petals yellow, 2–3 inches long. **FLOWERING:** Mar–Apr. **RANGE:** FL west to AL, north to VA. **NOTES:** In late summer, Yellow Pitcherplants growing in shady sites produce flat leaves called phyllodia. Yellow Pitcherplant is listed as Unusual by the state of Georgia. **SIMILAR TO** Green Pitcherplant (*Sarracenia oreophila*).

Hooded Pitcherplant

Sarracenia minor Walter
Pitcherplant Family (Sarraceniaceae)

PERENNIAL HERB found in wet pine savannas and flatwoods, seepage slopes, and bogs in the Coastal Plain. **LEAVES:** Pitchers up to 18 inches tall, green tinged with red, and with white patches on the strongly curved hood and upper pitcher; a wing runs the length of the pitcher. **FLOWERS:** Petals yellow, 1–2 inches long; the flower stalk is shorter than the pitchers. **FLOWERING:** Apr–May. **RANGE:** FL, GA, SC, and NC. **NOTES:** Hooded Pitcherplant is listed as Unusual by the state of Georgia. Hooded Pitcherplants in the Okefenokee Swamp, known as variety *okefenokeensis*, form huge clumps with pitchers up to 4 feet tall. **SIMILAR TO** Sweet Pitcherplant (*Sarracenia rubra*), a Georgia Threatened species that occurs in the Fall Line and Coastal Plain counties; its red pitchers lack white patches, and its flowers have red petals.

Parrot Pitcherplant

Sarracenia psittacina Michaux
Pitcherplant Family (Sarraceniaceae)

PERENNIAL HERB found in wet pine savannas and flatwoods, seepage slopes, and bogs in the Coastal Plain. **LEAVES:** Narrow tubes, 3–12 inches long, patterned in green and red with many white patches; the pitchers lie on the ground or nearly so, with a broad, curved wing on the upper side and a rounded hood ("parrot head") nearly closing the opening. **FLOWERS:** Petals maroon, 1–2 inches long. **FLOWERING:** Mar–Jun. **RANGE:** FL, GA, west to LA. Parrot Pitcherplant is listed as Threatened by the state of Georgia. **SIMILAR TO** Purple Pitcherplant (*Sarracenia purpurea*), a Georgia Endangered species that occurs in Blue Ridge mountain bogs and Coastal Plain seepage slopes and bogs; its red-veined pitchers are inflated and have erect, wavy-edged hoods that do not cover the opening.

Lizard's Tail

Saururus cernuus Linnaeus
Lizard's Tail Family (Saururaceae)

PERENNIAL HERB found in swamps, floodplains, and other wetlands throughout Georgia. **STEMS:** 1–4 feet tall, forming large colonies by the spread of underground stems. **LEAVES:** Up to 6¾ inches long and 4 inches wide, alternate, heart-shaped, smooth. **FLOWER CLUSTERS:** 2–14 inches long, erect and spikelike, nodding at the tip, white, fragrant; flowers open from the bottom to the top of the spike. **FLOWERS:** Consist of 6 white stamens, 3 or 4 white pistils, and a tiny white bract; there are no petals or sepals. **FLOWERING:** May–Jul. **RANGE:** FL west to TX, north to MI, ON, and CT. The shape of the flower cluster is **SIMILAR TO** Devil's Bit (*Chamaelirium luteum*), page 393.

Brook Saxifrage

Boykinia aconitifolia Nuttall
Saxifrage Family (Saxifragaceae)

PERENNIAL HERB found on rocky stream banks, wet cliffs, and seepages in the north Georgia mountains. **STEMS:** 6–32 inches tall, branched at the top, covered with gland-tipped hairs. **BASAL LEAVES:** Usually up to 3 inches long and 5 inches wide, sometimes larger, kidney-shaped, with 3–7 toothed lobes and stalks up to 7 inches long. **STEM LEAVES:** Similar to basal leaves but much smaller. **FLOWERS:** In branched clusters at the top of the stem, flowers up to ½ inch wide, with 5 white petals, 5 green or reddish sepals, and 5 stamens; flower stalks are glandular-hairy. **FLOWERING:** Jun–Jul. **RANGE:** GA and AL north to KY, WV, and VA. Leaves are **SIMILAR TO** those of Tassel-rue (*Trautvetteria caroliniensis*), page 308, which often occurs in wet, rocky habitats but has very different flowers.

Hairy Alumroot

Heuchera villosa Michaux
Saxifrage Family (Saxifragaceae)

PERENNIAL HERB found on shaded boulders, cliff ledges, and rock outcrops in the north Georgia mountains. **LEAVES:** 1–10 inches long and wide, maple leaf–shaped with 5–9 sharply pointed, toothed lobes, smooth or covered with gland-tipped hairs. **FLOWERS:** In a large cluster at the top of a tall, hairy stalk; flowers often nodding, each with 5 tiny, white, coiled petals, a tiny, white or pink, hairy, cup-shaped calyx, and 5 long, orange-tipped stamens. **FLOWERING:** Jun–Sep. **RANGE:** GA west to LA, north to MO and PA. **SIMILAR TO** American Alumroot (*Heuchera americana*), which occurs in north Georgia and the upper Coastal Plain on rock outcrops; its leaves have bluntly rounded lobes, and it flowers Apr–Jun. Cave Alumroot (*H. parviflora*) occurs in north Georgia in cave openings; it has nearly round leaves and flowers Jul–Sep.

Cliff Saxifrage, Michaux's Saxifrage

Hydatica petiolaris (Rafinesque) Small
SYNONYMS: *Saxifraga michauxii* Britton, *Micranthes petiolaris* (Rafinesque) Bush
Saxifrage Family (Saxifragaceae)

PERENNIAL HERB found on wet rock outcrops in the Blue Ridge. **LEAVES:** In a basal rosette, up to 6 inches long and 3 inches wide, spatula-shaped with a tapering base, coarsely and unevenly toothed, hairy, often tinged with red. **FLOWERS:** About ½ inch wide, with 5 white petals and 5 orange-tipped stamens; the 3 largest petals have yellow or green spots, the 2 smallest petals are pure white. **FLOWERING:** Jun–Aug. **RANGE:** GA north to KY and MD. **SIMILAR TO** Brook Lettuce (*Micranthes micranthidifolia*, synonym *Saxifraga micranthidifolia*), which occurs in creek beds and rocky seepages in the Blue Ridge; its leaves are smooth, oblong, and up to 14 inches long. Also see Early Saxifrage (*Micranthes virginiensis*), page 326.

Early Saxifrage

Micranthes virginiensis (Michaux) Small
SYNONYM: *Saxifraga virginiensis* Michaux
Saxifrage Family (Saxifragaceae)

PERENNIAL HERB found on rock outcrops and rocky stream banks and in moist forests in north Georgia. **STEMS:** Up to 20 inches tall, sticky, covered with gland-tipped hairs. **LEAVES:** In a flattened basal rosette, each leaf up to 3 inches long and 1 inch wide, spoon-shaped, coarsely toothed; overwintering leaves turn red. **FLOWERS:** About ⅓ inch wide, with 5 white, oval petals and 10 short, yellow-tipped stamens. **FLOWERING:** Mar–May. **RANGE:** GA west to LA, north to MB and NB. **NOTES:** The sticky stem prevents crawling insects such as ants from reaching the flowers and helps preserve the nectar for bees, which are more effective pollinators. **SIMILAR TO** Cliff Saxifrage (*Hydatica petiolaris*), page 325.

Foamflower

Tiarella cordifolia Linnaeus
Saxifrage Family (Saxifragaceae)

PERENNIAL HERB found in moist forests and on creek banks and seepy rock outcrops in north and southwest Georgia. **LEAVES:** Up to 5 inches long and 4 inches wide, more or less heart-shaped with 3–7 shallow, toothed lobes; hairy on both surfaces, sometimes mottled pale green or brown. **FLOWER CLUSTERS:** Spikelike, 2–16 inches tall, at the top of a leafless stalk (sometimes with 1 leaf), flowers opening from the bottom upward, the unopened buds often pink. **FLOWERS:** About ⅓ inch wide, with 5 white or pinkish, pointed petals. **FLOWERING:** Apr–Jun. **RANGE:** GA west to MS, north to MN and NS. Foamflower leaves are **SIMILAR TO** those of Miterwort (*Mitella diphylla*), which occurs in moist, rocky mountain forests. Miterwort flowers are small, white cups with 5 deeply fringed petals; its flowering stem has a pair of leaves about halfway up.

Snapdragon Family (Scrophulariaceae)

The Snapdragon Family has long been familiar to wildflower lovers for the showy flowers of turtleheads and monkey-flowers and to gardeners and herbalists for Snapdragon and Foxglove, among many others. About 15 years ago, molecular botanists began to take apart this long-recognized family, familiarly called the "scrophs," and moved its species to various other families. This guide reflects some of those changes. For example, the hemiparasitic scrophs, such as the false foxgloves, are now included in the Broomrape Family (Orobanchaceae) of parasitic plants (pp. 268–275). Some of the recent changes to the scroph family, such as the assignment of many species to the Plantain Family (Plantaginaceae) (pp. 280–281), involve both molecular analyses and rules of botanical nomenclature and are more controversial. In this guide, these latter species and their close relatives are grouped as "scrophs," with their new family designation included in the heading.

Blue Water-Hyssop

Bacopa caroliniana (Walter) B. L. Robinson
Snapdragon (Scrophulariaceae) or Plantain (Plantaginaceae) Family

PERENNIAL HERB found in shallow water of ponds, streams, marshes, and ditches in the Coastal Plain and a few Piedmont counties. **STEMS:** Sprawling, forming large mats by rooting at the leaf nodes, sending up erect, hairy stems to 12 inches high. **LEAVES:** Up to ¾ inch long and ⅓ inch wide, opposite, oval, widest at the base, clasping the stem. **FLOWERS:** Up to ½ inch long, blue, tubular with 4 spreading lobes, 1 of them notched. **FLOWERING:** May–Sep. **RANGE:** FL west to TX, north to MD. **NOTES:** Plants have a strong lemon scent when crushed. **SIMILAR TO** Monnier's Water-hyssop (*Bacopa monnieri*), which occurs in similar habitats; its flowers are white or pink with 5 distinct lobes, and the leaves are widest at the rounded tip, tapering to a narrow base; the plants are not fragrant.

White Turtlehead

Chelone glabra Linnaeus
Snapdragon (Scrophulariaceae) or Plantain (Plantaginaceae) Family

PERENNIAL HERB found on stream banks and in seepages, bogs, and floodplains in north Georgia. **STEMS:** 2–5 feet tall, 4-angled, smooth. **LEAVES:** 4–7 inches long and ½–2 inches wide, opposite, lance-shaped, toothed. **FLOWER CLUSTERS:** A compact spike at the top of the stem, flowers opening from the bottom upward. **FLOWERS:** 1–1½ inches long, white with a pink tip, tubular, inflated, 2-lipped, the upper lip hoodlike, the lower lip with 2 or 3 lobes. **FLOWERING:** Aug–Oct. **RANGE:** GA west to AR, north to MB and NL. **NOTES:** In profile the flower resembles a turtle's head. **SIMILAR TO** Cuthbert's Turtlehead (*Chelone cuthbertii*) and Purple Turtlehead (*C. lyonii*), both of which are rare in Georgia; they have dark pink or purple flowers.

Virginia Hedge-Hyssop

Gratiola virginiana Linnaeus
Snapdragon (Scrophulariaceae) or Plantain (Plantaginaceae) Family

PERENNIAL HERB found in swamps, edges of streams, and wet ditches throughout Georgia. **STEMS:** 4–16 inches long, sprawling and rooting at the leaf nodes, sending up erect branches, smooth or slightly glandular-hairy. **LEAVES:** Up to 2 inches long and 1 inch wide, opposite, oval or lance-shaped, unevenly toothed, tapering to the base, not clasping the stem. **FLOWERS:** About ½ inch long, tubular with 4 spreading, notched lobes, pale yellow or pinkish, with purple veins, on thick, ¼-inch-long stalks. **FLOWERING:** Mar–May. **RANGE:** FL west to TX, north to KS, MI, and RI. **SIMILAR TO** Branched Hedge-hyssop (*Gratiola ramosa*), which occurs in wetlands in the Coastal Plain; its leaves clasp the stem, and the flowers are white.

Flatrock Pimpernel

Lindernia monticola Muhlenberg ex Nuttall
Snapdragon (Scrophulariaceae) or False-Pimpernel (Linderniaceae) Family

PERENNIAL HERB found in seepages on granite outcrops in the Piedmont and low, wet areas in the Coastal Plain. **STEMS:** 4–8 inches tall. **LEAVES:** Mostly in a small basal rosette, each leaf up to 1 inch long and ⅓ inch wide, oval; stem leaves few, opposite, similar to basal leaves but smaller. **FLOWERS:** About ⅓ inch long, on long, slender stalks, tubular with a notched upper lip and a 3-lobed lower lip; blue, lavender, or white, with purple streaks and spots. **FLOWERING:** Apr–Sep. **RANGE:** FL west to AL, north to NC. **SIMILAR TO** False Pimpernel (*Lindernia dubia*), which occurs throughout Georgia in swamps and floodplains, pond edges, and other wet, sandy or muddy areas; all of its leaves are on the stem and the flower lacks the dark spots and streaks.

Japanese Mazus

Mazus pumilus (Burmann f.) Steenis
SYNONYM: *Mazus japonicus* (Thunberg) Kuntze
Snapdragon (Scrophulariaceae) or Mazus (Mazaceae) Family

PERENNIAL HERB found in lawns and other moist to wet disturbed areas throughout Georgia. **STEMS:** 2–8 inches long, sprawling, often very short due to mowing. **LEAVES:** Mostly in a basal rosette, up to 1¼ inches long and ½ inch wide, oval or spoon-shaped, coarsely and unevenly toothed, tapering to the base. **FLOWERS:** Less than ½ inch long, with a blue or purple tube, a small upper lip, and a large, white or lavender, 3-lobed lower lip; the lower lip has 2 white, yellow-dotted ridges. **FLOWERING:** Nearly all year. **RANGE:** FL west to TX, north to WI and MA; CA, OR, WA; native to East Asia. **SIMILAR TO** no other short, weedy species in Georgia.

Common Axil-Flower

Mecardonia acuminata (Walter) Small
Snapdragon (Scrophulariaceae) or Plantain (Plantaginaceae) Family

PERENNIAL HERB found in pine savannas, marshes, floodplains, pond margins, and wet ditches throughout Georgia. **STEMS:** 4–20 inches tall, branched, angled, and winged. **LEAVES:** Up to 2 inches long and ½ inch wide, opposite, elliptic to lance-shaped, tapering to a narrow base, toothed on the upper half. **FLOWERS:** Solitary at the tip of ¾-inch stalks arising from the leaf axils, about ⅓ inch long, white with purple veins, tubular with 5 spreading lobes, the 3 lower lobes slightly larger, the 2 upper lobes bearded at the throat; 5 narrow, pointed sepals almost equal to the length of the flower. **FLOWERING:** Jul–Sep. **RANGE:** FL west to TX, north to KS and DE. **SIMILAR TO** Winged Monkey-flower (*Mimulus alatus*), below.

Winged Monkey-Flower

Mimulus alatus Aiton
Snapdragon (Scrophulariaceae) or Lopseed (Phrymaceae) Family

PERENNIAL HERB found in bottomlands, marshes, and wet ditches in north Georgia and a few southwest Georgia counties. **STEMS:** 1½–4 feet tall, branched, angled, winged, smooth. **LEAVES:** Up to 6 inches long and 2 inches wide, opposite, oval to lance-shaped, toothed, with a ½-inch stalk. **FLOWERS:** Solitary at the tip of short stalks arising from the leaf axils, up to 1½ inches long, blue-violet or pink, the upper lip erect and notched, the lower lip 3-lobed with 2 white-and-yellow patches; the calyx is 5-angled and winged. **FLOWERING:** Jul–Nov. **RANGE:** FL west to TX, north to NE, ON, and MA. **SIMILAR TO** Monkey-flower (*Mimulus ringens*), which occurs in wetlands in north Georgia; its stems are angled but not winged, and its leaves have no stalks.

Toad-Flax

Nuttallanthus canadensis (Linnaeus) D. A. Sutton
SYNONYM: *Linaria canadensis* (Linnaeus) Dumortier
Snapdragon (Scrophulariaceae) or Plantain (Plantaginaceae) Family

ANNUAL OR BIENNIAL HERB found on rock outcrops and in disturbed areas throughout Georgia. **STEMS:** 6–28 inches tall, rising from a rosette of sprawling, nonflowering stems. **LEAVES:** Up to 1 inch long and ⅛ inch wide, alternate on the erect stems, opposite on the sprawling stems, linear, smooth. **FLOWERS:** About ½ inch long (including the spur), blue or purple, 2-lipped, the upper lip divided into 2 lobes, the lower lip larger and divided into 3 lobes with 2 white ridges; a narrow, pointed spur extends behind the flower. **FLOWERING:** Mar–Jul. **RANGE:** FL west to TX, north to ND and NS; CA, OR, WA. **SIMILAR TO** two rare species found in dry Coastal Plain habitats: Florida Toad-flax (*N. floridanus*) has a zigzag flower cluster; the flower of Texas Toad-flax (*N. texanus*) is ½–1 inch long.

Southern Beard-Tongue

Penstemon australis Small
Snapdragon (Scrophulariaceae) or Plantain (Plantaginaceae) Family

PERENNIAL HERB found in sandhills, dry woodlands, and roadsides throughout Georgia. **STEMS:** 8–28 inches tall, hairy. **LEAVES:** Up to 4 inches long and 1 inch wide, opposite, lance-shaped, coarsely toothed. **FLOWER CLUSTERS:** At the top of the stem, with opposite branches angled upward and held close to the main stem. **FLOWERS:** Up to 1 inch long, tubular with 2 lips, the upper lip 2-lobed, the lower lip 3-lobed; the outside of the tube is pink, lavender, or purple, the inside is white with purple lines; the "beard-tongue" (a sterile stamen) extends past the opening of the flower. **FLOWERING:** May–Jul. **RANGE:** FL west to MS, north to KY and VA. **SIMILAR TO** White Beard-tongue (*Penstemon pallidus*), which occurs in northwest Georgia in dry, rocky woodlands and clearings; its flowers are white with purple lines.

Appalachian Beard-Tongue

Penstemon canescens (Britton) Britton
Snapdragon (Scrophulariaceae) or Plantain (Plantaginaceae) Family

PERENNIAL HERB found in dry, rocky woodlands and forest edges and on roadsides in the mountains and upper Piedmont of Georgia. **STEMS:** 1–2 feet tall, very hairy. **LEAVES:** Up to 5½ inches long and 1¼ inches wide, opposite, lance-shaped, toothed. **FLOWERS:** Held at the tips of glandular-hairy branches, each about ½–1¼ inches long, tubular with 2 lips, the upper lip 2-lobed, the lower lip 3-lobed; the outside of the tube is pink or lavender, the inside is white with purple lines. **FLOWERING:** May–Jul. **RANGE:** GA, AL, north to IL and PA. **SIMILAR TO** Eastern Beard-tongue (*Penstemon laevigatus*), which occurs throughout Georgia in bottomlands and moist forests. Its flowers are less than ¾ inch long and are held on smooth branches; the stem below the flower cluster is only slightly hairy, with hairs in 2 lines.

Many-Flowered Beard-Tongue

Penstemon multiflorus Chapman ex Bentham
Snapdragon (Scrophulariaceae) or Plantain (Plantaginaceae) Family

PERENNIAL HERB found in sandhills, dry flatwoods, and sandy roadsides in the Coastal Plain. **STEMS:** 2½–5 feet tall, smooth, reddish. **LEAVES:** Up to 6 inches long, opposite, lance-shaped, gland-dotted, with wavy or slightly toothed margins. **FLOWER CLUSTERS:** With opposite branches and up to 30 flowers (most beard-tongues have only 10). **FLOWERS:** ¾ inch long, white (rarely pale pink), unlined, tubular with 5 spreading lobes and a wide opening; the white "beard-tongue" extends to the opening of the flower. **FLOWERING:** Apr–Sep. **RANGE:** GA, FL, and AL. With its many-flowered cluster, gaping, pure white flowers, and smooth, reddish stems, Many-flowered Beard-tongue is only slightly **SIMILAR TO** other beard-tongues.

Figwort, Carpenter's Square

Scrophularia marilandica Linnaeus
Snapdragon Family (Scrophulariaceae)

PERENNIAL HERB found in rich, moist forests, especially over mafic bedrock, in the Blue Ridge and adjacent Piedmont. **STEMS:** Up to 8 feet tall, 4-sided, grooved, smooth or hairy. **LEAVES:** 4–10 inches long and 1–4 inches wide, opposite, each pair at right angles to the next, mostly smooth, oval to lance-shaped, with rounded bases and toothed margins. **FLOWER CLUSTERS:** Held at the top of the stem, up to 1 foot high, with short, glandular-hairy branches. **FLOWERS:** About ⅓ inch long, reddish-green on the outside, maroon on the inside, with a yellowish-green, downcurved lower lip and a longer, 2-lobed, maroon upper lip; 4 blunt stamens and a downcurved style extend beyond the opening of the flower. **FLOWERING:** Jul–Oct. **RANGE:** FL west to TX, north to SD, QC, and ME. **SIMILAR TO** no other wildflower in Georgia.

Shaggy Hedge-Hyssop

Sophronanthe pilosa (Michaux) Small
SYNONYM: *Gratiola pilosa* Michaux
Snapdragon (Scrophulariaceae) or
Plantain (Plantaginaceae) Family

PERENNIAL HERB found in marshes, wet pine savannas, and other wetlands throughout Georgia (except mountains). **STEMS:** 4–28 inches tall, unbranched. **LEAVES:** Up to ¾ inch long and ½ inch wide, opposite, oval, toothed toward the tip, pebbly on the upper surface, without a stalk. **FLOWERS:** About ¼ inch long, with a yellowish tube and 5 white, spreading lobes veined with purple; 5 green, pointed sepals, also about ¼ inch long, surround the tube. **FLOWERING:** Jun–Sep. **RANGE:** FL west to TX, north to OK and NJ. **NOTES:** Long, spreading hairs cover the entire plant. **SIMILAR TO** Pineland Hedge-hyssop (*Sophronanthe hispida*, synonym *Gratiola hispida*), which occurs in sandhills, flatwoods, and dunes in southeast Georgia; its leaves are narrow and pointed, and the flowers are about twice as long as the sepals.

Moth Mullein

Verbascum blattaria Linnaeus
Snapdragon Family (Scrophulariaceae)

BIENNIAL HERB found in disturbed areas in north Georgia. **STEMS:** Emerging from a basal leaf rosette during the second year and growing to 5 feet tall, ribbed, smooth. **BASAL LEAVES:** First-year leaves in a large basal rosette, each leaf up to 8 inches long and 2 inches wide, deeply toothed, smooth. **STEM LEAVES:** Up to 6 inches long and 2 inches wide, smaller up the stem, alternate, toothed, clasping the stem, mostly smooth. **FLOWER CLUSTERS:** Spikelike, up to 2 feet high at the top of the stem, with glandular-hairy branches. **FLOWERS:** Up to 1½ inches wide, yellow or white, often tinged with purple, with 5 spreading petals, 5 purple-hairy stamens with orange anthers, and 5 glandular-hairy sepals. **FLOWERING:** May–Sep. **RANGE:** Native to Eurasia, now widespread in North America. **SIMILAR TO** Twiggy Mullein (*Verbascum virgatum*), an uncommon European species with hairy leaves and glandular-hairy stems.

Woolly Mullein

Verbascum thapsus Linnaeus
Snapdragon Family (Scrophulariaceae)

BIENNIAL HERB found in disturbed areas throughout Georgia. **STEMS:** Emerging from a basal leaf rosette during the second year, growing to 6 feet tall, winged, densely covered with long, white hairs. **LEAVES:** Up to 15 inches long and 4 inches wide, smaller up the stem, oval, alternate, entire or slightly toothed, densely covered with long, white hairs. **FLOWER CLUSTERS:** Spikelike, up to 2 feet high at the top of the stem, tightly packed with flowers. **FLOWERS:** About ¾ inch wide, yellow, with 5 spreading petals and 5 hairy stamens. **FLOWERING:** May–Sep. **RANGE:** Native to Europe, now widespread in North America. **SIMILAR TO** no other wildflower in Georgia.

Corn Speedwell

Veronica arvensis Linnaeus
Snapdragon (Scrophulariaceae) or Plantain (Plantaginaceae) Family

ANNUAL HERB found in lawns and other disturbed areas throughout Georgia. **STEMS:** Up to 8 inches long, sprawling or erect, hairy. **LEAVES:** Up to ½ inch long and ⅓ inch wide, oval to nearly round, mostly opposite, hairy, with rounded teeth; just beneath the flowers, the leaves are shorter, narrower, lance-shaped, and alternate. **FLOWERS:** Less than ¼ inch wide, 4-lobed, blue with darker lines and a pale "eye." **FLOWERING:** Mar–Sep. **RANGE:** Native to Eurasia, now widespread in North America. **SIMILAR TO** three other Eurasian species found in disturbed areas throughout Georgia: Bird's-eye Speedwell (*Veronica persica*) flowers are ¼–½ inch wide, and all of its leaves are oval; Ivy-leaf Speedwell (*Veronica hederifolia*) leaves are wider than they are long and have 3–5 lobes; Purslane Speedwell (*Veronica peregrina*) has white flowers, and its leaves are much longer than wide.

Culver's Root

Veronicastrum virginicum (Linnaeus) Farwell
Snapdragon (Scrophulariaceae) or Plantain (Plantaginaceae) Family

PERENNIAL HERB found in prairies, wet meadows, and rich woods, and on stream banks in north Georgia. **STEMS:** Up to 5 feet tall, hairy or smooth. **LEAVES:** Up to 6 inches long and 1¼ inches wide, whorled, lance-shaped, toothed, smooth on the upper surface, very hairy on the lower surface. **FLOWER CLUSTERS:** Several erect spikes, up to 8 inches tall, form a candelabra at the top of the stem, with the flowers opening from the bottom upward. **FLOWERS:** Up to ¼ inch long, white (rarely pinkish), tubular, with 4 tiny lobes and 2 protruding stamens with orange tips. **FLOWERING:** Jul–Sep. **RANGE:** FL west to TX, north to MB and NS. **SIMILAR TO** no other wildflower in Georgia.

Jimson Weed

Datura stramonium Linnaeus
Nightshade Family (Solanaceae)

ANNUAL HERB found in pastures, roadsides, and other disturbed areas throughout Georgia. **STEMS:** 2–5 feet tall, stout, purplish, with forking branches. **LEAVES:** 3–8 inches long and 1–5 inches wide, alternate, lance-shaped, coarsely and irregularly toothed, smooth. **FLOWERS:** Held in the forks, up to 4 inches long, white or light purple, funnel-shaped, partially closed and pleated lengthwise, with 5 pointed lobes, fragrant; the tubular calyx is angled and winged. **FRUIT:** A spiny, oval capsule up to 3 inches long, splitting into 4 sections with many black seeds. **FLOWERING:** Jul–Sep. **RANGE:** Native to western North America, now spread worldwide. **NOTES:** The flowers, which open at night and close in the morning, are pollinated by sphinx moths. All parts of Jimson Weed contain fatally toxic compounds. **SIMILAR TO** no other wildflower in Georgia.

Virginia Ground Cherry

Physalis virginiana Nuttall
Nightshade Family (Solanaceae)

PERENNIAL HERB found in dry woodlands and disturbed areas in north Georgia. **STEMS:** Usually 4–16 inches tall, sometimes taller, hairy, branched. **LEAVES:** Up to 4 inches long and 1½ inches wide, alternate, oval to lance-shaped with a tapering base, irregularly toothed, hairy. **FLOWERS:** About ¾ inch wide, with 5 dark spots at the center. **FRUIT:** A small, round berry completely enclosed in a papery husk; when ripe, the berry is yellow and the husk yellowish-green. **FLOWERING:** Apr–May. **RANGE:** FL west to NM, north to MB and QC. **SIMILAR TO** other ground cherries; see page 337.

Ground Cherries

Ground cherries, in the genus *Physalis*, are sometimes called Japanese Lanterns or Husk-tomatoes, both names referring to the papery husk that develops from the calyx and encloses the fruit. The leaves and unripe berries are poisonous, but the ripe fruits (usually red or orange) were eaten by Native Americans. They are all herbs, some annual, some perennial. All have nodding, yellow, bell-shaped flowers, usually with 5 dark spots at the center, and yellow or blue anthers tipping the stamens. Eight species occur in Georgia; the four described below and Virginia Ground Cherry (p. 336) are the most common. All occur in dry woodlands and disturbed areas; Dune Ground Cherry also occurs near beaches.

SMOOTH GROUND CHERRY (*PHYSALIS ANGULATA*), an annual, has smooth stems and leaves that are oval or lance-shaped, smooth, and usually toothed. The flowers are solid yellow, without dark spots, and the stamens are tipped with blue anthers. The inflated husk is round or 10-angled. It occurs throughout Georgia.

CLAMMY GROUND CHERRY (*PHYSALIS HETEROPHYLLA*) is a rhizomatous perennial covered with sticky hairs. Its leaves are oval to lance-shaped and coarsely toothed. The calyx is ¼–½ inch long with triangular lobes. Its anthers are yellow (rarely bluish), with purple stalks. Its berry is red-orange when mature and enclosed in a 10-angled, heart-shaped husk. It occurs mostly in north Georgia.

DOWNY GROUND CHERRY (*PHYSALIS PUBESCENS*) is a taprooted annual that is usually covered with sticky hairs. Its leaves are oval and usually toothed (rarely entire). Its calyx is less than ¼ inch long and has narrow, pointed lobes. Its stamens are tipped with blue anthers. Its berry is green when mature and enclosed in a strongly 5-angled, inflated husk. It occurs infrequently throughout Georgia.

DUNE GROUND CHERRY (*PHYSALIS WALTERI*) is a rhizomatous perennial with hairy stems and leaves, the leaves oval and entire. The calyx is ⅓–½ inch long. The flowers may or may not have dark spots at the center. The anthers are yellow, and the mature fruits are orange. It occurs only in the Coastal Plain, often on beach dunes.

American Black Nightshade

Solanum americanum P. Miller
Nightshade Family (Solanaceae)

ANNUAL HERB found in woodland margins and disturbed areas throughout Georgia. **STEMS:** Up to 3 feet tall, green or purplish, mostly smooth. **LEAVES:** Up to 7 inches long and 3½ inches wide, alternate, oval, mostly smooth, margins wavy or with a few coarse teeth, stalks sometimes winged. **FLOWERS:** About ¼ inch wide, white, with 5 pointed, spreading lobes. **FRUIT:** A round, shiny, black, ¼-inch berry; the calyx persists on the fruit, its lobes bent strongly backward from the berry. **FLOWERING:** May–Nov. **RANGE:** Most of southern and western North America. **NOTES:** All parts of the plant are deadly poisonous to humans. **SIMILAR TO** Deadly Nightshade (*Solanum ptychanthum*), which occurs in disturbed areas in a few north Georgia counties; its berry may be dull or shiny; the calyx lobes persist on the fruit but do not bend backward.

Carolina Horse-Nettle

Solanum carolinense Linnaeus
Nightshade Family (Solanaceae)

PERENNIAL HERB found in dry woodlands, maritime forests, pastures, and other disturbed areas throughout Georgia. **STEMS:** Up to 3 feet tall, hairy, spiny, greenish-purple. **LEAVES:** Up to 8 inches long and 2¾ inches wide, alternate, oval or lance-shaped, deeply toothed or lobed, hairy, with stout spines on the veins. **FLOWERS:** About ¾ inch wide, white or purple, with 5 pointed lobes, 5 long, yellow anthers, and a smooth, spineless calyx. **FRUIT:** A round, green, ½-inch berry, ripening to orange. **FLOWERING:** May–Oct. **RANGE:** Most of North America. **NOTES:** All parts of the plant, especially the berries, are deadly poisonous to humans. **SIMILAR TO** Sticky Nightshade (*Solanum sisymbriifolium*), a South American plant that occurs in disturbed areas in the Coastal Plain. Its leaves are deeply lobed and spiny, and the calyx is spiny and nearly encloses the red berry.

Piriqueta, Pitted Stripeseed

Piriqueta caroliniana (Walter) Urban
SYNONYM: *Piriqueta cistoides* (Linnaeus) Grisebach ssp. *caroliniana* (Walter) Urban
Turnera Family (Turneraceae)

PERENNIAL HERB found in dry, sandy woodlands and disturbed areas in the Coastal Plain. **STEMS:** 6–20 inches tall, densely hairy, often in colonies formed by root sprouts. **LEAVES:** Up to 2 inches long and ¾ inch wide, alternate, oblong to lance-shaped, hairy on both surfaces, with shallowly toothed or scalloped margins and short or no stalks; there is no basal rosette. **FLOWERS:** Up to 1½ inches wide, with 5 bright yellow petals, 5 stamens, and 5 green sepals; they open only on sunny days and drop their petals by the end of the day. **FLOWERING:** May–Sep. **RANGE:** GA, FL, AL, and SC. **SIMILAR TO** Carolina Sun-rose (*Crocanthemum carolinianum*), page 156.

False Nettle

Boehmeria cylindrica (Linnaeus) Swartz
Nettle Family (Urticaceae)

PERENNIAL HERB found in floodplains, marshes, and other wetlands throughout Georgia. **STEMS:** Up to 5 feet tall. **LEAVES:** Up to 6 inches long and 2¾ inches wide, opposite (sometimes alternate near the top of the stem), broadly oval or lance-shaped, sharply toothed, with 3 prominent veins and no stinging hairs; the stalk is about the same length as the blade. **FLOWER CLUSTERS:** Erect spikes arising from the leaf axils, the flowers in small, dense, round balls. **FLOWERS:** Tiny, with greenish or reddish sepals and no petals; female and male flowers may be in the same or different clusters. **FLOWERING:** Jun–Aug. **RANGE:** FL west to AZ, north to UT, MN, and NB. Plants have no stinging hairs, otherwise **SIMILAR TO** Wood Nettle (*Laportea canadensis*), page 340.

Stinging Nettles (Urticaceae)

The leaves and stems of stinging nettles are covered with hollow, brittle hairs with pointed tips. The tip easily breaks off when touched, and the remaining hair acts as a hypodermic that injects several irritating chemicals—histamine, acetylcholine, and formic acid, among others—into the skin, causing a painful rash. Browsing and grazing animals quickly learn to give stinging nettles a wide berth. As a result, nettles provide shelter for butterfly and moth larvae and other insects that are able to move among the hairs without breaking off the tips. Paradoxically, nettles have been used for centuries to *treat* painful conditions such as arthritis, gout, urinary tract infections, enlarged prostate, and eczema. They contain compounds that reduce inflammation and interfere with the body's pain signals. This is small comfort, however, when your ankles are on fire after a walk through the woods. Home remedies may help alleviate the pain. Because at least some of the pain is caused by acidic compounds, apply basic substances such as baking soda, vinegar, hydrogen peroxide, or soap to the irritated area. Or try removing the stinging hairs with sticky tape. Better yet, add stinging nettles to your life list of easily recognized wildflowers and walk around—not through—the nettle patch.

Wood Nettle

Laportea canadensis (Linnaeus) Weddell
Nettle Family (Urticaceae)

PERENNIAL HERB found in moist, rich forests in the Piedmont and mountains. **STEMS:** 1–5 feet tall, covered with stinging hairs. **LEAVES:** 2½–8 inches long and 1½–5 inches wide, alternate, oval to lance-shaped, with stinging hairs, toothed margins, and long stalks. **FLOWER CLUSTERS:** About 4 inches long with spreading branches, arising from the leaf axils; female-flowered clusters are near the top of the plant, male-flowered clusters are lower. **FLOWERS:** Female flowers have 4 tiny, greenish-white sepals, male flowers have 5; neither have petals. **FLOWERING:** Late Jun–Aug. **RANGE:** FL west to LA, north to MB and NS. **NOTES:** Tips of the brittle hairs break off when even lightly brushed, releasing irritating compounds. **SIMILAR TO** Dwarf Stinging Nettle (*Urtica chamaedryoides*), which occurs infrequently in the Coastal Plain in floodplains; it has opposite leaves and compact, round flower clusters.

Clearweed

Pilea pumila (Linnaeus) A. Gray
Nettle Family (Urticaceae)

ANNUAL HERB found in swamps and bottomlands in north Georgia and a few Coastal Plain counties. **STEMS:** 3–28 inches tall, smooth, translucent, pale green. **LEAVES:** 1–5 inches long and ½–3½ inches wide, opposite, oval, smooth, shiny, lower surface pale, with coarsely toothed margins and 3 prominent veins; the stalk is about half the length of the blade. **FLOWER CLUSTERS:** About 1 inch long, arising from the axils of the upper leaves. **FLOWERS:** Tiny, green, female and male flowers in the same cluster, the female flowers with 3 sepals, the male flowers with 4; neither with petals. **FLOWERING:** Aug–Sep. **RANGE:** FL west to TX, north to ND and QC. **NOTES:** Clearweed has no stinging hairs. **SIMILAR TO** Florida Pellitory (*Parietaria floridana*), which occurs on beaches, shell mounds, and disturbed areas near the coast; it has much smaller, alternate leaves.

Corn Salad

Valerianella radiata (Linnaeus) Dufresne
Valerian Family (Valerianaceae)

ANNUAL HERB found in moist forests, bottomlands, and disturbed areas throughout Georgia. **STEMS:** 4–18 inches tall, forking in the upper half, 4-angled with hairs on the angles. **LEAVES:** Up to 3 inches long and ¾ inch wide, opposite, oblong or spoon-shaped with rounded tips, usually clasping the stem, often toothed near the base of the leaf. **FLOWER CLUSTERS:** Less than 1 inch across, flat-topped, surrounded by lance-shaped bracts with hairy margins. **FLOWERS:** Tiny, white, with 5 lobes and 3 long, white stamens. **FLOWERING:** Apr–May. **RANGE:** FL west to TX, north to KS and CT. **NOTES:** Plants germinate in the fall and overwinter as a leaf rosette that sends up a stem in the spring. **SIMILAR TO** European Corn Salad (*Valerianella locusta*), a garden escape found in a few counties throughout the state; its flowers are usually light blue.

American Beauty-Berry

Callicarpa americana Linnaeus
Verbena (Verbenaceae) or Lamiaceae Family

PERENNIAL SHRUB found in woodlands, upper floodplains, maritime forests, and disturbed areas throughout Georgia. **STEMS:** Up to 8 feet tall, 4-angled, covered with white, star-shaped hairs (use 10× lens). **LEAVES:** 2¾–6 inches long and 1¼–4 inches wide, opposite, broadly oval, tapering at both ends, toothed, the lower surface covered with white, star-shaped hairs. **FLOWERS:** In rounded clusters arising from the leaf axils, each flower about ¼ inch long, lavender or pink, with 5 lobes. **FRUIT:** A small, round, magenta berry in compact, rounded, showy clusters. **FLOWERING:** Jun–Jul. **FRUITING:** Aug–Nov. **RANGE:** FL west to TX, north to OK and MD. **NOTES:** Stems and leaves have a rank odor. **SIMILAR TO** no other flowering shrub in Georgia.

Rose Verbena, Rose Vervain

Glandularia canadensis (Linnaeus) Nuttall
SYNONYM: *Verbena canadensis* Linnaeus
Verbena Family (Verbenaceae)

PERENNIAL HERB found in sandhills and dry, sandy, disturbed areas in the Coastal Plain and a few Piedmont counties. **STEMS:** Up to 2 feet tall, several arising from a woody crown, erect or sprawling, 4-angled, covered with long, spreading hairs. **LEAVES:** Up to 3 inches long and 1½ inches wide, opposite, hairy, divided into many lobed and toothed segments. **FLOWERS:** About ½ inch long and wide, pink to purple, with a hairy tube and 5 spreading, notched lobes, in rounded clusters at the tip of the stems. **FLOWERING:** Mar–May. **RANGE:** FL west to NM, north to MN and CT. **SIMILAR TO** Rough Verbena (*Verbena rigida*), a South American species found in disturbed areas throughout Georgia; its leaves are lance-shaped, coarsely toothed, and clasp the stem.

Moss Verbena, Moss Vervain

Glandularia pulchella (Sweet) Troncoso
SYNONYM: *Verbena tenuisecta* Briquet
Verbena Family (Verbenaceae)

ANNUAL OR PERENNIAL HERB found on roadsides and other disturbed areas in the Coastal Plain and a few Piedmont counties. **STEMS:** Up to 12 inches long, several stems arising from a woody crown, sprawling, rooting at the nodes and forming mats, slightly hairy. **LEAVES:** About 1 inch long and wide, opposite, divided into many very narrow segments. **FLOWER CLUSTERS:** A densely flowered spike up to 4 inches long, held at the tips of the stems. **FLOWERS:** Up to 1 inch long and ½ inch wide; purple, pink, or white, tubular, with 5 spreading, notched lobes and a white "eye." **FLOWERING:** Mar–Nov. **RANGE:** Native to South America, now widespread across the southern and western United States. **SIMILAR TO** Rose Verbena (*Glandularia canadensis*), page 342.

Frog-Fruit

Phyla nodiflora (Linnaeus) Greene
SYNONYM: *Lippia nodiflora* (Linnaeus) Michaux
Verbena Family (Verbenaceae)

PERENNIAL HERB found on back dunes, sandy lawns, roadsides, and other disturbed areas in the Coastal Plain and a few Piedmont counties. **STEMS:** Spreading, rooting at the nodes, flowering branches up to 4 inches high. **LEAVES:** Up to 2½ inches long and 1 inch wide, opposite, oval to wedge-shaped, widest above the middle, with 5 teeth per side. **FLOWER CLUSTERS:** A round or cylindrical head about ⅓ inch long (longer in fruit), held on long stalks arising from the leaf axils. **FLOWERS:** Tiny, 2-lipped, pink, purple, or white, surrounded by dark bracts. **FLOWERING:** Mar–Nov. **RANGE:** FL west to CA, north to OR and PA. **SIMILAR TO** Marsh Frog-fruit (*Phyla lanceolata*), which is rare in Georgia; it occurs in marshes, and its leaves are lance-shaped with 7 or more teeth per side.

Brazilian Vervain

Verbena brasiliensis da Conceição Vellozo
Verbena Family (Verbenaceae)

ANNUAL OR PERENNIAL HERB found in disturbed areas throughout Georgia. **STEMS:** Up to 8 feet tall, sharply 4-angled and grooved, roughly hairy on the angles. **LEAVES:** Up to 4 inches long and 1 inch wide, opposite, elliptic, tapering to but not clasping the stem, coarsely toothed on the upper half, with bristly veins on the lower surface. **FLOWER CLUSTERS:** Dense, cylinder-shaped spikes up to 2 inches high. **FLOWERS:** Tiny, pink or purple, with 5 spreading lobes, only a few flowers in each spike open at one time. **FLOWERING:** Apr–Nov. **RANGE:** FL west to TX, north to MO and VA; CA, OR; native to South America. **SIMILAR TO** Cluster-top Vervain (*Verbena bonariensis*), a South American species that occurs in disturbed areas throughout Georgia, mostly in the Coastal Plain; its flowers are in flat- or round-topped clusters, and the leaves clasp the stem.

Carolina Vervain

Verbena carnea Medikus
SYNONYM: *Stylodon carneus* (Medikus) Moldenke
Verbena Family (Verbenaceae)

PERENNIAL HERB found in sandhills, woodlands, and savannas in the Coastal Plain. **STEMS:** Up to 3 feet tall, 4-angled. **LEAVES:** Up to 3½ inches long and 1¼ inches wide, smaller up the stem, opposite, oval, tapered to the base, rounded at the tip, toothed. **FLOWER CLUSTERS:** 1–5 elongated spikes up to 16 inches high at the top of the stem, with widely separated flowers and fruits. **FLOWERS:** About ¼ inch wide, with a short tube and 5 spreading lobes, pink, lavender, or white. **FLOWERING:** Apr–Jul. **RANGE:** FL west to TX, north to VA. **SIMILAR TO** Hoary Vervain (*Verbena stricta*), which occurs infrequently in north Georgia in glades, pastures, and on roadsides; its flowers are blue to lavender.

Texas Vervain

Verbena halei Small
SYNONYM: *Verbena officinalis* Linnaeus ssp. *halei* (Small) S. C. Barber
Verbena Family (Verbenaceae)

PERENNIAL HERB found in dry woodlands and pastures and on roadsides in the Coastal Plain. **STEMS:** 1–2½ feet tall, with erect branches. **LEAVES:** Up to 3 inches long and 1½ inches wide, opposite, spoon-shaped, deeply to shallowly toothed and divided, tapering to the base, thick-textured. **FLOWER CLUSTERS:** Several elongated spikes 2½–16 inches tall at the top of the stem. **FLOWERS:** About ¼ inch wide, pale blue to lavender, with a ¼-inch tube and 5 spreading lobes. **FLOWERING:** Apr–Jun. **RANGE:** FL west to AZ, north to NC. **SIMILAR TO** Narrow-leaf Vervain (*Verbena simplex*), which occurs in glades and woodlands over mafic and limestone bedrock, mostly in northwest Georgia. It has only 1 flower spike per plant; its leaves are toothed, not deeply divided, and much longer than wide.

White Vervain

Verbena urticifolia Linnaeus
Verbena Family (Verbenaceae)

PERENNIAL HERB found in marshes, moist woodlands, pastures, and other disturbed areas in north Georgia and a few Coastal Plain counties. **STEMS:** Up to 8 feet tall, 4-angled, branching in the upper half, covered with long, spreading hairs. **LEAVES:** Up to 6 inches long and 2¾ inches wide, opposite, oval to lance-shaped, toothed, rough-hairy, with long stalks. **FLOWER CLUSTERS:** Slender spikes 1–10 inches long, arising from the leaf axils near the top of the stem; only a few flowers are open in a spike at one time. **FLOWERS:** About ⅛ inch wide, white, with 5 spreading lobes. **FLOWERING:** May–Nov. **RANGE:** FL west to TX, north to SK and NB. **SIMILAR TO** no other wildflower in Georgia.

Violets (Violaceae)

Violets come in two types. Some have leafy aboveground stems with flowers on stalks arising from the leaf axils. Other violets have underground stems called rhizomes. Their leaves and flower stalks arise from the tip of the rhizome and seem to emerge directly from the soil. Violets with aboveground stems are called caulescent—meaning stemmed. Violets with only underground stems are called acaulescent—without stems. In the latter group, what appears to be the stem is actually a long flower stalk.

Violets have two kinds of flowers: large, open, colorful flowers that bloom in the spring, and small, closed, self-pollinating flowers that bloom later in the summer near the base of the plant. The colorful flowers are pollinated by butterflies, bees, and flies. They also have the ability to self-pollinate. Thus, if cross-pollination fails to occur, violet flowers have a two-part backup system to ensure the production of seeds. Violet fruits dry and split explosively into three sections, flinging the seeds several feet away. The seeds bear fatty handles called elaiosomes that appeal to ants (see "The Skinny on Fat Bodies—Elaiosomes," p. 44).

The showy kind of violet flower is bilaterally symmetrical with 5 petals, the largest of which curves downward and also extends behind the flower into a spur where nectar is produced. Native violets typically range in color from violet to blue to yellow and cream-colored, and are sometimes a mix of two or more of these. Dark lines on the lower petals guide pollinators toward the spur and its nectar. The length of the spur can be important to identification, as are the size and shape of the stipules—small, leafy structures at the base of the leaf stalks. There are 28 species of violets in the genus *Viola* in Georgia, plus Green Violet in the genus *Hybanthus*.

Green Violet

Hybanthus concolor (T. F. Forster) Sprengel
Violet Family (Violaceae)

PERENNIAL HERB found in rich, moist forests and bottomlands in north Georgia and a few Coastal Plain counties. **STEMS:** 1–3 feet tall, hairy or smooth. **LEAVES:** 3½–7 inches long and 1¼–2½ inches wide, alternate, elliptic, hairy or smooth, veiny. **FLOWERS:** Held in the axils of midstem leaves, drooping on ½-inch-long stalks, each flower about ¼ inch long, green, facing downward, with 5 petals with upturned tips and 5 narrow, arching sepals. **FLOWERING:** Late May–Jun. **RANGE:** FL west to OK, north to WI and VT. **SIMILAR TO** no other wildflower in Georgia.

American Wild Pansy

Viola bicolor Pursh
SYNONYM: *Viola rafinesquii* Greene
Violet Family (Violaceae)

ANNUAL HERB found in pastures, lawns, and other disturbed areas throughout Georgia. **STEMS:** 1–16 inches tall, slender, branched from the base, mostly smooth. **LEAVES:** About 1 inch long, alternate, spoon-shaped, smooth, entire or slightly toothed, with deeply segmented stipules at the base of the leaf stalk. **FLOWERS:** About ½ inch across, pale blue or white, with dark veins and a yellow patch in the center; the petals are twice as long as the sepals. **FLOWERING:** Mar–May. **RANGE:** Most of North America. **SIMILAR TO** two European species that also occur in disturbed areas. European Field Pansy (*Viola arvensis*) has white or cream-colored flowers with yellow centers; its roots smell of wintergreen. Johnny-jump-up (*Viola tricolor*) has yellow, white, and blue-purple flowers.

Sweet White Violet

Viola blanda Willdenow
Violet Family (Violaceae)

PERENNIAL HERB found in moist forests and along streams in the mountains, Ridge & Valley, and upper Piedmont. **STEMS:** There are no aboveground stems. **LEAVES:** 1–3½ inches long and wide, heart-shaped, dark green, hairy, with reddish stalks, toothed margins, and a satiny sheen on the upper surface. **FLOWERS:** About ½ inch wide, white, with dark veins at the base of the 3 lower petals, the upper 2 petals twisted backward, faintly fragrant; flower stalks up to 4 inches long and reddish. **FLOWERING:** Apr–Jun. **RANGE:** GA and AL, north to SK and NL. **SIMILAR TO** Small White Violet (*Viola macloskeyi*), which occurs in similar habitats; its leaves are mostly smooth with green, hairy stalks; its flowers are less than ½ inch wide with green stalks.

Canada Violet

Viola canadensis Linnaeus
Violet Family (Violaceae)

PERENNIAL HERB found in cove forests, rich slopes, and along streams in the north Georgia mountains. **STEMS:** 6–16 inches tall. **LEAVES:** 1–4¾ inches long and ½–2½ inches wide, heart-shaped with long, pointed tips, toothed, hairy or smooth; the stipules are lance-shaped and not toothed. **FLOWERS:** About 1 inch wide, bright white with a yellow center and purple veins on the 3 lower petals; the back side of the flower is tinted purple. **FLOWERING:** Apr–Jul. **RANGE:** Most of North America. **SIMILAR TO** Creamy Violet (*Viola striata*), which occurs in north Georgia in moist forests and floodplains; its flowers are creamy white with purple veins but are not purple on the back side; its stipules are deeply toothed.

Marsh Blue Violet

Viola cucullata Aiton
Violet Family (Violaceae)

PERENNIAL HERB found in marshes, bogs, swamps, and streamsides in northeast and southeast Georgia. **STEMS:** There are no aboveground stems. **LEAVES:** 1½–4 inches long and ¾–2 inches wide, heart-shaped to nearly round, smooth, with toothed margins and long stalks. **FLOWERS:** About ¾ inch wide, pale blue or blue-violet, darker toward the center, with a white, purple-veined throat; the hairs on the 2 side petals have swollen, knobby tips (use 10× lens); the flower stalks may be up to 7 inches long. **FLOWERING:** Apr–Jun. **RANGE:** GA west to AR, north to MN and NL. **SIMILAR TO** Common Blue Violet (*Viola sororia*), page 353, which has a pale throat and club-shaped hairs on the side petals. Most stemless blue violets have straight or club-shaped hairs on the side petals.

Silver-Leaf Violet

Viola hastata Michaux
Violet Family (Violaceae)

PERENNIAL HERB found in dry to moist forests in north Georgia. **STEMS:** Up to 12 inches tall, smooth. **LEAVES:** Up to 4 inches long and 2 inches wide, alternate, elongated heart-shaped, smooth, the upper surface usually mottled with silvery patches. **FLOWERS:** About ½ inch wide, bright yellow with purple veins. **FLOWERING:** Late Mar–May. **RANGE:** FL west to MS, north to OH and PA. **NOTES:** It is time to retire the old common name, Halberd-leaf Violet, which refers to the shape of a medieval weapon. **SIMILAR TO** Downy Yellow Violet (*Viola pubescens*), which occurs in rich deciduous forests in north Georgia; its leaves are heart-shaped, very hairy, and solid green. Smooth Yellow Violet (*Viola pensylvanica*) occurs in similar habitats; its leaves are heart-shaped, smooth, and solid green.

Lance-Leaf Violet

Viola lanceolata Linnaeus
Violet Family (Violaceae)

PERENNIAL HERB found in bogs, wet savannas and flatwoods, and edges of depression ponds in the Coastal Plain and one county in northwest Georgia. **STEMS:** There are no above-ground stems. **LEAVES:** 1–8 inches long and ¼–¾ inch wide, linear to narrowly lance-shaped with pointed tips and tapering bases, shallowly toothed, smooth, held erect on long, reddish, partially winged stalks. **FLOWERS:** About ½ inch wide, white with a greenish "eye" and purple veins on the lower petals; flower stalks reddish. **FLOWERING:** Mar–May. **RANGE:** FL west to TX, north to ON and NL. **SIMILAR TO** Primrose-leaf Violet (*Viola primulifolia*), page 351.

Wood Violet

Viola palmata Linnaeus
Violet Family (Violaceae)

PERENNIAL HERB found in moist forests, woodlands, and glades in north Georgia and several Coastal Plain counties. **STEMS:** There are no aboveground stems. **LEAVES:** 2–4 inches long and wide, early leaves round, oval, or heart-shaped; later leaves deeply divided into 3–11 lobes, hairy or smooth, with scalloped margins. **FLOWERS:** ½–1½ inches wide; blue, violet, white, or white streaked with violet, the 2 side petals hairy at the base; bright orange stamens are sometimes visible deep in the throat. **FLOWERING:** Late Mar–May. **RANGE:** GA west to LA, north to WI, ON, and ME. **SIMILAR TO** Bird's Foot Violet (*Viola pedata*), below, and Seven-lobed Violet (*Viola septemloba*), page 352.

Bird's Foot Violet

Viola pedata Linnaeus
Violet Family (Violaceae)

PERENNIAL HERB found in dry, rocky, or sandy woodlands and on dry road banks in north Georgia and the upper Coastal Plain. **STEMS:** There are no aboveground stems. **LEAVES:** 1–2 inches long and wide, deeply divided into 5–11 narrow lobes, smooth. **FLOWERS:** 1–1¾ inches wide; blue, violet, or both, usually with a white, purple-lined throat and orange stamens protruding from the throat. **FLOWERING:** Mar–May. **RANGE:** GA west to TX, north to ON and ME. **NOTES:** Compared with other violets, this species has a flatter "face" often held horizontally, providing a good landing surface for butterflies and skippers, which visit these flowers more than other violets. **SIMILAR TO** Wood Violet (*Viola palmata*), above, and Seven-lobed Violet (*Viola septemloba*), page 352.

Primrose-Leaf Violet

Viola primulifolia Linnaeus
Violet Family (Violaceae)

PERENNIAL HERB found in wetlands throughout Georgia. **STEMS:** There are no aboveground stems. **LEAVES:** ½–4 inches long and ¼–2 inches wide, oval to lance-shaped, entire or shallowly toothed, smooth or hairy, with blade tissue forming narrow wings on the stalk; stalks 1–7 inches long. **FLOWERS:** About ⅓ inch wide, white with purple lines on the lower petals. **FLOWERING:** Mar–May. **RANGE:** FL west to TX, north to ON and NL. **SIMILAR TO** Lance-leaf Violet (*Viola lanceolata*), page 349.

Long-Spurred Violet

Viola rostrata Pursh
Violet Family (Violaceae)

PERENNIAL HERB found in rich, moist forests in the north Georgia mountains. **STEMS:** Up to 10 inches tall, smooth. **LEAVES:** ¾–2¾ inches long and wide, alternate, heart-shaped, smooth except for hairs on the midvein, with scalloped margins. **FLOWERS:** ½–1 inch wide, lavender with dark purple streaks on the petals and a dark purple circle surrounding the pale throat; the 2 side petals lack hairs at the base, and the lowest petal has a slender ¾-inch spur. **FLOWERING:** Apr–May. **RANGE:** GA, AL, north to WI, ON, and QC. **SIMILAR TO** American Dog Violet (*Viola labradorica*, synonym *V. conspersa*), which occurs in moist forests in the foothills of the Blue Ridge; its flower has a white throat without the dark circle, and the 2 side petals have hairs at the base.

Round-Leaf Violet

Viola rotundifolia Michaux
Violet Family (Violaceae)

PERENNIAL HERB found in rich, moist forests in the north Georgia mountains. **STEMS:** There are no aboveground stems. **LEAVES:** About 1 inch long and wide at flowering time, up to 4¾ inches long by midsummer, lying nearly flat on the ground, round or oval, glossy, slightly succulent, hairy, with scalloped margins. **FLOWERS:** ½–1 inch wide, yellow, the 3 lower petals with purple veins, the 2 side petals with hairs at the base. **FLOWERING:** Mar–Apr. **RANGE:** GA, north to ON and NL. **NOTES:** This is the earliest-flowering of all our violets and the only yellow-flowered violet without aboveground stems. **SIMILAR TO** no other violets in Georgia.

Seven-Lobed Violet

Viola septemloba Le Conte
Violet Family (Violaceae)

PERENNIAL HERB found in wet to dry, sandy pinelands throughout Georgia. **STEMS:** There are no aboveground stems. **LEAVES:** Up to 2 inches long and wide, smooth; early leaves round, oval, or heart-shaped; mature leaves deeply divided into 3–11 narrow lobes. **FLOWERS:** ¾–1½ inches wide, purple with a white throat and darker purple veins, the lower 3 petals with patches of hair at the base, the stamens not visible. **FLOWERING:** Mar–May. **RANGE:** FL west to TX, north to TN and NC. **SIMILAR TO** Salad Violet (*Viola esculenta*), which occurs in floodplains and on banks of blackwater streams in the Coastal Plain; its mature leaves usually have 3–5 relatively wide lobes, and the 2 side petals have patches of hair while the lowest petal is smooth.

Common Blue Violet

Viola sororia Willdenow
Violet Family (Violaceae)

PERENNIAL HERB found in moist forests, bottomlands, lawns, and other disturbed areas throughout Georgia. **STEMS:** There are no aboveground stems. **LEAVES:** 2–4 inches long and wide, broadly heart-shaped, smooth, with toothed margins. **FLOWERS:** Up to 1 inch wide; blue, violet, or purple (rarely white) with a white or yellowish throat and darker veins on the petals; the 2 side petals have club-shaped hairs at the base, and the lowest petal is smooth. **FLOWERING:** Mar–Jun. **RANGE:** FL west to TX, north to MB and NL. **SIMILAR TO** Arrow-leaf Violet (*Viola sagittata*), which occurs infrequently in dry woodlands and forests in north Georgia; its leaves are oblong to lance-shaped with a few coarse teeth or lobes at the base; all 3 of the lower petals have hairs at the base.

Three-Parted Yellow Violet

Viola tripartita Elliott
Violet Family (Violaceae)

PERENNIAL HERB found in moist forests and bottomlands in north Georgia and a few southwest Georgia counties, especially over mafic or limestone bedrock. **STEMS:** Up to 12 inches tall, hairy or smooth. **LEAVES:** 1½–4¾ inches long, alternate, deeply divided into 3 lance-shaped segments (variety *tripartita*) *or* not divided and oval to heart-shaped (variety *glaberrima*). **FLOWERS:** Up to ¾ inch wide, yellow with purple veins on the 3 lower petals; the 2 side petals have hairs at the base, the lowest petal is smooth. **FLOWERING:** Late Mar–May. **RANGE:** FL west to MS, north to OH and PA. Plants with divided leaves (variety *tripartita*) are **SIMILAR TO** no other wildflower in our area—it is the only stemmed violet east of the Rocky Mountains with divided leaves. Plants with undivided leaves (variety *glaberrima*) are **SIMILAR TO** Silver-leaf Violet (*Viola hastata*) and Downy Yellow Violet (*Viola pubescens*), page 349.

Walter's Violet

Viola walteri House
Violet Family (Violaceae)

PERENNIAL HERB found in moist forests and bottomlands in the Piedmont and Coastal Plain, especially over mafic or limestone bedrock. **STEMS:** Up to 4 inches long, trailing along the ground and rooting at the nodes. **LEAVES:** 1–2 inches long and wide, in a basal rosette and also along the stems, heart-shaped with rounded tips, finely toothed, with dark green or purplish veins on the upper surface and usually solid purple on the lower surface. **FLOWERS:** Up to 1 inch wide, blue or violet with a white throat. **FLOWERING:** Mar–May. **RANGE:** FL west to TX, north to OH and VA. **NOTES:** The earliest flowers arise from the basal leaf rosette, creating the appearance of a stemless violet. With its trailing stems and purple-backed leaves, Walter's Violet is similar to no other violet.

MONOCOTS

Rattlesnake Master, False Aloe

Manfreda virginica (Linnaeus) Salisbury ex Rose
SYNONYMS: *Agave virginica* Linnaeus, *Polianthes virginica* (Linnaeus) Shinners
Agave Family (Agavaceae)

PERENNIAL HERB found in shallow soils around granite outcrops and limestone cedar glades, and in open woodlands over mafic bedrock throughout Georgia. **STEMS:** 1½–4½ feet tall, stout, nearly leafless. **LEAVES:** In a basal rosette, up to 20 inches long and 3½ inches wide, fleshy, somewhat folded lengthwise, often with a few spiny teeth, usually blue-green with maroon speckles. **FLOWERS:** About 1 inch long, pale green or tan, tubular with 6 short lobes; 6 purple stamens with long, yellow anthers extend about ½ inch from the tip of the flower. **FRUIT:** A round, 3-lobed capsule about ½ inch long. **FLOWERING:** May–Aug. **RANGE:** FL west to TX, north to MO, OH, and NC. **NOTES:** The flowers have a sweet, fruity fragrance and are pollinated by moths. **SIMILAR TO** no other wildflower in Georgia.

Georgia Bear-Grass

Nolina georgiana Michaux
Agave (Agavaceae) or Ruscus (Ruscaceae) Family

PERENNIAL HERB found in Longleaf Pine and Turkey Oak sandhills and other habitats with very dry, sandy soils in the eastern Coastal Plain and Fall Line counties. **STEMS:** 1–5 feet tall, waxy blue-green, with a large, much-branched flower cluster at the top. **LEAVES:** In a grasslike clump, blades 12–50 inches long and about ⅓ inch wide, waxy blue-green, with tiny teeth along the margins. **FLOWERS:** Evenly spaced along the flower cluster branches, on stalks less than ⅓ inch long, each flower about ¼ inch wide, with 6 white, curved tepals and 6 long stamens. **FRUIT:** A papery, 3-lobed capsule about ¼ inch long. **FLOWERING:** May–Jun. **RANGE:** GA and SC. **SIMILAR TO** Eastern Turkeybeard (*Xerophyllum asphodeloides*), a Georgia Special Concern species; its long-stalked flowers are in dense, somewhat cylinder-shaped clusters.

Spanish Dagger

Yucca aloifolia Linnaeus
Agave Family (Agavaceae)

PERENNIAL SHRUBLIKE HERB found on dunes and other coastal habitats, cultivated inland. **STEMS:** 5–23 feet tall and 3–5 inches in diameter, usually branched at least once, often toppled over. **LEAVES:** 5–24 inches long and 1–2½ inches wide, evergreen, thick and rigid with a sharp point and tiny, cutting teeth along the margins, dark green, pale only at the base. **FLOWER CLUSTERS:** Up to 2 feet tall, the lower fourth hidden by leaves. **FLOWERS:** About 2 inches long, nodding, bell-shaped with 6 white tepals. **FLOWERING:** May–Jul. **RANGE:** FL west to TX, north to VA; Central America, West Indies. **NOTES:** Old, withered leaves persist, forming a "grass skirt." **SIMILAR TO** Spanish Bayonet (*Yucca gloriosa*), which occurs in the same habitats; it forms mounds without tall trunks and has grayish-green leaves. Its flower cluster is held well above the leaves.

Curly-Leaf Yucca

Yucca filamentosa Linnaeus
Agave Family (Agavaceae)

PERENNIAL SHRUBLIKE HERB found in dry woodlands and disturbed areas throughout Georgia. **LEAVES:** 12–30 inches long and 1–2 inches wide, evergreen, thick and rigid with a sharp point and curling fibers along the margins. **FLOWER CLUSTERS:** 2½–5 feet tall, at the top of a stout stalk 3–10 feet tall. **FLOWERS:** About 2 inches long, nodding, bell-shaped with 6 white tepals. **FRUIT:** An erect, oblong, 3-lobed capsule, 1½–2 inches long, green becoming brown and woody. **FLOWERING:** Apr–Jun. **RANGE:** FL west to LA, north to TN and NJ. **NOTES:** After flowering and fruiting, the leaves and the stalk slowly disintegrate during the winter; buds on the underground stem form new plants the following spring. **SIMILAR TO** Adam's Needle (*Yucca flaccida*), which occurs throughout Georgia in similar habitats; it has smaller flowers and narrower, floppy leaves.

Yucca Pollination

Insect pollination can be a casual relationship, with insects drifting from one type of flower to another. Or it can be a tightly choreographed dance, with the insect and plant partners locked in mutual dependency, as in the case of the yuccas and yucca moths. Yucca flowers are visited by many nectar-seeking insects throughout the day, but only yucca moths pollinate the flowers. In the hours between sunset and midnight, female and male yucca moths mate on the petals of yucca's flowers. The female then goes to work, gathering pollen from the stamens. She tucks the pollen under her chin, forming a ball up to three times the size of her head. She then visits the flowers of another yucca plant, searching for one that has not been visited by another female, which she can determine through smell. She lays a few eggs (too many would abort the flower) in the ovary of the chosen flower and then carefully deposits part of the pollen ball on the stigma at the top of the ovary. She then moves on, in search of another flower and repeats the process. The pollen does its usual thing, sending a tube down into the ovary and releasing sperm cells that fertilize the flower's egg cells. Thus fertilized, the yucca ovary develops into a fruit with enough seeds to both nourish the moth larvae and guarantee another generation of yucca plants. In a few weeks, the larvae hatch, fall to the ground, and form an underground cocoon. The following spring (or sometimes the spring after that), the adult moths emerge from the cocoon and go in search of yucca flowers, beginning the cycle all over again.

Curly-leaf Yucca flowers with yucca moths (*Tegeticula yuccasella*)

Southern Water Plantain

Alisma subcordatum Rafinesque
Water Plantain Family (Alismataceae)

PERENNIAL HERB found in open wetlands in north Georgia. **LEAVES:** Rising in a cluster from an underground stem, each up to 6 inches long and 4 inches wide, oval to nearly heart-shaped with a sharp tip, entire, smooth, fleshy, with long, 3-sided stalks. **FLOWER CLUSTERS:** Up to 3 feet tall, 1 per plant, erect, open, with several whorls of branches. **FLOWERS:** Less than ¼ inch wide, with 3 white, rounded petals, 3 green sepals visible between the petals, and a ring of green ovaries in the center. **FLOWERING:** Apr–Nov. **RANGE:** GA west to TX, north to ND and NB. **SIMILAR TO** Creeping Burr-head (*Echinodorus cordifolius*), below. Also see Duck Potato (*Sagittaria latifolia*), page 360.

Creeping Burr-head

Echinodorus cordifolius (Linnaeus) Grisebach
Water Plantain Family (Alismataceae)

PERENNIAL HERB found in swamps, wet ditches, streams, and marshes in the Coastal Plain. **LEAVES:** Clustered, rising above the water from an underground stem, blades up to 12½ inches long and 7½ inches wide, oval to heart-shaped, entire, main veins curved, cross-veins straight and parallel; held above the water on long, spongy, grooved stalks. **FLOWER CLUSTERS:** Up to 2 feet long, with several widely spaced whorls of flowers, arching and rooting at the nodes and tips. **FLOWERS:** Up to 1 inch wide, with 3 white, rounded petals. **FLOWERING:** Jun–Nov. **RANGE:** GA west to TX, north to KS and VA; Mexico, West Indies, South America. **SIMILAR TO** Tall Burr-head (*Echinodorus berteroi*), which occurs in similar habitats in southwest Georgia; its flower clusters are held rigidly erect. Also see Duck Potato (*Sagittaria latifolia*), page 360.

Grass-Leaved Arrowhead

Sagittaria graminea Michaux
Water Plantain Family (Alismataceae)

PERENNIAL HERB found in ponds and wetlands in Georgia's Coastal Plain and a few north Georgia counties. **LEAVES:** Two types: lance-shaped blades up to 7 inches long and 1½ inches wide with stalks up to 7 inches long; *and* narrow, grasslike blades up to 14 inches long. **FLOWER CLUSTERS:** Held at the top of a stalk; female flowers, with many green ovaries in the center, are usually at the bottom of the cluster; male flowers, with many yellow-tipped stamens, are at the top. **FLOWERS:** Up to 1 inch wide, with 3 white, rounded petals. **FLOWERING:** May–Nov. **RANGE:** FL west to TX, north to SD and NL; West Indies. **SIMILAR TO** Lance-leaved Arrowhead (*Sagittaria lancifolia*), which occurs in wetlands in the Coastal Plain; it has lance-shaped leaf blades up to 16 inches long with stout, spongy stalks up to 32 inches long.

Duck Potato, Broad-Leaved Arrowhead

Sagittaria latifolia Willdenow
Water Plantain Family (Alismataceae)

PERENNIAL HERB found in ponds and marshes throughout Georgia. **LEAVES:** Up to 12 inches long and 7 inches wide, arrowhead-shaped with spongy stalks raising the leaf blades above water level. **FLOWER CLUSTERS:** Held at the top of a stalk about as long as the leaves; female flowers, with many green ovaries in the center, are usually at the bottom of the cluster; male flowers, with many yellow-tipped stamens, are at the top. **FLOWERS:** Up to 1½ inches wide, with 3 white, rounded petals and very short bracts at the base of the flower stalk. **FLOWERING:** Jun–Sep. **RANGE:** Eastern and central North America, south to tropical America. **SIMILAR TO** Engelmann's Arrowhead (*Sagittaria engelmanniana*), which occurs in blackwater swamps in the Coastal Plain; bracts at the base of its flower stalks are as long as or longer than the flower stalks.

Wild Onion, Wild Garlic

Allium canadense Linnaeus
Onion (Alliaceae), Amaryllis (Amaryllidaceae), or Lily (Liliaceae) Family

PERENNIAL HERB with a strong onion odor found in woodlands and disturbed areas throughout Georgia. **STEMS:** Up to 2 feet tall, round, smooth. **LEAVES:** Up to 20 inches long and ¼ inch wide, flat, with a groove running the entire length. **FLOWER CLUSTERS:** At the top of the stem, with a mix of flowers and bulblets (variety *canadense*) or with flowers only (variety *mobilense*). **FLOWERS:** About ½ inch wide, with 6 pink or white, spreading tepals; stamens pink; ovary green, the upper surface smooth, not bumpy. **FLOWERING:** Apr–May. **RANGE:** FL west to TX, north to MT and NB. **SIMILAR TO** Cuthbert's Onion (*Allium cuthbertii*), which occurs on rock outcrops throughout Georgia (except the mountains); the upper surface of its ovary is covered with bumps. Also see False Garlic (*Nothoscordum bivalve*), below.

False Garlic

Nothoscordum bivalve (Linnaeus) Britton
SYNONYM: *Allium bivalve* (Linnaeus) Kuntze
Onion (Alliaceae), Amaryllis (Amaryllidaceae), or Lily (Liliaceae) Family

PERENNIAL HERB found on rock outcrops and in disturbed areas throughout Georgia. **STEMS:** 4–16 inches tall, round, hollow, smooth. **LEAVES:** Up to 12 inches long and ¼ inch wide, flat, smooth. **FLOWERS:** Held in a cluster at the top of the stem, not fragrant, each about 1 inch wide, with 6 white tepals marked with green at the base, a greenish-yellow ovary, and 6 yellow stamens. **FLOWERING:** Mar–May, Sep–Dec. **RANGE:** FL west to AZ, north to KS and VA; South America. **NOTES:** The entire plant lacks any odor of garlic or onion. **SIMILAR TO** the lawn weed, Star-of-Bethlehem (*Ornithogalum umbellatum*), which has a wide, green stripe on the lower surface of the tepals and white, flattened stamens. Also see Wild Onion (*Allium canadense*), above.

Hammock Spider-Lily

Hymenocallis occidentalis (J. Le Conte) Kunth

Amaryllis Family (Amaryllidaceae)

PERENNIAL HERB found in moist slope forests, bottomlands, calcareous glades, and upland flatwoods throughout Georgia. **LEAVES:** 14–24 inches long and up to 2½ inches wide, waxy, bluish-green, arising from a bulb. **FLOWER CLUSTERS:** At the top of a waxy, bluish-green, 2-edged stalk up to 2 feet tall. **FLOWERS:** Up to 8 inches wide, with 6 narrow, white tepals radiating from the top of a green tube, a central white cup (the corona) less than 2 inches across, and 6 stamens rising from the margin of the corona. **FLOWERING:** Summer. **RANGE:** FL west to LA, north to IL and NC. **NOTES:** Flowers open late in the day and wither the following afternoon; they are pollinated by sphinx moths. **SIMILAR TO** Coastal Spider-lily (*Hymenocallis crassifolia*), which grows on the banks of blackwater streams and in freshwater tidal marshes near the coast.

Atamasco-Lily, Rain-Lily

Zephyranthes atamasca (Linnaeus) Herbert

Amaryllis Family (Amaryllidaceae)

PERENNIAL HERB found in moist forests, bottomlands, seepy rock outcrops, and roadsides in the Piedmont and Coastal Plain. **LEAVES:** 6–18 inches long and about ¼ inch wide, grasslike, sprawling, somewhat succulent, shiny green, forming large patches. **FLOWERS:** 2–4 inches across, funnel-shaped, with 6 white tepals spreading from the top of a slender, green tube, fragrant, turning pink with age; held at the top of a hollow stalk 4–12 inches tall. **FLOWERING:** Mar–Apr. **RANGE:** FL west to MS and north to MD. **SIMILAR TO** Treat's Zephyr-lily (*Zephyranthes treatiae*), which occurs in wet pine flatwoods and roadside ditches in Coastal Plain counties near the Florida border; it does not form large patches, and its leaves are narrower than ¼ inch and have a groove running the length of the upper surface.

Arum Family (Araceae)

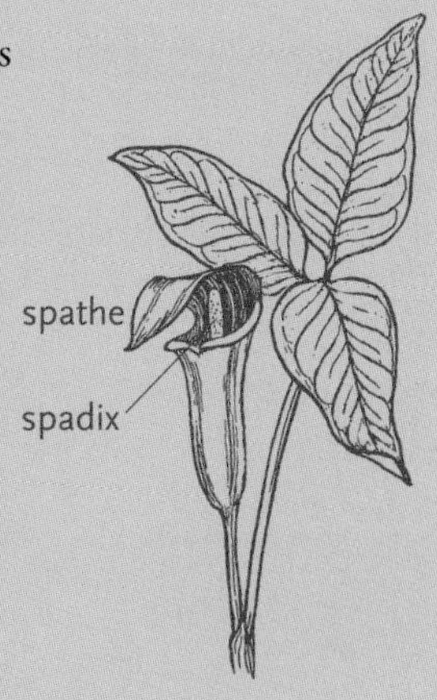

The arums are largely tropical, with only seven species in Georgia, but their peculiar flowers are well known to spring wildflower explorers. The "flower" is actually an inflorescence containing many flowers held on a fleshy column called a spadix. The spadix is usually surrounded by a colorful, leaflike spathe. In Jack-in-the-pulpit, our best-known arum, the Jack (or Jill!) is the green or purple spadix standing inside the pulpit-like spathe. The female flowers are held at the base of the spadix; they are tiny green balls with a tuft of fuzzy stigmas. The male flowers are above the female flowers and open later to prevent self-pollination. Jack-in-the-pulpit plants may have female flowers or male flowers or both. The flowers lack petals and sepals and are pollinated mostly by flies. Some commonly cultivated arums include Peace Lily and Elephant-ear. Skunk Cabbage is known from the Blue Ridge and farther north but does not quite make it south into Georgia.

Green Dragon

Arisaema dracontium (Linnaeus) Schott
Arum Family (Araceae)

PERENNIAL HERB found in moist, hardwood forests and bottomlands throughout Georgia. **LEAVES:** 1 per plant, up to 2 feet tall (including the stout stalk) and wide, with 7–13 elliptic leaflets forming a semicircle at the top of the stalk. **FLOWER CLUSTERS:** Up to 1 foot tall, including the stalk; a pale green, waxy spathe encloses the base of a fleshy, elongated, pointed spadix that extends up to 10 inches beyond the spathe. The base of the spadix holds many tiny flowers that have neither petals nor sepals; flowers on a single plant may be all female, all male, or a mix, with female flowers closest to the base of the spadix. **FRUIT:** Orange-red berries in a cluster at the top of a stalk. **FLOWERING:** May–Jun. **RANGE:** FL west to TX, north to MN and QC. **SIMILAR TO** Jack-in-the-pulpit (*Arisaema triphyllum*), page 364.

Jack-in-the-Pulpit, Jill-in-the-Pulpit

Arisaema triphyllum (Linnaeus) Schott
Arum Family (Araceae)

PERENNIAL HERB found in moist, hardwood forests and near streams throughout Georgia. **LEAVES:** 1 or 2 per plant, up to 32 inches tall (including the long stalk), with 3–5 oval or lance-shaped leaflets with a vein that runs parallel to the margins; the overlapping leaf bases look like a stem. **FLOWER CLUSTERS:** 8–32 inches tall, with a green, white- or purple-striped spathe enclosing a spadix. **FRUIT:** A cluster of bright red berries. **FLOWERING:** Mar–Apr. **RANGE:** FL west to TX, north to MN and NS. **NOTES:** Female-flowered Jills and bisexual plants are larger, have 2 leaves, and produce fruits at the base of the spadix. Male Jacks are smaller with a single leaf and produce pollen. The leaves are **SIMILAR TO** those of trilliums but have a vein running parallel to the leaf margins.

Golden Club, Never-Wet

Orontium aquaticum Linnaeus
Arum Family (Araceae)

PERENNIAL AQUATIC HERB found in swamps, marshes, beaver ponds, and blackwater streams in the Coastal Plain and a few north Georgia counties. **LEAVES:** 4–12 inches long, 2–4 inches wide, held above the water on stalks up to 16 inches long, elliptic, blue-green, velvety, coated with a waxy substance that repels water. **FLOWERS:** Tiny, golden yellow, at the tip of a fleshy, yellow-and-white "club" (the spadix) with a long, green stalk; there is no spathe. **FRUIT:** Round, about ½ inch wide, blue-green, embedded in the spadix. **FLOWERING:** Mar–Apr. **RANGE:** FL west to TX, north to TN and MA. **NOTES:** The leaves are buoyant and will pop back up, shedding droplets, if held underwater. **SIMILAR TO** no other wildflower in Georgia.

Arrow-Arum

Peltandra virginica (Linnaeus) Schott
Arum Family (Araceae)

PERENNIAL HERB found in swamps, marshes, slow-moving streams, and wet ditches throughout Georgia. **LEAVES:** Blades up to 20 inches long and 12 inches wide, arrowhead-shaped, smooth, glossy, with a vein running parallel to the leaf margins. **FLOWER CLUSTERS:** A green, leaflike spathe partly encloses the spadix, a column of tightly packed flowers; flat, yellow male flowers are held above the round, green female flowers on the lower, more enclosed portion of the spadix. **FRUIT:** Green to purplish-black berries in a cluster partly enclosed by the base of the spathe. **FLOWERING:** May–Jun. **RANGE:** FL west to TX, north to MN, ON, and ME. **SIMILAR TO** Spoon-flower (*Peltandra sagittifolia*), a Georgia Special Concern species that occurs in wetlands in southeast Georgia; its spathe is white, and its berries are red. Also see Duck Potato (*Sagittaria latifolia*), page 360.

Canada May-Flower

Maianthemum canadense Desfontaines
Asparagus (Asparagaceae), Ruscus (Ruscaceae), or Lily (Liliaceae) Family

PERENNIAL HERB found in moist, high-elevation forests in the Blue Ridge. **STEMS:** 1½–7 inches tall, forming colonies. **LEAVES:** 1–4 inches long and up to 2 inches wide, 1–3 per plant, alternate, oval to lance-shaped, glossy, smooth. **FLOWER CLUSTERS:** 1 per plant, held at the tip of the stem, each cluster up to 1½ inches high with 12–25 flowers. **FLOWERS:** Tiny, white, with 4 pointed, short-lived tepals and 4 white, persistent stamens. **FRUIT:** A round berry less than ¼ inch wide, mottled red-and-green when young, bright red when mature. **FLOWERING:** May–Jun. **RANGE:** GA north and west across Canada. Canada May-flower is a miniature version of Solomon's Plume (*Maianthemum racemosum*), page 366; otherwise it is **SIMILAR TO** no other wildflower in Georgia.

Solomon's Plume, False Solomon's Seal

Maianthemum racemosum (Linnaeus) Link

SYNONYM: *Smilacina racemosa* (Linnaeus) Desfontaines

Asparagus (Asparagaceae), Ruscus (Ruscaceae), or Lily (Liliaceae) Family

PERENNIAL HERB found in moist forests in north Georgia and a few Coastal Plain counties. **STEMS:** Up to 4 feet tall, solitary, erect or arching, slightly zigzag. **LEAVES:** 2¾–7 inches long and 1–3 inches wide, alternate, nearly clasping, oval or elliptic, with 3–5 main parallel veins. **FLOWER CLUSTERS:** Up to 3 inches long, held at the tip of the stem, plumelike, with many tiny flowers. **FLOWERS:** Tiny, white, with 6 pointed tepals and 6 flat, petal-like stamens. **FRUIT:** A round berry less than ¼ inch wide, coppery green when young, bright red when mature. **FLOWERING:** Apr–Jun. **RANGE:** FL west to TX, north to OK, MB, and NL. Stems and leaves are **SIMILAR TO** Solomon's Seal (*Polygonatum biflorum*), below.

Solomon's Seal

Polygonatum biflorum (Walter) Elliott

Asparagus (Asparagaceae), Ruscus (Ruscaceae), or Lily (Liliaceae) Family

PERENNIAL HERB found in moist forests in north and southwest Georgia. **STEMS:** 1–6½ feet tall, solitary, erect or arching, smooth, waxy, bluish-green. **LEAVES:** 2–8 inches long and ½–3½ inches wide, alternate, elliptic to lance-shaped, lower surface waxy white. **FLOWERS:** Dangling from the leaf axils on slender stalks, each flower ½–¾ inch long, pale yellow, tubular with 6 spreading tips. **FRUIT:** A round, blue-black berry about ⅓ inch wide. **FLOWERING:** Apr–Jun. **RANGE:** FL west to AZ, north to MB and ME. **SIMILAR TO** Hairy Solomon's Seal (*Polygonatum pubescens*), which occurs in moist coves in the Blue Ridge; its leaves are hairy on the lower surface, and its flowers are less than ½ inch long. Stems and leaves of Solomon's Seal are also **SIMILAR TO** Solomon's Plume (*Maianthemum racemosum*), above.

Yellow Sunny-Bells

Schoenolirion croceum (Michaux) Wood
Asparagus (Asparagaceae), Agave (Agavaceae), or Lily (Liliaceae) Family

PERENNIAL HERB found in wet depressions on Piedmont granite outcrops and on Altamaha Grit outcrops in the Coastal Plain; also in wet prairies in the Coosa River valley. **STEMS:** 8–14 inches tall. **LEAVES:** 8–13 inches long and about ¼ inch wide, grasslike but fleshy, smooth, arising from a bulb. **FLOWERS:** Less than ½ inch wide, bright yellow, with 6 tepals, in a spikelike cluster at the top of the stem. **FLOWERING:** Mar–Apr. **RANGE:** FL west to TX, north to TN and NC. **NOTES:** Sunny-bells survives the harsh summer conditions on rock outcrops by going dormant by June. It has contractile roots that pull the bulb deeper into the soil. **SIMILAR TO** Yellow Star-grass (*Hypoxis hirsuta*), page 382.

Spanish Moss

Tillandsia usneoides (Linnaeus) Linnaeus
Pineapple Family (Bromeliaceae)

PERENNIAL HERB that droops from the limbs and trunks of trees and utility lines in the Coastal Plain and lower Piedmont. **STEMS:** Up to 10 feet long, stringlike, silvery gray. **LEAVES:** Up to 1¼ inches long, very narrow, twisted and curled, silvery gray, nearly indistinguishable from the stems. **FLOWERS:** About ⅓ inch long, tubular, with 3 spreading, yellow-green petals. **FRUIT:** A narrow, tan, inch-long cylinder that splits into 3 parts to release plumed seeds. **FLOWERING:** Jun–Aug. **RANGE:** FL west to TX, north to VA; Central and South America, West Indies. **NOTES:** Spanish Moss is not a parasite and uses trees only for support; it is covered with gray scales that absorb water from the air and nutrients from airborne dust. Spanish Moss reproduces by the spread of stem fragments and wind-blown seeds. **SIMILAR TO** no other plant in Georgia.

Nodding Nixie

Apteria aphylla (Nuttall) Barnhart ex Small

Burmannia Family (Burmanniaceae)

PERENNIAL HERB found in swamps, bogs, and rich, moist, upland forests in the Coastal Plain. **STEMS:** 1–11 inches tall, delicate, often branched, purple. **LEAVES:** Scale-like, tiny, alternate, purple-brown. **FLOWERS:** Nodding at the top of the stem, about ⅓ inch long, purple or white with purple stripes inside, tubular with 6 erect lobes. **FLOWERING:** Aug–Oct. **RANGE:** GA west to TX, south to South America and West Indies. **NOTES:** Nodding Nixie lacks chlorophyll and does not photosynthesize; it derives carbohydrates from an underground fungus that extracts nutrients from tree roots (see "A Forest Menage à Trois," p. 168). **SIMILAR TO** no other wildflower in Georgia.

Northern Blue Thread, Violet Burmannia

Burmannia biflora Linnaeus

Burmannia Family (Burmanniaceae)

ANNUAL HERB found in swamps, bogs, and wet savannas and flatwoods in the Coastal Plain. **STEMS:** 1–7 inches tall, threadlike, green. **LEAVES:** Few, scalelike, tiny, alternate, green. **FLOWERS:** Solitary or in a small cluster at the top of the stem, less than ¼ inch long and wide, violet, tubular with 3 rounded wings and 6 tiny, pointed lobes at the top. **FLOWERING:** Aug–Nov. **RANGE:** FL west to TX, north to VA. **NOTES:** All species in the Burmannia Family, including those like Northern Blue Thread that can photosynthesize to some extent, are dependent on an underground fungus for at least some of their nutrients (see "A Forest Menage à Trois," p. 168). **SIMILAR TO** no other wildflower in Georgia.

Yellow Mandarin

Prosartes lanuginosa (Michaux) D. Don
SYNONYM: *Disporum lanuginosum* (Michaux) Nicholson
Meadow Saffron (Colchicaceae) or Lily (Liliaceae) Family

PERENNIAL HERB found in rich, moist cove forests in the north Georgia mountains. **STEMS:** 16–30 inches tall, softly hairy, with forking branches. **LEAVES:** 2–6 inches long and up to 2 inches wide, alternate, oval or lance-shaped, softly hairy. **FLOWERS:** 1–1¾ inches wide, with 6 pale yellow or greenish-yellow, spreading, pointed, unspotted tepals; usually in groups of 2 or 3, dangling from leaf axils. **FRUIT:** A bright orange or red, shiny, oval berry about ½ inch long. **FLOWERING:** Apr–May. **RANGE:** GA west to AK, north to NY, in the mountains. **SIMILAR TO** Spotted Mandarin (*Prosartes maculata*), which occurs in similar habitats; its flowers are cream-colored or white with purple spots, and its fruit is yellowish-white, 3-lobed, and hairy.

Perfoliate Bellwort

Uvularia perfoliata Linnaeus
Meadow Saffron (Colchicaceae) or Lily (Liliaceae) Family

PERENNIAL HERB found in dry to moist, hardwood forests in north Georgia and a few Coastal Plain counties. **STEMS:** 6–20 inches tall, smooth and waxy, branched, with 2–4 leaves below the lowest branch. **LEAVES:** Up to 4 inches long and 2½ inches wide, alternate, completely surrounding the stem, elliptic, smooth. **FLOWERS:** About 1 inch long, dangling at the tips of downcurved stalks, with 6 pale yellow, pointed tepals, each with tiny, orange bumps on the inside. **FLOWERING:** Apr–early May. **RANGE:** FL west to TX, north to AR and ME. **SIMILAR TO** Large-flowered Bellwort (*Uvularia grandiflora*), which occurs in rich, moist forests in northwest Georgia; its leaves are hairy on the lower surface, and the tepals are twisted, golden yellow, and do not have orange bumps on the inner surface.

Wild Oats, Sessile-Leaf Bellwort

Uvularia sessilifolia Linnaeus
Meadow Saffron (Colchicaceae) or Lily (Liliaceae) Family

PERENNIAL HERB found in moist coves and bottomlands in north Georgia. **STEMS:** 4–18 inches tall, smooth, forked once, slightly zigzag, in colonies of widely spaced plants (5–10 inches apart). **LEAVES:** Up to 3 inches long and 1½ inches wide, alternate, elliptic, pointed, smooth, without stalks. **FLOWERS:** About 1 inch long, solitary, dangling at the tip of a short, leafless stalk, with 6 pale yellow, pointed tepals. **FLOWERING:** Late Mar–early May. **RANGE:** FL west to TX, north to ND and NS. **SIMILAR TO** Appalachian Bellwort (*Uvularia puberula*), which forms colonies of closely spaced stems in hardwood forests in north Georgia; its upper stems and leaves are usually finely hairy. Florida Bellwort (*U. floridana*) occurs in moist forests in the Coastal Plain; its flower stalk has a leafy bract below the flower.

Spring Ephemerals: Here Today, Gone Tomorrow

Spring ephemerals are wildflowers that emerge in early spring, complete their life cycle, and go dormant by late spring. This rapid dash through the life cycle takes advantage of the sun shining through the still leafless trees and warming the soil. Depending on the latitude and altitude, the cycle may begin as early as Feb ruary or as late as May. And it is indeed a race to grow to maturity, flower, set fruit, and manufacture and store carbohydrates before the expanding tree canopy blocks the sun and the plants disappear underground. Spring ephemerals survive the environmental stresses of summer, fall, and winter as underground organs—bulbs, corms, or rhizomes—and reemerge the following spring to begin their brief life cycle all over again. Ants disperse the seeds of many spring ephemerals (see "The Skinny on Fat Bodies—Elaiosomes," p. 44). Among the spring ephemerals found in Georgia are:

Bellwort
Blue Cohosh
Crane's-bill
Dutchman's Britches
Early Saxifrage
Eastern Columbine
Mandarin
May-apple
Phacelia
Rue-anemone
Shooting Star
Spring-beauty
Squirrel Corn
Toothwort
Trout Lily
Twinleaf
Wild Geranium
Wood Anemone

Erect Dayflower, Slender Dayflower

Commelina erecta Linnaeus
Spiderwort Family (Commelinacaceae)

PERENNIAL HERB found in dry woodlands, sandhills, rock outcrops, and sand dunes in the Piedmont and Coastal Plain. **STEMS:** Up to 2 feet tall, erect, slender, succulent, usually hairy. **LEAVES:** Up to 6 inches long and 1½ inches wide, alternate, smooth; the base of the leaf encloses the stem in a small sheath fringed with white hairs. **FLOWERS:** Up to 1½ inches wide, with 2 blue petals, 1 very small white petal, 3 purple stamens, and 3 yellow stamens; flowers open in the early morning and wither by afternoon. **FLOWERING:** Jun–Oct. **RANGE:** FL west to AZ, north to WY, MN, and NY. **SIMILAR TO** Common Dayflower (*Commelina communis*), an Old World species that occurs in moist disturbed areas throughout Georgia; its sprawling stems form mats by rooting at the leaf nodes.

Virginia Dayflower

Commelina virginica Linnaeus
Spiderwort Family (Commelinacaceae)

PERENNIAL HERB found in floodplains and disturbed wetlands throughout Georgia. **STEMS:** Up to 3 feet tall, stout, smooth, often forming patches. **LEAVES:** Up to 8 inches long and 2 inches wide, alternate, smooth; the base of the leaf encloses the stem in a hairy, tubular sheath fringed at the top with brownish-red bristles. **FLOWERS:** About ¾ inch wide, with 3 blue petals, one only slightly smaller than the others; the stamens are not hairy. **FLOWERING:** Jul–Oct. **RANGE:** FL west to TX, north to IL and NJ. **SIMILAR TO** Creeping Dayflower (*Commelina diffusa*), an Old World species with weak, sprawling stems that root at the leaf nodes; it occurs in disturbed areas throughout Georgia.

Grass-Leaf Roseling

Cuthbertia graminea Small
SYNONYMS: *Callisia graminea* (Small) G. Tucker, *Tradescantia rosea* Ventenat var. *graminea* (Small) E. S. Anderson & Woodson
Spiderwort Family (Commelinaceae)

PERENNIAL HERB found in dry, sandy woodlands in the Coastal Plain. **STEMS:** Up to 16 inches tall, smooth. **LEAVES:** Up to 7 inches long and about ⅛ inch wide, forming clumps of nearly erect leaves; the base of the leaf encloses the stem in a tubular sheath that, if slit open and flattened, is wider than the leaf blade. **FLOWER CLUSTERS:** 1 or 2 flowers open and wither each day; there are no long, leafy bracts beneath the flowers. **FLOWERS:** About ¾ inch wide, with 3 pink petals and 6 hairy, pink stamens. **FLOWERING:** May–Jul. **RANGE:** FL north to VA. **SIMILAR TO** Common Roseling (*Cuthbertia rosea*), which occurs in similar habitats; its leaves are loosely and widely spreading and ¼–½ inch wide.

Hairy Spiderwort

Tradescantia hirsuticaulis Small
Spiderwort Family (Commelinaceae)

PERENNIAL HERB found in rocky woodlands and on granite outcrops in the Piedmont and a few Coastal Plain counties. **STEMS:** Up to 16 inches tall, hairy. **LEAVES:** Up to 12 inches long and 1½ inches wide, very hairy; the leaf base encloses the stem in a tubular sheath that, if slit open and flattened, is as wide as or wider than the leaf blade. **FLOWER CLUSTERS:** Long, leaflike bracts spread beneath each cluster. **FLOWERS:** About 1 inch wide, with 3 blue, purple, pink, or white petals and 6 hairy stamens; sepals and flower stalks have long, gland-tipped hairs. **FLOWERING:** Apr–Jun. **RANGE:** GA west to OK, north to NC. **SIMILAR TO** Sandhill Spiderwort (*Tradescantia hirsutiflora*), which occurs in dry woodlands in the Coastal Plain; its stems, sepals, and flower stalks are very hairy but have few or no gland-tipped hairs.

Smooth Spiderwort

Tradescantia ohiensis Rafinesque
Spiderwort Family
(Commelinaceae)

PERENNIAL HERB found in forests and disturbed areas throughout Georgia. **STEMS:** Up to 4 feet tall, with a waxy, white coating. **LEAVES:** Up to 18 inches long and 1¾ inches wide, alternate, smooth; the base of the leaf encloses the stem in a tubular sheath that, if slit open and flattened, is as wide as or wider than the leaf blade. **FLOWER CLUSTERS:** 2 leaflike bracts spread beneath each cluster. **FLOWERS:** Up to 1½ inches wide, on hairless stalks, with 3 blue, rose, or white petals and 6 hairy stamens; sepals smooth or with a small tuft of hairs at the tip. **FLOWERING:** Apr–Jul. **RANGE:** FL west to TX, north to NE, MN, and MA. **SIMILAR TO** Virginia Spiderwort (*Tradescantia virginiana*), which occurs in similar habitats; its stems are smooth, and its leaf sheaths, flower stalks, and sepals are hairy (but not glandular-hairy).

Zigzag or Wide-Leaved Spiderwort

Tradescantia subaspera Ker-Gawler
Spiderwort Family
(Commelinaceae)

PERENNIAL HERB found in moist to dry forests in north Georgia. **STEMS:** Up to 3 feet tall, smooth or hairy, somewhat zigzag. **LEAVES:** Up to 12 inches long and 2½ inches wide, alternate, smooth or hairy, upper surface dark green, lower surface paler; the base of the leaf encloses the stem in a tubular sheath that, if slit open and flattened, is narrower than the leaf blade. **FLOWER CLUSTERS:** 2 leaflike bracts spread beneath each cluster. **FLOWERS:** About 1 inch wide, with 3 blue or purple petals and 6 hairy, blue stamens. **FLOWERING:** May–Sep. **RANGE:** FL west to LA, north to MO and VA. **SIMILAR TO** other spiderworts but distinguished by its zigzag stems, wide leaves with narrower leaf bases, and pale green lower leaf surfaces.

Sedge Family (Cyperaceae)

The Sedge Family is huge, with about 5,500 species worldwide and more than 300 species in Georgia. Sedges are ecologically important as major components of marshes and fens, and their seeds support many animals. Florida's famous "River of Grass"—the Everglades—is actually a river of Sawgrass, a member of the Sedge Family. Sedges make up the first part of the handy jingle used to distinguish the three common grasslike plant families: "Sedges have edges, rushes are round, and grasses are jointed all down to the ground." Sedges have edges because most of them have 3-sided stems. Roll the stem lightly between your fingers and the edges are obvious. (There are, of course, exceptions.) Like grasses, sedges have tiny, wind-pollinated flowers that lack showy petals and sepals for attracting pollinators. Also like grasses, the flowers are held in clusters called spikelets. In the largest genus of sedges, *Carex*, the female flowers are enveloped by tiny sacs and the male flowers occur in separate, slender spikelets. In another large genus, *Cyperus*, the flat-sedges, the flowers are usually held in flattened spikelets. Learning to identify sedges can be difficult because of the small size of the flowers and the specialized vocabulary used to describe them. The species presented here are all relatively easy to recognize without resort to technical keys. "An Introduction to the Sedges of Georgia," by Dr. Richard Carter, published in *Tipularia—the Journal of the Georgia Botanical Society* in 2005, is a good way to get started with this family. That issue can be ordered by e-mailing gabotany@comcast.net.

Cherokee Sedge

Carex cherokeensis Schweinitz
Sedge Family (Cyperaceae)

PERENNIAL HERB found in floodplains and moist forests throughout Georgia, especially over mafic or limestone bedrock. **STEMS:** 1–3 feet tall, 3-angled, with old, dried leaves clustered at the base; often in large colonies. **LEAVES:** Forming clumps, grasslike, up to 3 feet long and ¼ inch wide, flat, channeled only near the base, with rough margins; the leaf base encloses the stem in a reddish-brown sheath. **FEMALE FLOWER CLUSTERS:** 3–6 spikes per stem, up to 1¾ inches long and ⅓ inch wide, cylinder-shaped, most of them drooping on long, slender stalks. **MALE FLOWER CLUSTERS:** 1–4 slender, erect spikes at the top of the stem, up to 2¼ inches long. **FLOWERING:** May–Jun. **RANGE:** FL west to TX, north to MO, OH, and VA. **SIMILAR TO** other sedges, but distinguished by the old leaf remains and the slender, drooping flower clusters.

Fringed Sedge

Carex crinita Lamarck
Sedge Family (Cyperaceae)

PERENNIAL HERB found in floodplains, beaver swamps, pond edges, and other wetlands in north Georgia. **STEMS:** Up to 5 feet tall, 3-angled. **LEAVES:** Long and grasslike, up to ½ inch wide, with rough margins and a groove running down the center; the leaf base encloses the stem in a reddish-brown sheath. **FEMALE FLOWER CLUSTERS:** 2–6 spikes per stem, up to 4 inches long and ¼ inch wide, covered with stiff bristles, drooping on long, slender stalks. **MALE FLOWER CLUSTERS:** 1–3 spikes at the top of the stem, up to 2½ inches long, slender, drooping, soft, not bristly. **FLOWERING:** May–Jul. **RANGE:** FL west to TX, north to MB and NS. **SIMILAR TO** Bearded or Bristly Sedge (*Carex comosa*), which also occurs in wetlands; its female spikes are shorter (less than 2½ inches long) and wider (about ½ inch), erect or drooping on short stalks.

Blue Sedge, Southern Waxy Sedge

Carex glaucescens Elliott
Sedge Family (Cyperaceae)

PERENNIAL HERB found in wetlands in the Ridge & Valley and Coastal Plain. **STEMS:** Up to 4 feet tall, 3-angled. **LEAVES:** Up to 30 inches long and ⅓ inch wide. **FEMALE FLOWER CLUSTERS:** 3–6 spikes per stem, each spike up to 4 inches long and ¼ inch wide, bristly, drooping on long, slender stalks when mature; a few male flowers may occur at the tips. **MALE FLOWER CLUSTERS:** 1 spike at the top of the stem, up to 2½ inches long, erect, not bristly. **FLOWERING:** Jul–Sep. **RANGE:** FL west to TX, north to TN and VA. **NOTES:** All parts of the plant have a white, waxy coating that gives them a blue-green cast. **SIMILAR TO** Warty Sedge (*Carex verrucosa*), which occurs in wetlands mostly in the Coastal Plain; its female spikes are held erect.

Lurid Sedge

Carex lurida Wahlenberg
Sedge Family (Cyperaceae)

PERENNIAL HERB found in wetlands throughout Georgia. **STEMS:** Up to 4 feet tall, sharply 3-angled, the base enclosed in a reddish-purple leaf sheath. **LEAVES:** About 12 inches long and ⅓ inch wide, grooved down the center, usually in large, dense clumps. **FEMALE FLOWER CLUSTERS:** 1–4 spikes per stem, each up to 1½ inches long and ½ inch wide, yellow-green, cylindrical, bristly, closely packed with pointed fruit sacs about ⅓ inch long. **MALE FLOWER CLUSTERS:** In 1 slender spike at the top of the stem, less than 2 inches long, erect, not bristly. **FLOWERING:** Jun–Sep. **RANGE:** FL west to TX, north to WI and NL. **SIMILAR TO** Hop Sedge (*Carex lupulina*) and Giant Sedge (*C. gigantea*), which also occur in wetlands throughout Georgia; both have fruit sacs longer than ½ inch.

Baldwin's Flat-Sedge

Cyperus croceus Vahl
SYNONYM: *Cyperus globulosus* Aublet
Sedge Family (Cyperaceae)

PERENNIAL HERB found in savannas, flatwoods, sandy woodlands, and open, disturbed areas throughout Georgia (except the mountains). **STEMS:** Up to 18 inches tall, 3-angled, smooth. **LEAVES:** Up to 17 inches long and ¼ inch wide, smooth, clustered at the base of the stem. **FLOWER HEADS:** About ½ inch wide, round, with short, loosely spreading branchlets; each head is held at the tip of a slender stalk that rises from the top of the stem—except for the central head, which is stalkless; several leaflike bracts spread from the base of the cluster. **FLOWERS:** Enclosed in the overlapping scales on the branchlets. **FLOWERING:** Jul–Oct. **RANGE:** FL west to TX, north to OK and NJ. **SIMILAR TO** Globe Flatsedge (*Cyperus echinatus*), which occurs in dry uplands throughout Georgia; it has dense, compact flower clusters with 50–100 branches.

Fragrant Flat-Sedge

Cyperus odoratus Linnaeus
Sedge Family (Cyperaceae)

ANNUAL OR PERENNIAL HERB found in wetlands in the Coastal Plain and a few Piedmont counties. **STEMS:** Usually ½–2 feet tall, 3-angled, purplish at the base, roots white. **LEAVES:** 2–24 inches long and ⅛–½ inch wide, clustered at the base of the stem. **FLOWER HEADS:** ½–1½ inches long, cylindrical, with many short, spreading branchlets; 1–3 heads are held at the tip of a slender stalk that rises from the top of the stem; 3–10 leaflike bracts spread from the base of the flower cluster. **FLOWERING:** Jul–Sep. **RANGE:** Throughout the tropics and most of North America. **NOTES:** When the seeds drop, the flowering branchlets shatter apart. **SIMILAR TO** Red-root Flat-sedge (*Cyperus erythrorhizos*); its roots and lower stems are red; when its seeds drop, the branchlets are left intact in the flower head.

Plukenet's Flat-Sedge

Cyperus plukenetii Vahl
Sedge Family (Cyperaceae)

PERENNIAL HERB found in sandhills and other dry, sandy woodlands and clearings in the Coastal Plain and a few Piedmont counties. **STEMS:** 1–3 feet tall, round or slightly 3-angled, rough-hairy, with 3–10 leaflike bracts at the top. **LEAVES:** Up to 28 inches long and ¼ inch wide, hairy, clustered at the base of the stem. **FLOWER CLUSTERS:** With 6–12 rough-hairy branches radiating from the top of the stem, each bearing at its tip an oval, burlike head composed of many slender, down-pointing spikelets. **FLOWERING:** Jul–Oct. **RANGE:** FL west to TX, north to MO and NJ. **SIMILAR TO** Rough Flat-sedge (*Cyperus retrofractus*), which occurs in dry, sandy or rocky uplands in the Piedmont and a few Coastal Plain counties; its stems are sharply 3-angled and the flower cluster branches are mostly smooth.

Three-Way Sedge

Dulichium arundinaceum (Linnaeus) Britton
Sedge Family (Cyperaceae)

PERENNIAL HERB found in marshes, stream and lake edges, bogs, and ditches in the Coastal Plain and a few north Georgia counties. **STEMS:** Up to 3 feet tall, round or slightly 3-angled in cross-section, jointed, encased by leaf sheaths, spreading by rhizomes and forming large colonies. **LEAVES:** Blades up to 3 inches long and ⅓ inch wide, spreading horizontally from 3 sides of the stem (when seen from above), linear, pointed, flat, with rough margins. **FLOWER CLUSTERS:** Spreading from the axils of the upper leaves, up to 2 inches long, narrow, pointed, brownish-green, with 3–10 tiny, overlapping scales, each enclosing a tiny flower. **FLOWERING:** Jul–Oct. **RANGE:** Most of North America. **SIMILAR TO** no other wildflower in Georgia.

White-Top Sedge

Rhynchospora colorata (Linnaeus) H. Pfeiffer
SYNONYM: *Dichromena colorata* (Linnaeus) Hitchcock
Sedge Family (Cyperaceae)

PERENNIAL HERB found in wet savannas, dune swales, marshes, and wet ditches in the Coastal Plain. **STEMS:** Up to 2 feet tall, 3-angled, leafy on the lower half. **LEAVES:** Up to 16 inches long, narrow and grasslike, alternate. **FLOWER CLUSTERS:** With 3–7 narrow, spreading and drooping, 4-inch-long, green-and-white bracts surrounding a group of yellowish-white spikelets; the white part of the bract is less than 1 inch long and gradually tapers into the green part. **FLOWERING:** May–Sep. **RANGE:** FL west to TX, north to AR and VA; Central America, West Indies. **SIMILAR TO** Broad-leaf White-top Sedge (*Rhynchospora latifolia*), which occurs in similar habitats; it is a larger plant with 6–10 bracts, the white part 1–2 inches long, abruptly tapering into the green part.

Woolly Bulrush

Scirpus cyperinus (Linnaeus) Kunth
Sedge Family (Cyperaceae)

PERENNIAL HERB found in wetlands and wet ditches throughout Georgia. **STEMS:** Up to 6 feet tall, stout, 3-angled, leafy. **LEAVES:** Up to 32 inches long, narrow and grasslike, alternate, with rough margins and a channel the length of the leaf. **FLOWER CLUSTERS:** Held at the top of the stem, 3–6 inches long, with many drooping branches and 2 or 3 leafy bracts beneath. There are 200–500 flowers and fruits per cluster, each surrounded by reddish-brown scales and 6 long, curly bristles, giving the whole cluster a rusty, woolly look. **FLOWERING:** Jul–Sep. **RANGE:** FL west to TX, north to BC and NL. **SIMILAR TO** Leafy Bulrush (*Scirpus polyphyllus*), which occurs in wetlands in north Georgia; it is a smaller plant with its flower clusters on erect branches.

Wild Yam

Dioscorea villosa Linnaeus
Yam Family (Dioscoreaceae)

PERENNIAL HERBACEOUS VINE found in moist, hardwood forests throughout Georgia. **STEMS:** Up to 23 feet long, twining (without tendrils) over other plants, slender, angled. **LEAVES:** Up to 5 inches long and wide with 3-inch stalks; alternate, opposite, or whorled; heart-shaped, entire, smooth, with 7–11 curved main veins. **FLOWERS:** Less than ¼ inch wide, greenish-white, with 6 tepals; female and male flowers are on separate plants, held on long, drooping stalks. **FRUIT:** A 3-winged, golden green capsule about 1 inch wide, best seen in winter when the tan, papery capsules persist on dried vines. **FLOWERING:** Apr–Jun. **RANGE:** FL west to TX, north to MN, ON, and MA. **SIMILAR TO** Cinnamon Vine (*Dioscorea polystachya*), an invasive pest plant from China; its leaves are arrowhead-shaped and its stems are round, often bearing small "potatoes." It should be eradicated wherever it is found. Also see Smooth Carrion Flower (*Smilax herbacea*), page 418.

Flattened Pipewort

Eriocaulon compressum Lamarck
Pipewort Family (Eriocaulaceae)

PERENNIAL HERB found in ponds, bogs, wet savannas and flatwoods, and wet ditches in the Coastal Plain. **STEMS:** Up to 28 inches tall, ridged, slightly twisted, smooth. **BASAL LEAVES:** 2–12 inches long, tapering from a 1½-inch-wide base to a narrowly pointed tip. **FLOWER HEADS:** Solitary at the top of the stem, about ½ inch wide, domed, round in outline, soft and easily compressed, tightly packed with tiny, hairy, white flowers. **FLOWERING:** Mar–Oct. **RANGE:** FL west to TX, north to NJ. **NOTES:** The black tips of bracts and stamens look like black pepper scattered across the flower head. **SIMILAR TO** Ten-angled Pipewort (*Eriocaulon decangulare*), which occurs in similar habitats; it has hard heads that do not easily compress, and its leaves have bluntly pointed or rounded tips; it flowers Jun–Oct. Also see White Bog Buttons (*Lachnocaulon anceps*), below.

White Bog Buttons

Lachnocaulon anceps (Walter) Morong
Pipewort Family (Eriocaulaceae)

PERENNIAL HERB found in pine flatwoods, bogs, and wet clearings in the Coastal Plain and a few north Georgia counties. **ROOTS:** Slender, fibrous, dark brown, branched. **STEMS:** 6–16 inches tall, very thin, twisted, covered with long, spreading hairs. **LEAVES:** In mounded clumps, each leaf 1–2⅜ inches long and about ¼ inch wide at the base, tapering to a point, smooth or hairy. **FLOWER HEADS:** Solitary at the top of the stem, about ¼ inch wide, round, domed or slightly flattened, white or grayish-white, tightly packed with tiny, hairy flowers. **FLOWERING:** May–Oct. **RANGE:** FL west to TX, north to TN and NJ; Cuba. **SIMILAR TO** Small's Bog Buttons (*Lachnocaulon minus*), which occurs in similar habitats; its heads are grayish-brown. Yellow Hatpins (*Syngonanthus flavidulus*) has thick, white, spongy, unbranched roots, and its heads are yellowish-white.

Redroot

Lachnanthes caroliniana (Lamarck) Dandy

Bloodwort Family (Haemodoraceae)

PERENNIAL HERB found in wet savannas, pine flatwoods, and disturbed wetlands in the Coastal Plain. **STEMS:** 1–3 feet tall, hairy only near the top. **BASAL LEAVES:** Up to 18 inches long and ¾ inch wide, linear and pointed, overlapping at their bases; stem leaves few, smaller, alternate, clasping the stem. **FLOWERS:** About ⅓ inch long, with 6 tepals that are yellow on the inner surface, silvery green and very hairy on the outer surface; a style and 3 stamens extend well beyond the flower. **FLOWERING:** Jun–Sep. **RANGE:** FL west to LA, north to NS. **NOTES:** The roots and rhizomes have red sap. **SIMILAR TO** Golden Crest (*Lophiola americana*), a Georgia Special Concern species that occurs in the same habitats; its stem is completely hairy, and the flowers have 6 stamens and a puff of yellow hairs in the center.

Orange Day-Lily

Hemerocallis fulva (Linnaeus) Linnaeus

Day-Lily Family (Hemerocallidaceae)

PERENNIAL HERB found on old homesites, roadsides, and other disturbed areas throughout Georgia. **STEMS:** Up to 5 feet tall, smooth, stout. **BASAL LEAVES:** Up to 3 feet long and 1 inch wide, arching downward; stem leaves few and much smaller. **FLOWERS:** Up to 6 inches wide, orange with a yellow throat, with 6 spreading and downcurved tepals; the edges of the tepals are crimped. **FLOWERING:** May–Jul. **RANGE:** Native to Asia, Orange Day-lily was brought to North America in the 17th century and has since escaped from cultivation and spread throughout much of the continent. **NOTES:** Each flower lasts only a single day. **SIMILAR TO** another, less common Asian garden escape, Yellow Day-lily (*Hemerocallis lilioasphodelus*), which has yellow flowers.

Yellow Star-Grass

Hypoxis hirsuta (Linnaeus) Coville
Star-Grass Family (Hypoxidaceae)

PERENNIAL HERB found in upland forests, meadows, and disturbed areas in north Georgia and a few Coastal Plain counties. **STEMS:** Up to 8 inches tall, hairy, with a cluster of 2–7 flowers at the top. **LEAVES:** 4–20 inches long and up to ⅓ inch wide, grasslike, clustered at the base of the stem, usually firm and nearly erect, usually hairy, sometimes smooth, the midvein slightly off-center. **FLOWERS:** Up to ¾ inch across, with 6 yellow, spreading tepals that are longer than the ovary. **FLOWERING:** Mar–Jun. **RANGE:** FL west to NM, north to MN, ON, and ME. **SIMILAR TO** Swamp Star-grass (*Hypoxis curtissii*), which occurs in swamp forests and floodplains in the Coastal Plain; its leaves are soft and flexible and mostly smooth, and its tepals are shorter than the ovary. Also see Yellow Sunny-bells (*Schoenolirion croceum*), page 367.

Fringed Star-Grass

Hypoxis juncea J. E. Smith
Star-Grass Family (Hypoxidaceae)

PERENNIAL HERB found in wet pine savannas and flatwoods in the Coastal Plain. **STEMS:** Up to 8 inches tall, hairy, with a cluster of 2–7 flowers at the top. **LEAVES:** 2–7½ inches long, very narrow, stiff, resembling pine needles, U-shaped in cross-section, smooth or hairy. **FLOWERS:** ½–1 inch wide, with 6 yellow, spreading tepals that are longer than the ovary and shorter than the flower stalk. **FLOWERING:** All year, especially in response to fire, but usually in the spring. **RANGE:** FL west to AL, north to NC. **SIMILAR TO** Savanna Star-grass (*Hypoxis rigida*), which occurs in similar habitats; its tepals are longer than the flower stalk; the leaves are very narrow but not U-shaped in cross-section, and old leaf bases persist as fibrous bristles.

Irises (Iridaceae)

The striking beauty of iris flowers is due to their vivid colors and to the unusual shapes and sizes of the flower parts. An outer whorl of 3 spreading or down-curved sepals (called "falls") is usually the showiest part of the flower. The colorful patch or crest on each sepal is a nectar guide directing insects to the nectar and to the pollen-bearing stamens hidden between the falls and the petal-like stigmas that curve over the falls. Three small, more or less erect petals are in the center of the flower. (An exception to this typical pattern is Blackberry-lily, *Iris domestica*, p. 384). Iris leaves are also distinctive. They arise in a group from the tip of an underground stem (rhizome), overlap at their bases, and spread into a fan shape, an arrangement called "equitant." Twelve species of iris grow wild in Georgia, including a few that have escaped from cultivation, such as Blackberry-lily, Bearded Iris (*Iris germanica*), and Sweet Iris (*Iris pallida*).

Blue-eyed grasses, in the genus *Sisyrinchium*, are also members of the Iris Family and are definitely not grasses. Their flowers, with six radiating tepals, could hardly look more different from the irises. There are about 12 species of blue-eyed grass in Georgia, and most are difficult to identify to species. The species described here are the most common and the easiest to recognize.

Crested Iris

Iris cristata Aiton
Iris Family (Iridaceae)

PERENNIAL HERB found in moist to dry, upland, hardwood forests, mostly in the upper Piedmont and mountains of north Georgia; the plants are connected by rhizomes on the soil surface and form colonies. **LEAVES:** Up to 6 inches long and 1 inch wide, curved, with overlapping bases. **FLOWERS:** Up to 3 inches high and 5 inches across, bluish-purple, each fall with a large, white patch outlined in purple and crested with 3 toothed, yellow ridges; not fragrant. **FLOWERING:** Apr–May. **RANGE:** GA west to OK, north to MO and MA. **SIMILAR TO** Dwarf Iris (*Iris verna*), which occurs in dry woodlands in the Coastal Plain and mountains (rarely in the Piedmont); its flowers are very fragrant, its falls have an orange-and-white patch but no crest, the leaves are more or less straight, and the rhizomes are not visible at the soil surface.

Blackberry-Lily

Iris domestica (Linnaeus) Goldblatt & Mabberley
SYNONYM: *Belamcanda chinensis* (Linnaeus) A. P. de Candolle
Iris Family (Iridaceae)

PERENNIAL HERB found in dry, upland forests and in woodlands around granite outcrops in the eastern Piedmont of Georgia. **STEMS:** Up to 3 feet tall, leafless, with 3–5 short branches near the top. **LEAVES:** 1–2 feet long and ¾ inch wide, with a whitish, waxy coating; arising from the tip of an underground stem (rhizome) and spreading into a fan shape. **FLOWERS:** Up to 2 inches wide, with 6 spreading, orange or yellow tepals with purple spots. **FRUIT:** An oval, 1-inch capsule, splitting to reveal several shiny, black seeds in a blackberry-like cluster. **FLOWERING:** Jun–Aug. **RANGE:** FL west to TX, north to MN and VT; native to China and widely escaped from cultivation. **SIMILAR TO** no other wildflower in Georgia.

Yellow Flag

Iris pseudacorus Linnaeus
Iris Family (Iridaceae)

PERENNIAL HERB found in swamps, marshes, wet ditches, and edges of lakes, ponds, and streams in north Georgia. **STEMS:** Up to 4 feet tall. **LEAVES:** Up to 3 feet long and 1 inch wide, dark green, smooth, entire. **FLOWERS:** About 3 inches high and wide, yellow, with brown lines and speckles on the falls. **FLOWERING:** May–Jun. **RANGE:** Native to Eurasia and Africa, now escaped from cultivation and widely naturalized, sometimes aggressively so, in wetlands across most of North America. In size and shape, Yellow Flag is **SIMILAR TO** the blue-flowered wetland irises, but it is the only yellow-flowered iris occurring in Georgia's natural wetlands.

Blue Flag

Iris virginica Linnaeus
Iris Family (Iridaceae)

PERENNIAL HERB found in swamps, wet ditches, and edges of lakes, ponds, and streams in the Coastal Plain and a few north Georgia counties. **STEMS:** Up to 3 feet tall, often fallen over after flowering. **BASAL LEAVES:** Up to 3 feet long and 1½ inches wide, smooth, entire, often bluish-green. **STEM LEAVES:** Similar to basal leaves but smaller, alternate. **FLOWERS:** About 4 inches high and wide; blue, violet, or purple; the falls have a central white-and-yellow patch veined with dark purple, and the petals are marked with white. **FLOWERING:** May–Jul. **RANGE:** FL west to TX, north to MN and NS. **SIMILAR TO** two species that are rare in Georgia: Lamance Iris (*Iris brevicaulis*), which has a flower stalk never more than 1½ feet tall; and Anglepod Blue Flag (*I. hexagona*), which has a zigzag flower stalk.

Atlantic Blue-Eyed Grass

Sisyrinchium atlanticum P. Miller
SYNONYM: *Sisyrinchium mucronatum* var. *atlanticum* (E. P. Bicknell) H. E. Ahles
Iris Family (Iridaceae)

PERENNIAL HERB found in dry, sandy or rocky woods and clearings throughout Georgia. **STEMS:** Up to 18 inches tall and 1/10 inch wide, flat and narrowly winged, branched only at the top, without old, dried leaf bases at the base. **LEAVES:** Up to 14 inches long, less than ⅛ inch wide, alternate, overlapping at the base of the stem, smooth. **FLOWERS:** ½–¾ inch wide, with 6 blue, purple, or white tepals, the tips rounded or notched, tipped with a tiny, pointed tooth. **FLOWERING:** Mar–Jun. **RANGE:** FL west to LA, north to WI and NS. **SIMILAR TO** other blue-eyed grasses, which are hard to tell apart. Narrow-leaf Blue-eyed Grass (*Sisyrinchium angustifolium*) is common in moist areas in north Georgia; its flowers are less than ½ inch wide and are blue with purple stripes.

Nash's Blue-Eyed Grass

Sisyrinchium nashii E. P. Bicknell
SYNONYM: *Sisyrinchium fibrosum* E. P. Bicknell
Iris Family (Iridaceae)

PERENNIAL HERB found in dry to moist woodlands throughout Georgia. **STEMS:** Up to 19 inches tall and 1⁄10 inch wide, branched, flattened and narrowly winged, smooth on the margins, dull green, with tufts of old, dried leaves at the base. **LEAVES:** Up to 12 inches long, narrow and grasslike, alternate, overlapping at the base of the stem, smooth on the margins, dull green. **FLOWERS:** ½–1 inch wide, with 6 blue or purple tepals and a yellow "eye," the tepals rounded or notched, tipped with a tiny, pointed tooth. **FLOWERING:** Apr–Jun. **RANGE:** FL west to MS, north to TN and NC. **SIMILAR TO** other blue-eyed grasses, which are hard to tell apart. Coastal Plain Blue-eyed Grass (*Sisyrinchium fuscatum*) occurs in Coastal Plain sandhills; its stems and branches are rough on the margins.

Annual Blue-Eyed Grass

Sisyrinchium rosulatum E. P. Bicknell
Iris Family (Iridaceae)

ANNUAL HERB found in lawns, roadsides, and other clearings throughout Georgia. **STEMS:** Up to 14 inches tall but usually much shorter due to mowing; there is no clump of old, dried leaf bases at the base of the stem. **LEAVES:** Grasslike, up to 14 inches long. **FLOWERS:** Up to ¾ inch across (usually much smaller), with 6 white, yellow, or lavender-pink tepals and a yellow "eye" encircled by a purplish ring; the tepal tips are pointed, rounded, or notched, with a tiny tooth. **FLOWERING:** Apr–May. **RANGE:** FL west to AR and TX, north to VA, possibly introduced. **SIMILAR TO** other blue-eyed grasses except for flower color; this is the easiest blue-eyed grass to identify.

Soft Rush

Juncus effusus Linnaeus
Rush Family (Juncaceae)

PERENNIAL HERB found in freshwater wetlands throughout Georgia. **STEMS:** Up to 5 feet tall, bright green, round in cross-section, forming clumps. **LEAVES:** None; a brown, papery, bristle-tipped sheath encloses the base of each stem. **FLOWER CLUSTER:** Appears to emerge from the side of the stem several inches below the tip; actually the cluster is at the top of the stem, and a pointed, green bract identical with the stem extends a few inches above the cluster. **FLOWERS:** Tiny, golden tan, with 6 tepals, held singly at the tips of short stalks. **FLOWERING:** Jun–Sep. **RANGE:** Most of North America; Mexico. **SIMILAR TO** Leathery Rush (*Juncus coriaceus*), which occurs throughout Georgia in similar habitats; the sheaths at the base of its stems occasionally have blades instead of bristle tips, and the bract at the top of the stem (above the flower cluster) is grooved.

Hedgehog Wood Rush

Luzula echinata (Small) Hermann
Rush Family (Juncaceae)

PERENNIAL HERB found in floodplains and moist upland forests in north Georgia and a few Coastal Plain counties. **STEMS:** Up to 18 inches tall, smooth, loosely clustered. **LEAVES:** Up to 6 inches long and ¼ inch wide, flat, grasslike, often reddish, clustered at the base of the plant and alternate on the stem, with a stiff, pointed tip and hairy margins. **FLOWER CLUSTERS:** At the top of the stem, with 4–15 round heads, each about ⅓ inch wide, on long, widely spreading stalks. **FLOWERS:** Tiny, brown, with 6 tepals. **FRUIT:** Tiny, round, brown, 3-lobed capsules. **FLOWERING:** Mar–Aug. **RANGE:** GA west to TX, north to IA and MA. **SIMILAR TO** Common Wood Rush (*Luzula multiflora*), which has cylindrical heads on erect stalks. Hairy Wood Rush (*L. acuminata*) has individual flowers instead of heads at the tips of stalks.

Lily Family (Liliaceae)

The Lily Family (Liliaceae) was once recognized as one of the largest plant families, with 4,000 species, and included such wildly different plants as trilliums, trout lilies, and greenbriers. Twenty-first-century botanists have used molecular genetics to divide the broadly defined Lily Family into as many as 30 families or as few as 15, depending on the researcher. This much-needed reclassification is a work-in-progress and will be refined in the coming years. For this guide, species in the broadly defined Lily Family, as presented in the *Manual of the Vascular Flora of the Carolinas*, are placed in the following families, according to the Angiosperm Phylogeny Group III system (Stevens 2001) and/or the *Flora of the Southern and Mid-Atlantic States* (Weakley 2014).

Alliaceae—Onion Family, p. 361
Allium, Onion
Nothoscordum, False Garlic

Asparagaceae—Asparagus Family, pp. 365–367
Maianthemum, *Smilacina*, May-flower, Solomon's Plume
Polygonatum, Solomon's Seal
Schoenolirion, Sunny-bells

Colchicaceae—Meadow Saffron Family, pp. 369–370
Prosartes (*Disporum*), Mandarin
Uvularia, Bellwort

Hemerocallidaceae—Day-Lily Family, p. 381
Hemerocallis, Orange Day-lily

Hypoxidaceae—Star-Grass Family, p. 382
Hypoxis, Star-grass

Liliaceae—True Lilies, pp. 389–391
Clintonia, Wood Lily
Erythronium, Trout Lily
Lilium, Lilies
Medeola, Indian Cucumber Root

Melanthiaceae—Bunchflower Family, pp. 392–394
Amianthium, Fly Poison
Chamaelirium, Devil's Bit
Stenanthium, Feather-bells and others
Veratrum, *Melanthium*, Bunchflower

Nartheciaceae—Bog Asphodel Family, p. 395
Aletris, Colic Root

Smilacaceae—Smilax Family, pp. 418–420
Smilax, Greenbrier and others

Tofieldiaceae—False Asphodel Family, p. 420
Triantha, *Tofieldia*, Bog Asphodel

Trilliaceae—Trillium Family, pp. 421–426
Trillium, Trillium

Speckled Wood-Lily

Clintonia umbellulata (Michaux) Morong
Lily Family (Liliaceae)

PERENNIAL HERB found in moist, hardwood forests in the north Georgia mountains. **LEAVES:** 6–12 inches long and 1–3½ inches wide, in a basal rosette, oblong to elliptic, dark green, entire, with a deeply inset, hairy midvein and long hairs on the margins. **FLOWER CLUSTERS:** Rounded, held at the top of a slender stalk 8–24 inches tall, with 6–25 flowers. **FLOWERS:** About ½ inch wide, with 6 spreading, white tepals usually speckled with purple. **FRUIT:** A round, black berry about ¼ inch wide. **FLOWERING:** May–Jun. **RANGE:** Found only in the Appalachian Mountains from GA north to NY. **SIMILAR TO** Blue-bead Lily (*Clintonia borealis*), a Georgia Special Concern species found in high-elevation forests; its flowers are yellow, the berries are blue, and the leaves lack hairs on the margins and midvein.

Dimpled Trout Lily

Erythronium umbilicatum Parks & Hardin
Lily Family (Liliaceae)

PERENNIAL HERB found in moist forests, bottomlands, and granite outcrop seepages in north and southwest Georgia. **LEAVES:** Up to 7 inches long and 3 inches wide, 2 leaves per flowering plant, oval to lance-shaped, entire, fleshy, mottled purplish-brown. **FLOWERS:** Up to 1 inch long, with 6 yellow, purple-speckled tepals and 6 dangling stamens with long, brown or purple anthers. **FRUIT:** Oval, green capsules up to 1 inch long, usually lying on the ground, the broad tip slightly indented. **FLOWERING:** Feb–Apr. **RANGE:** FL west to AL, north to KY and MD. **NOTES:** The tepals close at night and then curve strongly backward in the morning. **SIMILAR TO** American Trout Lily (*Erythronium americanum*), which occurs in moist forests in north Georgia; it has yellow anthers, and each tepal has small "ears" or pockets at its base. The fruit is rounded or pointed at the tip and held above the ground.

Pine Lily, Catesby's Lily

Lilium catesbaei Walter
Lily Family (Liliaceae)

PERENNIAL HERB found in wet pine savannas and pitcherplant bogs in the Coastal Plain. **STEMS:** 1–3 feet tall. **LEAVES:** 1–3 inches long and up to ½ inch wide, alternate, narrowly elliptic, smooth, entire, closely pressed against the stem. **FLOWERS:** 6–9 inches across, held erect at the top of the stem and facing upward, the 6 tepals spreading and curved at the tips, orange with a yellow, maroon-spotted blotch above a narrow, green base. **FLOWERING:** Jun–Oct. **RANGE:** FL west to LA, north to VA. **NOTES:** Pine Lily has the largest flower of any North American lily. **SIMILAR TO** no other wildflower in Georgia's Coastal Plain.

Carolina Lily

Lilium michauxii Poiret
Lily Family (Liliaceae)

PERENNIAL HERB found in dry to moist forests in north Georgia and a few Coastal Plain counties. **STEMS:** 1–3 feet tall. **LEAVES:** Up to 4½ inches long and 1½ inches wide, whorled and alternate on the same plant, tapering at both ends, widest above the middle, smooth, entire, fleshy, paler on the lower surface. **FLOWERS:** 1–4 per plant, nodding and facing downward, the 6 tepals strongly curved upward, orange toward the tips, yellow (never green) and maroon-spotted at the base; 6 stamens with rust-colored anthers dangle from the flower. **FLOWERING:** Jul–Aug. **RANGE:** FL west to TX, north to VA. **NOTES:** Carolina Lily is the only fragrant lily east of the Rocky Mountains. Its flowers are pollinated by swallowtail butterflies. **SIMILAR TO** Turk's-cap Lily (*Lilium superbum*), page 391.

Turk's-Cap Lily

Lilium superbum Linnaeus
Lily Family (Liliaceae)

PERENNIAL HERB found in cove forests, boulderfields, moist slopes, and streamsides in the north Georgia mountains and a few Piedmont and Coastal Plain counties. **STEMS:** 4–9 feet tall. **LEAVES:** Up to 10 inches long and 1 inch wide, in whorls (occasionally single and alternate), lance-shaped or narrowly elliptic, widest at or below the middle, entire, smooth. **FLOWER CLUSTERS:** Large, branched, held at the top of the stem, with 1–22 flowers. **FLOWERS:** 5–8 inches across, nodding downward from the tips of long stalks, the 6 tepals strongly curved upward, orange with maroon spots, with green nectaries forming a star in the center; 6 stamens with rust-colored anthers dangle from the flower. **FLOWERING:** Jul–Aug. **RANGE:** FL west to AR, north to MO and NH. **SIMILAR TO** Carolina Lily (*Lilium michauxii*), page 390.

Indian Cucumber-Root

Medeola virginiana Linnaeus
Lily Family (Liliaceae)

PERENNIAL HERB found in acidic pine-hardwood forests in north Georgia and a few Coastal Plain counties. **STEMS:** 8–30 inches tall, wiry, covered with white, cobwebby hairs. **LEAVES:** 5–9 oval or lance-shaped leaves up to 6⅓ inches long and 2 inches wide, whorled; flowering plants have an upper whorl of 3–5 similar but smaller leaves. **FLOWERS:** Dangling below the upper whorl of leaves, each with 6 strongly curved, yellowish-green tepals, 6 dangling stamens, and a 3-branched style. **FRUITS:** Purple-black berries held above the upper leaves. **FLOWERING:** Mid-Apr–mid-Jun. **RANGE:** FL west to LA, north to QC and NS. **NOTES:** Upper leaves turn red when the plant is in fruit, attracting animals that may disperse the seeds. The leaves are **SIMILAR TO** those of Large Whorled Pogonia (*Isotria verticillata*), page 402.

Bog Moss

Mayaca fluviatilis Aublet
Bog Moss Family (Mayacaceae)

PERENNIAL HERB found in marshes, spring runs, seepages, slow-moving streams, and edges of ponds and streams in the Coastal Plain. **STEMS:** Up to 2 feet long, usually much shorter, floating in water or matted and mosslike on the shore. **LEAVES:** About ½ inch long, spiraled and very crowded around the stem. **FLOWERS:** Less than ½ inch wide, with 3 oval, pink-and-white petals, 3 pointed, green sepals, and 3 yellow-tipped stamens. **FLOWERING:** May–Jul. **RANGE:** FL west to TX, north to NC; West Indies, Central and South America. **SIMILAR TO** several other aquatic plants. Water-milfoil and parrot-feather (*Myriophyllum* spp.) leaves are whorled and finely dissected into 10 or more threadlike segments; their flowers are inconspicuous. Brazilian Waterweed, or Elodea (*Egeria densa*), an aquarium escapee, has inch-long leaves about ⅛ inch wide; its flowers are white.

Fly Poison

Amianthium muscitoxicum (Walter) A. Gray
Bunchflower (Melanthiaceae) or Lily (Liliaceae) Family

PERENNIAL HERB found in moist to dry forests, bogs, and pine savannas throughout Georgia. **STEMS:** 1–4 feet tall, slender, unbranched, smooth. **LEAVES:** Up to 1½ feet long and 1 inch wide, grasslike, deeply channeled, each with a purple sheath at the base; in large clumps of many leaves. **FLOWER CLUSTERS:** Up to 5 inches long, cylindrical or cone-shaped, with many flowers opening from the bottom upward. **FLOWERS:** About ½ inch wide, white, with 6 oval, blunt-tipped tepals that turn green with age. **FLOWERING:** May–Jul. **RANGE:** FL west to LA and OK, north to MO and NY. **NOTES:** Leaves and bulbs are extremely toxic. **SIMILAR TO** Crow Poison (*Stenanthium densum*), page 393.

Devil's Bit, Fairy Wand

Chamaelirium luteum Michaux
Bunchflower (Melanthiaceae), Heloniadaceae (Swamp-Pink), or Lily (Liliaceae) Family

PERENNIAL HERB found in acidic cove forests and pine woodlands in north and southwest Georgia. **STEMS:** ½–4 feet tall, smooth. **BASAL LEAVES:** 3–8 inches long, ½–2½ inches wide, broadly oval or spatula-shaped, evergreen, smooth, with 5–7 parallel veins and many smaller, netted veins. **STEM LEAVES:** Few, smaller than basal leaves, alternate. **FLOWER CLUSTERS:** Female-flowered plants have erect flower spikes up to 12 inches long; male-flowered plants, which are more common, have flower spikes up to 5 inches long that droop at the tip as they mature. **FLOWERS:** About ¼ inch wide, with 6 greenish-white (female) or white (male) tepals. **FLOWERING:** Mar–Jul. **RANGE:** FL west to LA, north to ON and MA. **SIMILAR TO** Galax (*Galax urceolata*), page 164.

Crow Poison, Osceola's Plume

Stenanthium densum (Desrousseaux) Zomlefer & Judd
SYNONYM: *Zigadenus densus* (Desrousseaux) Fernald
Bunchflower (Melanthiaceae) or Lily (Liliaceae) Family

PERENNIAL HERB found in pine savannas and flatwoods in the Coastal Plain. **STEMS:** 1–5 feet tall. **LEAVES:** 4–20 inches long and less than ½ inch wide, usually 3 or fewer (occasionally up to 10), grasslike, with no purple sheath at the base; stem leaves few, short, alternate. **FLOWER CLUSTERS:** At the top of the stem, up to 6 inches tall and 2 inches wide, densely flowered, flowers opening from the bottom upward, the lower flowers turning purplish with age while upper flowers are still in bud. **FLOWERS:** About ⅓ inch wide, white, with 6 oval, blunt-tipped tepals. **FLOWERING:** Mar–Jul. **RANGE:** FL west to TX, north to VA. **SIMILAR TO** Fly Poison (*Amianthium muscitoxicum*), page 392, which has a purple sheath at the base of the leaves.

Feather-Bells

Stenanthium gramineum (Ker-Gawler) Morong
Bunchflower (Melanthiaceae) or Lily (Liliaceae) Family

PERENNIAL HERB found in dry to moist forests and on stream banks in north Georgia and a few southwest Georgia counties. **STEMS:** 2–6½ feet tall, smooth. **BASAL LEAVES:** Few, not in clumps, 8–28 inches long and ¼–1 inch wide, grasslike, rising upward from an underground bulb then bending downward at midpoint, channeled along the midvein. **STEM LEAVES:** Few, smaller than basal leaves, alternate. **FLOWERS:** In a large, branched cluster at the top of the stem, each flower about ½ inch wide, with 6 white, narrow, pointed tepals. **FLOWERING:** Jul–early Sep. **RANGE:** FL west to TX, north to MO and PA. **SIMILAR TO** Fly Poison (*Amianthium muscitoxicum*), page 392, which forms large clumps of leaves and has a purple sheath at the base of each leaf.

Appalachian Bunchflower

Veratrum parviflorum Michaux
SYNONYM: *Melanthium parviflorum* (Michaux) S. Watson
Bunchflower (Melanthiaceae) or Lily (Liliaceae) Family

PERENNIAL HERB found on moist hardwood slopes in the north Georgia mountains. **STEMS:** Up to 5 feet tall, hairy, with branches for most of the length, seldom seen because the plants flower only every few years. **BASAL LEAVES:** 7–14 inches long and up to 5½ inches wide, broadly oval, pleated, smooth, entire. **FLOWERS:** About ½ inch wide, with 6 green, oval, hairless tepals. **FLOWERING:** Jul–early Sep. **RANGE:** GA and AL, north to WV and VA, in the Appalachian Mountains. **NOTES:** Hundreds of basal leaf rosettes may blanket a slope, few or none with a flower stalk. **SIMILAR TO** Corn Lily (*Veratrum viride*), a Georgia Special Concern species that occurs along seepages and creeks in northeast Georgia; its stems are leafy, and the tepals are hairy.

Northern White Colic Root

Aletris farinosa Linnaeus
Bog Asphodel (Nartheciaceae) or Lily (Liliaceae) Family

PERENNIAL HERB found in pine flatwoods and savannas, bogs, upland woodlands, meadows, and roadsides throughout Georgia. **STEMS:** Up to 4 feet tall, smooth, sticky. **LEAVES:** Up to 8 inches long and 1 inch wide, in basal rosettes. **FLOWER CLUSTERS:** 4–14 inches tall, spikelike, held at the top of the stem. **FLOWERS:** Almost ½ inch long, white, tubular with 6 pointed, spreading lobes at the tip; outer surface sticky and bumpy. **FLOWERING:** Apr–Jun. **RANGE:** FL west to TX, north to MN, ON, and ME. **SIMILAR TO** Southern White Colic Root (*Aletris obovata*), which occurs in pine savannas and flatwoods in Georgia's Coastal Plain. Its flowers are less than ¼ inch long and wide, with the tip of the flower rounded and nearly closed.

Yellow Colic Root

Aletris lutea Small
Bog Asphodel (Nartheciaceae) or Lily (Liliaceae) Family

PERENNIAL HERB found in bogs, wet pine savannas and flatwoods, and roadside ditches in the Coastal Plain. **STEMS:** Up to 3 feet tall, unbranched. **LEAVES:** 1½–6 inches long and up to ¾ inch wide, yellowish-green, elliptic or lance-shaped, in basal rosettes. **FLOWER CLUSTERS:** Spikelike, held at the top of the stem. **FLOWERS:** About ¾ inch long, yellow, tubular with 6 pointed, spreading lobes at the tip; outer surface grainy-textured. **FLOWERING:** Mar–May. **RANGE:** FL west to LA, north to GA. **NOTES:** Yellow Colic Root hybridizes with Southern White Colic Root (*Aletris obovata*), above, especially in disturbed areas, and produces a plant with pale yellow flowers. **SIMILAR TO** Golden Colic Root (*Aletris aurea*), which occurs in the same habitats; it blooms mid-May–Jul and has oval, golden yellow flowers that are nearly closed at the tip.

Orchid Family (Orchidaceae)

With their exquisite shapes and colors, unique pollination strategies, and endangered or remote habitats, orchids are often the first plants to lure people out of their cars and offices and into the woods. More than 30,000 species are found around the globe in mind-boggling diversity. Georgia has about 60 orchid species, ranging from large, showy lady's slippers to diminutive coral-roots. They occur in just about every natural community and habitat in Georgia, including disturbed areas.

Although orchids may appear quite different from one another, the flowers follow a basic, bilaterally symmetrical pattern (see "Regular or Irregular: The Basic Patterns of Flowers," p. 275). Orchid flowers have 3 colorful sepals, usually spreading or sometimes forming a hood over the top of the flower. Two of the petals sometimes join with a sepal to form a hood, or they may spread away from the center of the flower. A third petal, called a lip, curves down and forward from the center (in a few species, the lip is erect at the top of the flower). The lip is larger than the other petals and decorated with fringes, crests, and contrasting colors that direct insects and hummingbirds to the interior of the flower where pollination takes place.

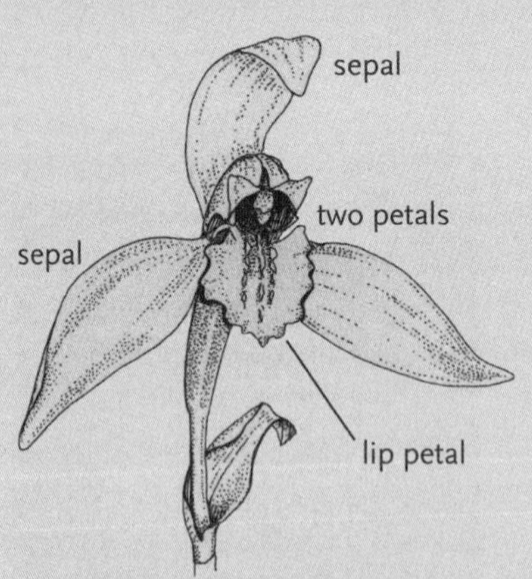

Each orchid species has a close relationship with its pollinator, an interdependence that has shaped both flower and pollinator. This relationship, first described by Charles Darwin in his book *Fertilisation of Orchids* (1862), is a textbook example of coevolution. Orchids are also famous for their deceptive pollination strategies—producing flowers that appear to offer nectar or mating opportunities that result in cross-pollination for the flower but disappointment for the pollinator.

Orchids are closely tied to another life form: fungi. Unless their seeds form a connection with certain soil fungi, they will not germinate. The fungi provide the orchid's tiny, dustlike seeds with the nutrients and moisture needed for germination and seedling growth, and continue to support the mature plant, to one degree or another, throughout its life. Without its special fungus a transplanted orchid will die, which explains why most wild-collected (poached) native orchids die in gardens within a year or two. Take-home message: never dig up wild orchids!

Puttyroot, Adam-and-Eve Orchid

Aplectrum hyemale (Muhlenberg ex Willdenow) Torrey
Orchid Family (Orchidaceae)

PERENNIAL HERB found in rich forests in ravines and bottomlands in a few north Georgia counties. **STEMS:** 7–20 inches tall, leafless, emerging in spring. **LEAVES:** 1 leaf per plant, resting on the ground, up to 8 inches long and 3 inches wide, oval, dark bluish-green with raised, white, parallel veins; leaves emerge in the autumn, persist through the winter, and wither the following spring. **FLOWERS:** Widely spaced at the top of the stem, each about 1 inch wide, with 3 spreading, greenish-yellow and magenta sepals, 2 similarly colored petals arching over the center of the flower, and a white, purple-spotted lip. **FLOWERING:** Late Apr–Jun. **RANGE:** GA west to OK, north to MN and QC. **NOTES:** Puttyroot's bulblike corms contain a putty-like substance once used to mend pottery. **SIMILAR TO** no other wildflower in Georgia.

Bearded Grass-Pink

Calopogon barbatus (Walter) Ames
Orchid Family (Orchidaceae)

PERENNIAL HERB found in wet flatwoods and savannas, seepage slopes, and bogs in the Coastal Plain. **STEMS:** Up to 1 foot tall while flowering, up to 2 feet in fruit. **LEAVES:** 1 or 2 per plant, up to 8 inches long and ½ inch wide, pleated lengthwise, erect, arising from the base of the stem. **FLOWER CLUSTERS:** At the top of the stem, with 3–7 closely spaced flowers opening almost simultaneously. **FLOWERS:** Up to 1 inch across, dark pink (rarely white), with 3 sharply pointed sepals, 2 blunt-tipped petals, and an upright lip petal bearing a crest of golden hairs; not fragrant. **FLOWERING:** Mar–May. **RANGE:** FL west to LA, north to NC. **SIMILAR TO** other grass-pinks but distinguished by the combination of small overall size, closely spaced flowers, and the nearly simultaneous opening of the flowers.

Tuberous Grass-Pink

Calopogon tuberosus (Linnaeus) B.S.P.
Orchid Family (Orchidaceae)

PERENNIAL HERB found in wet flatwoods and savannas, seepage slopes, and bogs, mostly in the Coastal Plain, rarely in north Georgia. **STEMS:** 1–4 feet tall, smooth. **LEAVES:** Up to 20 inches long and 2 inches wide, pleated lengthwise. **FLOWER CLUSTERS:** At the top of the stem, with up to 25 buds and flowers but only a few open at the same time; lowest buds open first. **FLOWERS:** About 1½ inches across, dark pink (rarely pale pink or white), with 3 oval sepals, 2 oblong petals, and an upright, T-shaped lip petal with a crest of golden hairs; faintly fragrant. **FLOWERING:** Apr–Jul. **RANGE:** FL west to TX, north to MB and NL. **SIMILAR TO** Pale Grass-pink (*Calopogon pallidus*), which occurs in similar habitats and has pale pink to white flowers; its 2 side petals are strongly curved and arched backward.

Large Spreading Pogonia

Cleistesiopsis divaricata (Linnaeus) Pansarin & F. Barros
SYNONYM: *Cleistes divaricata* (Linnaeus) Ames
Orchid Family (Orchidaceae)

PERENNIAL HERB found in wet flatwoods and savannas, seepage slopes, and bogs in the Coastal Plain. **STEMS:** 6–30 inches tall, waxy blue-green. **LEAVES:** 1 (rarely 2) at midstem, up to 8 inches long and 1 inch wide, elliptic, entire, smooth, waxy blue-green; a smaller leafy bract rises behind the flower. **FLOWERS:** 1–3 per stem, with 3 narrow, spreading, maroon sepals up to 2½ inches long, and a pink tube 1–2¼ inches long, the lip with scalloped edges and purple speckles inside; smells like daffodils. **FLOWERING:** May–Jun. **RANGE:** FL north to NJ. **SIMILAR TO** Small Spreading Pogonia (*Cleistesiopsis bifaria*), a Georgia Special Concern species that occurs in mountain bogs; it has smaller, odorless flowers, and the lip and sepals are usually less than 1½ inches long.

Spring Coral-Root

Corallorhiza wisteriana Conrad
Orchid Family (Orchidaceae)

PERENNIAL HERB found infrequently in moist forests throughout Georgia. **STEMS:** 4–17 inches tall, slender, yellowish-brown to reddish-purple. **LEAVES:** None. **FLOWERS:** About ⅓ inch long, with reddish-brown sepals and petals forming a hood over a white, purple-spotted lip. **FLOWERING:** Apr–May. **RANGE:** FL west to AZ, north to OR and CT; Mexico. **NOTES:** Lacking chlorophyll, coral-roots cannot photosynthesize; instead they derive nutrients from a three-way relationship with a fungus and another plant (see "A Forest Menage à Trois," p. 168). **SIMILAR TO** Autumn Coral-root (*Corallorhiza odontorhiza*), which occurs in forests in north Georgia and flowers Aug–Oct; its flowers are smaller, less than ¼ inch long.

Pink Lady's Slipper

Cypripedium acaule Aiton
Orchid Family (Orchidaceae)

PERENNIAL HERB found in pine- or oak-dominated forests in north Georgia. **STEMS:** 6–24 inches tall. **LEAVES:** In a basal rosette, each leaf up to 12 inches long and 6 inches wide, elliptic to oblong, hairy with raised, parallel veins. **FLOWERS:** Up to 2½ inches long, with a showy, pink, pouch-shaped lip and 2 narrow, spreading, twisted, reddish-brown or green petals; a large, green sepal curves over the top of the flower. **FLOWERING:** Apr–Jun. **RANGE:** GA, north to AB and NL. **SIMILAR TO** Yellow Lady's Slipper (*Cypripedium parviflorum*), a Georgia Special Concern species with yellow flowers and leafy stems.

Showy Orchis

Galearis spectabilis (Linnaeus) Rafinesque
SYNONYM: *Orchis spectabilis* Linnaeus
Orchid Family (Orchidaceae)

PERENNIAL HERB found in cove forests and on moist slopes in Georgia's Blue Ridge. **STEMS:** 2–8 inches tall, smooth, succulent. **LEAVES:** 2 per plant, up to 8 inches long and 4 inches wide, oval to oblong, fleshy, smooth, shiny, slightly folded up along the midvein. **FLOWER CLUSTERS:** At the top of the stem, with 3–12 flowers, each flower with a leafy bract beneath. **FLOWERS:** About 1 inch long, with lavender-pink petals and sepals forming a hood over a white, spurred lip. **FLOWERING:** Apr–Jun. **RANGE:** GA west to OK, north to MN and NB. **SIMILAR TO** no other wildflower in Georgia.

Downy Rattlesnake-Orchid

Goodyera pubescens (Willdenow) R. Brown
Orchid Family (Orchidaceae)

PERENNIAL HERB found in oak-pine and hemlock forests in north Georgia. **STEMS:** 4–18 inches tall, very hairy. **LEAVES:** Several leaves form a rosette at the base of the stem, each up to 2½ inches long and 1¼ inches wide, evergreen, oval to elliptic, dark bluish-green with a bright white midvein and the side veins traced in white. **FLOWER CLUSTERS:** Spikelike, with 10 or more flowers evenly distributed around the upper stem. **FLOWERS:** About ¼ inch long, white, hairy, with 2 petals and 1 sepal forming a hood over a pouch formed by 1 petal and 2 sepals. **FLOWERING:** Jul–Aug. **RANGE:** FL west to AR, north to MN and NS. The flower spike is **SIMILAR TO** those of Ladies' Tresses (pp. 405–407), but the white-patterned leaves are unique.

Water-Spider Orchid

Habenaria repens Nuttall
Orchid Family (Orchidaceae)

PERENNIAL HERB found in marshes, swamps, ponds, and wet ditches in the Coastal Plain, often in floating mats. **STEMS:** Up to 3 feet tall, leafy, smooth. **LEAVES:** Up to 10 inches long and 1¾ inches wide, lance-shaped, clasping the stem, prominently veined. **FLOWER CLUSTERS:** Spikelike, at the top of the stem, with many crowded flowers. **FLOWERS:** About ½ inch across, green, the petals and lip divided into very narrow segments, giving the flower a spidery look; a slender spur less than ½ inch long curves below each flower. **FLOWERING:** Apr–Nov. **RANGE:** FL west to TX, north to NC; West Indies, Latin America. **SIMILAR TO** Michaux's Orchid (*Habenaria quinqueseta*), a Georgia Threatened species that occurs in the Coastal Plain; its flowers are greenish-white with a spur 1½–4 inches long.

Crested Coral-Root

Hexalectris spicata (Walter) Barnhart
Orchid Family (Orchidaceae)

PERENNIAL HERB found in moist to dry forests, often over mafic or limestone bedrock, with red cedars, pines, or oaks, in the Piedmont and a few Coastal Plain counties. **STEMS:** 1–2½ feet tall, leafless; yellow, brown, or reddish. **LEAVES:** None. **FLOWERS:** About 1 inch wide, yellow, tan, or greenish with purple stripes, sepals and petals spreading, the downcurved lip with 5–7 purple ridges (crests). **FLOWERING:** Apr–Aug. **RANGE:** FL west to AZ, north to MO and MD. **NOTES:** Lacking chlorophyll, Crested Coral-root cannot photosynthesize; it derives nutrients from a three-way relationship with a fungus and another plant (see "Menage à Trois," p. 168). **SIMILAR TO** no other wildflower in Georgia.

Large Whorled Pogonia

Isotria verticillata (Muhlenberg ex Willdenow) Rafinesque
Orchid Family (Orchidaceae)

PERENNIAL HERB found in acidic soils of wetlands and upland forests, mostly in north Georgia. STEMS: Up to 16 inches tall, smooth, purplish. LEAVES: 5 or 6 in a whorl at the top of the stem, each leaf up to 4 inches long and 2 inches wide during flowering, enlarging as the fruit matures, oval, lance-shaped, or elliptic. FLOWERS: 1 (rarely 2), up to 2 inches wide, with 3 widely spreading, purplish-green sepals, 2 yellow petals, and a white or yellowish lip edged and streaked with purple. FLOWERING: Apr–Jun. RANGE: FL west to TX, north to MO, MI, and ME. SIMILAR TO Small Whorled Pogonia (*Isotria medeoloides*), a federally Threatened species that occurs in pine-oak woodlands in the Blue Ridge; it has smaller flowers and its stem has a waxy, whitish coating. Also see Indian Cucumber-root (*Medeola virginiana*), page 391.

Lily-Leaved Tway-Blade

Liparis liliifolia (Linnaeus) L. C. Richard ex Ker-Gawler
Orchid Family (Orchidaceae)

PERENNIAL HERB found in moist, acidic forests in the Blue Ridge. STEMS: Up to 1 foot tall, purplish. LEAVES: 2 at the base of the stem, up to 7 inches long and 4 inches wide, oval with a blunt tip, glossy, smooth, slightly succulent, folded upward along the midvein. FLOWER CLUSTERS: 4–10 inches high with up to 30 flowers opening from the bottom up. FLOWERS: About 1 inch across, on slender, purple, ¾-inch stalks, with 3 narrow, spreading, green sepals, 2 purple petals, and a large, oval, mauve lip. FLOWERING: May–Jul. RANGE: GA west to AR, north to MI and VT. SIMILAR TO Small's Tway-blade (*Listera smallii*), a Georgia Special Concern species that occurs in rhododendron thickets in the Blue Ridge; it has a large, mauve, notched lip, and a pair of small leaves at midstem.

Green Adder's-Mouth Orchid

Malaxis unifolia Michaux
Orchid Family (Orchidaceae)

PERENNIAL HERB found in moist to dry forests with acidic soils throughout Georgia, most common in the Piedmont. **STEMS:** 1–20 inches tall, smooth. **LEAVES:** 1 leaf about midway up the stem, up to 4 inches long and 2 inches wide, oval, glossy, clasping the stem. **FLOWER CLUSTERS:** Initially a compact, densely flowered cylinder, elongating as the flowers open from the bottom up. **FLOWERS:** About ¼ inch long, on a slender stalk about twice that long, sepals and petals narrow and curved backward, the lip broader and flat, the flower appearing "flat-faced." **FLOWERING:** May–Aug. **RANGE:** FL west to TX, north to MN, MB, and NL; Mexico, Cuba, West Indies, Central America. **SIMILAR TO** no other wildflower in Georgia; other green-flowered orchids in Georgia have 2 or more stem leaves.

Yellow Fringed Orchid

Platanthera ciliaris (Linnaeus) Lindley
SYNONYM: *Habenaria ciliaris* (Linnaeus) R. Brown
Orchid Family (Orchidaceae)

PERENNIAL HERB found in sunny, moist to wet habitats, including roadsides, infrequently throughout Georgia. **STEMS:** Up to 3 feet tall. **LEAVES:** 2–4 per stem, up to 16 inches long and 2½ inches wide, much smaller up the stem, lance-shaped, clasping the stem. **FLOWER CLUSTERS:** Up to 8 inches high and 3 inches wide, cylinder- or cone-shaped, with 25–115 flowers, the lower opening first. **FLOWERS:** About 1½ inches long with a downcurved, 1-inch spur, yellow or orange, with 2 petals and a sepal forming a hood, 2 winglike sepals, and a deeply fringed lip. **FLOWERING:** Jul–Sep. **RANGE:** FL west to TX, north to MI and NH. **SIMILAR TO** Crested Fringed Orchid (*Platanthera cristata*), which occurs in wet areas in the Coastal Plain; it is shorter, and its flower spur is less than ½ inch long.

Small Green Wood Orchid, Green Rein Orchid

Platanthera clavellata (Michaux) Luer
SYNONYM: *Habenaria clavellata* (Michaux) Sprengel
Orchid Family (Orchidaceae)

PERENNIAL HERB found in wetlands in north Georgia and a few southwest Georgia counties. **STEMS:** Up to 18 inches tall, angled and narrowly winged. **LEAVES:** 1 leaf near the base of the stem, lance-shaped, up to 7 inches long and 1½ inches wide, with small bracts above. **FLOWER CLUSTERS:** Up to 3 inches tall, with flowers twisted so that the spurs overlap. **FLOWERS:** About ½ inch long with a downcurved, ½-inch-long spur; pale green, yellowish, or white, with 2 petals and 1 sepal forming a hood, 2 winglike sepals, and a slightly lobed lip petal. **FLOWERING:** Jun–Sep. **RANGE:** FL west to TX, north to MN and NL. **SIMILAR TO** Southern Rein Orchid (*Platanthera flava*), which occurs in similar habitats; its lip petal has a round bump near its base.

Snowy Orchid

Platanthera nivea (Nuttall) Luer
SYNONYMS: *Habenaria nivea* (Nuttall) Sprengel, *Gymnadeniopsis nivea* (Nuttall) Rydberg
Orchid Family (Orchidaceae)

PERENNIAL HERB found in wet savannas, bogs, and cypress swamps in the Coastal Plain. **STEMS:** Up to 3 feet tall. **LEAVES:** 1–3 per stem, up to 12 inches long, narrowly lance-shaped, nearly erect, often withered at flowering time. **FLOWER CLUSTERS:** Up to 6 inches high, with 20–50 flowers, the lowest flowers opening first. **FLOWERS:** About ½ inch across with a ½-inch-long spur, bright white, with 2 spreading sepals, 2 spreading petals, and the lip petal topmost; none of the petals is fringed. **FLOWERING:** May–Sep. **RANGE:** FL west to TX, north to DE. **SIMILAR TO** no other wildflower in Georgia; other white-flowered *Platanthera* have a fringed lip at the bottom of the flower.

Rose Pogonia

Pogonia ophioglossoides (Linnaeus) Ker-Gawler
Orchid Family (Orchidaceae)

PERENNIAL HERB found in bogs, seepage slopes, and wet pine savannas and flatwoods in the Coastal Plain and a few north Georgia counties. **STEMS:** Up to 2 feet tall, smooth. **LEAVES:** 1 leaf about midway up the stem, up to 5 inches long and 1⅓ inches wide, lance-shaped, entire, smooth, clasping the stem; there may be 1 or 2 long-stalked leaves at the base of the stem. **FLOWERS:** 1–2 inches wide, dark pink to white, with spreading, oval petals and sepals, and a large lip with fringed edges and a showy crest of fleshy, yellow or white bristles; a green leafy bract stands erect behind the flower; smells like raspberries. **FLOWERING:** Mar–Jun. **RANGE:** FL west to TX, north to MB and NL. **SIMILAR TO** Grass-pinks, pages 397–398.

Nodding Ladies' Tresses

Spiranthes cernua (Linnaeus) L. C. Richard
Orchid Family (Orchidaceae)

PERENNIAL HERB found in pine flatwoods, woodlands, swamps, bogs, meadows, and roadsides throughout Georgia. **STEMS:** 4–20 inches tall. **LEAVES:** 3–5, basal, linear, up to 10 inches long and ¾ inch wide, present at flowering time. **FLOWER CLUSTERS:** A spike with 10–50 flowers in 2–4 overlapping spirals that make them appear randomly placed rather than tightly spiraled; gland-tipped hairs cover the bracts and stem within the spike. **FLOWERS:** About ⅓ inch long, creamy white, sometimes tinged pale yellow or pale green, tubular with 5 sepals and petals and a downcurved lip with a crinkled tip. **FLOWERING:** Jul–Nov. **RANGE:** FL west to TX, north to MN and NS. **SIMILAR TO** other ladies' tresses but distinguished by the combination of long, narrow leaves present at flowering, glandular-hairy spike, fragrant flowers, and fall flowering season.

Southern Slender Ladies' Tresses

Spiranthes lacera (Rafinesque) Rafinesque var. *gracilis* (Bigelow) Luer
SYNONYM: *Spiranthes gracilis* (Bigelow) Beck
Orchid Family (Orchidaceae)

PERENNIAL HERB found in prairies, woodland openings, roadsides, and other clearings throughout Georgia. **STEMS:** 6–26 inches tall, leafless, smooth. **LEAVES:** 2–4 in a basal cluster, up to 2 inches long, oval, smooth, usually withered by flowering time. **FLOWER CLUSTERS:** A narrow spike with 10–35 flowers tightly spiraled around the stem. **FLOWERS:** About ¼ inch long, tubular, with 5 white sepals and petals and a downcurved lip with crinkled edges and a green patch in the center. **FLOWERING:** Aug–Oct. **RANGE:** GA west to TX, north to ON and NS. **SIMILAR TO** other ladies' tresses but distinguished by the combination of green-throated flowers, leaves absent at flowering, and summer to fall flowering season.

Grass-Leaved Ladies' Tresses

Spiranthes praecox (Walter) S. Watson
Orchid Family (Orchidaceae)

PERENNIAL HERB found in pine savannas and flatwoods, swamps, bogs, fields, and roadsides throughout Georgia. **STEMS:** 8–30 inches tall, mostly leafless, smooth or slightly hairy near the top. **LEAVES:** 2–7, grasslike, up to 10 inches long and 2 inches wide, present during flowering. **FLOWER CLUSTERS:** A narrow spike with 10–40 flowers spiraled around a sparsely hairy stem. **FLOWERS:** About ⅓ inch long, tubular, with 5 white sepals and petals, and a white lip with crinkled edges and green stripes. **FLOWERING:** Mar–Jun. **RANGE:** FL west to TX, north to NJ. **SIMILAR TO** Spring Ladies' Tresses (*Spiranthes vernalis*), which occurs throughout Georgia in similar habitats; its flower spike is densely hairy, and the flowers are white with a yellow lip or occasionally cream-colored; it flowers Mar–Jun.

Little Ladies' Tresses

Spiranthes tuberosa Rafinesque
SYNONYM: *Spiranthes grayi* Ames
Orchid Family (Orchidaceae)

PERENNIAL HERB found in dry woodlands, old fields, cemeteries, and roadsides throughout Georgia. **STEMS:** 2–12 inches tall, smooth. **LEAVES:** 2–5 oval leaves in a basal rosette up to 4 inches wide; leaves overwinter but wither and disappear before flowering. **FLOWER CLUSTERS:** A narrow spike with 10–35 flowers spiraled around the stem. **FLOWERS:** Less than ¼ inch long, short-tubular, with 5 pure white sepals and petals and a broad, pure white lip with crinkled edges. **FLOWERING:** Jun–Sep. **RANGE:** FL west to TX, north to MI and MA. **SIMILAR TO** other ladies' tresses but distinguished by the combination of upland habitat, leaves absent at flowering time, pure white flowers with crinkled lips, and summer flowering season.

Crane-Fly Orchid

Tipularia discolor (Pursh) Nuttall
Orchid Family (Orchidaceae)

PERENNIAL HERB found in moist to dry forests throughout Georgia. **LEAVES:** 1 per plant, appearing in the fall, overwintering, and withering the next spring; up to 4 inches long and 2¾ inches wide, oval, the upper surface dull green with purple bumps, the lower surface glossy purple. **FLOWER CLUSTERS:** An elongated cluster of 20–40 flowers appears in early summer at the top of a leafless stalk 4–26 inches tall. **FLOWERS:** About 1 inch long, drooping; greenish, tan, or purplish, with 5 narrowly oblong sepals and petals, a 3-lobed lip, and a slender, ¾-inch-long spur. **FLOWERING:** Jul–Sep. **RANGE:** FL west to TX, north to MO and MA. **NOTES:** Night-flying moths probe the spur looking for nectar; pollen sticks to their *eyes* and is then deposited on the next Crane-fly flower they visit. **SIMILAR TO** no other wildflower in Georgia.

Grasses (Poaceae or Gramineae Family)

Grasses and their look-alike cousins, sedges and rushes, don't get a lot of love from most botanists, whether seasoned veterans or eager amateurs. Most of us have been heard to say at one time or another, "I don't *do* grasses." Why such disrespect for one of the world's largest and most economically important plant families? Though ranging in size from a few inches to many feet tall, the majority of grasses do tend to look alike, with their long, strappy leaves, slender stems, and open, airy flower and seed heads. Grasses are flowering plants, but their flowers have neither petals nor sepals. Since their copious pollen is spread by the wind (as allergy sufferers know), grass flowers don't need colorful petals and sweet fragrances to attract pollinators. Learning to identify grasses involves learning a whole new vocabulary that names the peculiar parts of a grass plant. But it's worth the effort to learn at least the most common and ecologically important grasses in our area. Grasses are keystone species in at least two of Georgia's most important plant communities: coastal salt marshes and Longleaf Pine–Wire Grass woodlands. They are basic to the process of secondary succession because they start the 100-year-long process that results in forest reestablishment on abandoned farmland. And far more effectively than the invasive plants, such as kudzu and sericea lespedeza, that were imported from Asia in the last century, grasses stop erosion with their deep, wide-spreading, fibrous root systems. Here are some grass basics to get you started:

- Grass stems (called culms) are usually hollow and are round or slightly flattened, but never 3-sided.
- Nodes (the point where a leaf attaches to the stem) are usually solid.
- Leaves consist of a blade, a sheath that wraps tightly around the stem, and a collar where blade and sheath meet.
- Flowers (called florets) have pistils and stamens like "normal" flowers, but instead of petals and sepals there are tiny, nested, canoe-shaped structures called paleas and lemmas.
- Florets are compressed into a compact structure called a spikelet with (usually) 2 bracts at the base called glumes. Spikelets are arranged into flower clusters (or seed heads) called inflorescences.

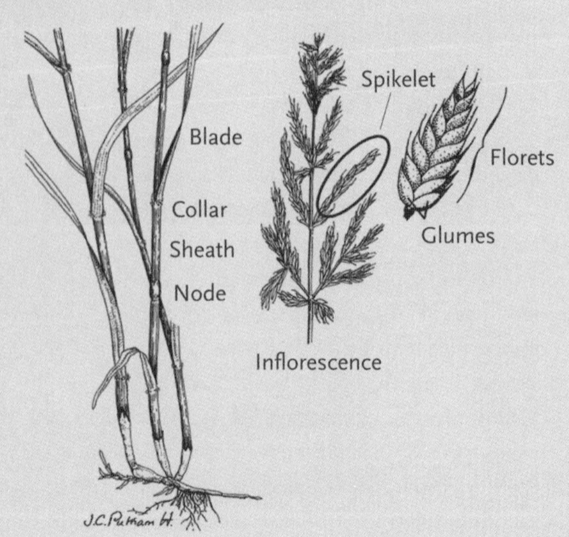

Big Bluestem, Turkey-Foot

Andropogon gerardii Vitman
Grass Family (Poaceae)

PERENNIAL GRASS found in prairies, sandhills, woodland borders, glades, rock outcrops, and clearings throughout Georgia. **STEMS:** Up to 8 feet tall, often waxy blue-green, forming large, leafy clumps. **LEAVES:** Blades up to 20 inches long and ⅓ inch wide, alternate, dull or bluish-green, with softly hairy sheaths that wrap around the stem. **FLOWER CLUSTERS:** Held at the top of the stem, with 2–6 branches (usually 3) emerging from a leaflike sheath and spreading to form a V-shaped "turkey foot"; each branch is 2–4 inches long, dull green to reddish brown, with scattered, single awns (bristles). **FLOWERING:** Jul–Oct. **RANGE:** FL west to AZ, north to SK and QC. **SIMILAR TO** Broom-sedge Grass (*Andropogon virginicus*), page 410, which is smaller and has tufts of white bristles on its flower heads.

Splitbeard Bluestem

Andropogon ternarius Michaux
Grass Family (Poaceae)

PERENNIAL GRASS found in woodlands, grasslands, and disturbed areas throughout Georgia. **STEMS:** 2–5 feet tall, reddish; blue-green leaf sheaths clasp the stem, alternating with lengths of red stem and creating a striped look. **LEAVES:** Blades up to 20 inches long and ¼ inch wide, alternate, usually hairy, copper-colored in the fall. **FLOWER CLUSTERS:** Divided into 2 hairy spikes, each up to 2 inches long, held at the top of a slender, erect stalk up to 3 inches long; a tuft of silvery white hair remains after the seeds disperse. **FLOWERING:** Aug–Oct. **RANGE:** FL west to TX, north to KS and DE. **SIMILAR TO** Little Bluestem (*Schizachyrium scoparium*), page 414, which also appears to have striped stems.

Broom-Sedge Grass

Andropogon virginicus Linnaeus
Grass Family (Poaceae)

PERENNIAL GRASS found in open woodlands, rock outcrops, and disturbed areas throughout Georgia. **STEMS:** Up to 4 feet tall, strawlike, with short, erect branches near the top. **LEAVES:** Blades up to 20 inches long and ¼ inch wide, alternate, bluish-green in summer, copper-colored in the fall. **FLOWER CLUSTERS:** Held at the top of the stem and partly concealed by several leaflike bracts, each cluster bearing several seeds with tufts of white, bristly hairs at their tips. **FLOWERING:** Sep–Oct. **RANGE:** FL west to TX, north to IA and MA. **SIMILAR TO** two common wetland grasses with large, bushy flower clusters. Bushy Bluestem (*Andropogon glomeratus*) occurs throughout Georgia; it has copper-colored stems and leaves in the fall. Chalky Bluestem (*A. glaucopsis*) has whitish-blue stems; it occurs in the Coastal Plain.

Wire Grass

Aristida stricta Michaux
SYNONYM: *Aristida beyrichiana* Trinius & Ruprecht
Grass Family (Poaceae)

PERENNIAL GRASS found in sandhills and dry to moist pinelands in the Coastal Plain. **STEMS:** Up to 3 feet tall, wiry. **LEAVES:** In dense clumps, the blades 1–1½ feet long, alternate, wiry, round in cross-section, with a tiny patch of white hairs at the base of each leaf blade. **FLOWER CLUSTERS:** Up to 12 inches long, erect, narrow, wispy, covered with ¼-inch-long seeds that are tipped with 3-parted, ½-inch-long bristles. **FLOWERING:** Sep–Nov. **RANGE:** FL west to MS, north to NC. **NOTES:** Wire Grass is the keystone species of the Longleaf Pine ecosystem, efficiently carrying fire through the understory; it quickly leafs out after a fire and will not flower and set seed unless burned. Its leaves are **SIMILAR TO** those of Muhly Grass (*Muhlenbergia capillaris*), page 413.

River Oats, Fish-on-a-Pole, Spangle Grass

Chasmanthium latifolium (Michaux) Yates
SYNONYM: *Uniola latifolia* Michaux
Grass Family (Poaceae)

PERENNIAL GRASS found on stream banks and in bottomlands throughout Georgia. **STEMS:** 1½–5 feet tall, smooth, round in cross-section. **LEAVES:** Blades 3–9 inches long, up to 1 inch wide, alternate, lance-shaped, blue-green, usually smooth or sometimes hairy on the lower surface. **FLOWER CLUSTERS:** Up to 14 inches long, nodding at the top of the stem, with many slender, drooping stalks ½–1 inch long, each bearing a flattened seed head up to 2 inches long at its tip. **FLOWERING:** Jun–Oct. **RANGE:** FL west to TX, north to IA and PA. **NOTES:** River Oats can form large colonies, spreading by underground stems. The dangling spikelets catch the wind, which disperses the pollen. **SIMILAR TO** Sea Oats (*Uniola paniculata*), page 416.

Wood Oats, Longleaf Spike Grass

Chasmanthium sessiliflorum (Poiret) Yates
SYNONYM: *Uniola sessiliflora* Poiret
Grass Family (Poaceae)

PERENNIAL GRASS found in moist, hardwood forests throughout Georgia. **STEMS:** 1½–5 feet tall, slender, wiry, unbranched, arching, leafy only on the lower half. **LEAVES:** Blades up to 20 inches long and ½ inch wide, alternate, smooth; long hairs cover the sheath and the collar (the back of the blade where it meets the sheath). **FLOWER CLUSTERS:** About ⅓ inch long, roughly V-shaped, stiff, scattered along the upper half of the stem. **FLOWERING:** Aug–Oct. **RANGE:** FL west to TX, north to OK and VA. **NOTES:** *Chasmanthium* species are among the few native grasses that thrive in shady forests. **SIMILAR TO** Slender Wood Oats (*Chasmanthium laxum*, synonym *Uniola laxa*), which occurs in similar habitats; its leaf sheaths and collars are smooth.

Toothache Grass

Ctenium aromaticum (Walter) Wood
Grass Family (Poaceae)

PERENNIAL GRASS found in wet savannas and flatwoods, bogs, and seepages in the Coastal Plain. **STEMS:** Up to 5 feet tall. **LEAVES:** Forming large clumps, the blades up to 18 inches long and ¼ inch wide, bluish-green on the upper surface, pale green on the lower. **FLOWER CLUSTERS:** A yellowish spike, 2–6 inches long, with 2 rows of erect, bristly spikelets on one side of the spike; as the spike dries, it coils and twists. **FLOWERING:** Jun–Aug, following a fire. **RANGE:** FL west to TX, north to VA. **NOTES:** A compound in the plant, especially in the rhizome, numbs the mouth. **SIMILAR TO** Florida Toothache Grass (*Ctenium floridanum*), a Georgia Special Concern species found in sandhills in southeast Georgia and northeast Florida; its stems arise from a long rhizome, in lines rather than in clumps.

Purple Love Grass

Eragrostis spectabilis (Pursh) Steudel
Grass Family (Poaceae)

PERENNIAL GRASS found in woodlands, fields, prairies, and roadsides throughout Georgia. **STEMS:** Up to 2 feet tall. **LEAVES:** Blades up to 13 inches long and ⅓ inch wide, alternate, smooth or hairy; long, white hairs mark the transition from blade to sheath; the sheaths are hairy only on the edges. **FLOWER CLUSTERS:** Up to 15 inches tall and wide, open and airy, with wiry, spreading, purple branches with tiny tufts of white hair at the base; the branchlets are tipped with flat, reddish-purple flower spikelets about ¼ inch long. **FLOWERING:** Aug–Oct. **RANGE:** FL west to NM, north to ND and ME. **NOTES:** The flower clusters form pink "clouds" on roadsides. **SIMILAR TO** Bigtop Love Grass (*Eragrostis hirsuta*), which occurs in disturbed areas throughout Georgia; it has densely hairy leaf sheaths, green branches, and green spikelets only slightly tinged with purple.

Muhly Grass

Muhlenbergia capillaris (Lamarck) Trinius
Grass Family (Poaceae)

PERENNIAL GRASS found in woodlands, prairies, sandhills, and pine savannas and flatwoods throughout Georgia, mostly in the Coastal Plain. **STEMS:** Up to 3½ feet tall, the upper half a much-branched flower cluster. **LEAVES:** In dense clumps, the blades up to 20 inches long, wiry, smooth. **FLOWER CLUSTERS:** Up to 20 inches tall, with delicate, dark pink branches, each tipped with a tiny flower bearing a ½-inch-long bristle. **FLOWERING:** Late Aug–Oct. **RANGE:** FL west to TX, north to KS and MA. **NOTES:** In full bloom, stands of Muhly Grass and Sweet Grass look like pink, low-lying clouds. **SIMILAR TO** Sweet Grass (*Muhlenbergia sericea*, synonym *M. capillaris* variety *filipes*), which occurs in dune swales, wet grasslands, and edges of marshes on barrier islands and is made into beautiful baskets by the Gullah and Geechee people who live there.

Eastern Needle Grass

Piptochaetium avenaceum (Linnaeus) Parodi
SYNONYM: *Stipa avenacea* Linnaeus
Grass Family (Poaceae)

PERENNIAL GRASS found in dry woodlands and forests, especially around rock outcrops, throughout Georgia. **STEMS:** Up to 3½ feet tall, smooth. **LEAVES:** Blades up to 1 foot long and ⅛ inch wide, rough. **FLOWER CLUSTERS:** Up to 12 inches high, open and airy, with long, wiry, nodding branches bearing black, 1-seeded spikelets, each spikelet with short barbs at its base and a bent, spirally twisted awn (bristle) up to 2½ inches long at its tip. **FLOWERING:** Apr–Jun. **RANGE:** FL west to TX, north to MO, ON, and MA. **NOTES:** The long bristles on the seed heads react to changes in humidity by twisting and boring the seeds into the ground. **SIMILAR TO** no other grass in Georgia.

Silver Plume Grass

Saccharum alopecuroides (Linnaeus) Nuttall
SYNONYM: *Erianthus alopecuroides* (Linnaeus) Elliott
Grass Family (Poaceae)

PERENNIAL GRASS found in upland forests and clearings in north Georgia and a few Coastal Plain counties. **STEMS:** 5–10 feet tall, stout, hairy just below the flower head. **LEAVES:** Blades up to 2½ feet long and 1 inch wide, flat, with a conspicuous white midvein and a patch of long hairs near the base. **FLOWER CLUSTERS:** An erect, densely flowered plume up to 12 inches high and 5 inches wide, silvery to pinkish before and during flowering, whitish and woolly after; covered with long, spirally twisted (but not bent) bristles. **FLOWERING:** Sep–Oct. **RANGE:** FL west to TX, north to IL and NJ. **SIMILAR TO** Sugarcane Plume Grass (*Saccharum giganteum*), which occurs in wetlands, mostly in the Coastal Plain. The bristles in its flower clusters are curved or slightly wavy, but not spirally twisted.

Little Bluestem

Schizachyrium scoparium (Michaux) Nash
Grass Family (Poaceae)

PERENNIAL GRASS found in woodlands, grasslands, and disturbed areas throughout Georgia. **STEMS:** Up to 5 feet tall, slightly zigzag, reddish; blue-green leaf sheaths alternate with lengths of red stem, creating a striped look. **LEAVES:** Blades up to 3 feet long and ¼ inch wide, alternate, smooth or hairy, blue-green, forming mounded clumps, and turning coppery brown in the fall. **FLOWER CLUSTERS:** 1–3 inches long, held singly at the tip of a long stalk, sparsely flowered, slightly zigzag between pairs of spikelets, covered with long, white hairs. **FLOWERING:** Aug–Oct. **RANGE:** FL west to TX, north to KS and DE. **SIMILAR TO** Splitbeard Bluestem (*Andropogon ternarius*), page 409, which also appears to have striped stems.

Yellow Indian Grass

Sorghastrum nutans (Linnaeus) Nash
Grass Family (Poaceae)

PERENNIAL GRASS found in sandhills, woodlands, prairies, roadsides, and utility rights-of-way throughout Georgia. **STEMS:** Usually 3–5 feet tall, smooth except at the nodes. **LEAVES:** Blades up to 3 feet long and ½ inch wide, alternate, smooth, with a conspicuous white midvein and a pair of pointed "ears" where the blade meets the sheath. **FLOWER CLUSTERS:** 4–30 inches long, erect, plumelike, yellow-brown at maturity, with the spikelets arranged on erect branches all around the stem (not 1-sided); each ¼-inch spikelet bears a once-bent, spirally twisted, brown bristle (awn) usually about ½ inch long. **FLOWERING:** Aug–Oct. **RANGE:** FL west to AZ, north to MB, QC, and ME. **SIMILAR TO** Slender Indian Grass (*Sorghastrum elliottii*), which occurs in similar habitats; its flower clusters are narrowly spikelike with long (up to 1¾ inches), twice-bent awns.

Lopsided Indian Grass

Sorghastrum secundum (Elliott) Nash
Grass Family (Poaceae)

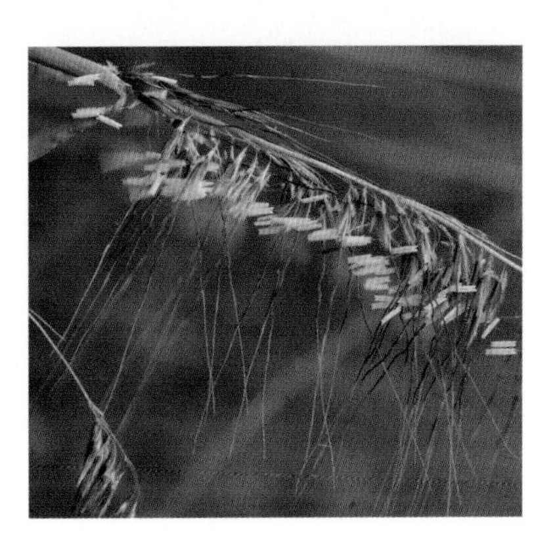

PERENNIAL GRASS found in sandhills and other dry woodlands in the Coastal Plain. **STEMS:** 3–6 feet tall, mostly smooth. **LEAVES:** Up to 20 inches long and ¼ inch wide, alternate, rough. **FLOWER CLUSTERS:** Up to 16 inches long, golden brown at maturity, with all of the spikelets drooping from one side of the cluster, each ¼-inch spikelet is densely covered with hairs and tipped with a twice-bent, spirally twisted, brown bristle (awn) ¾–1¾ inches long. **FLOWERING:** Sep–Oct. **RANGE:** FL west to MS, north to SC. **SIMILAR TO** Yellow Indian Grass (*Sorghastrum nutans*) and Slender Indian Grass (*S. elliottii*), which occur throughout Georgia in woodlands and clearings; they have erect (not 1-sided), plumelike, golden brown flower clusters. Yellow Indian Grass is also distinguished by its awns, which are less than ¾ inch long.

Purple Top, Greasy Grass

Tridens flavus (Linnaeus) A. S. Hitchcock
Grass Family (Poaceae)

PERENNIAL GRASS found in prairies, roadsides, and other disturbed areas throughout Georgia. **STEMS:** 2–5 feet tall, smooth. **LEAVES:** Blades up to 2 feet long and ½ inch wide, alternate, smooth or rough on the upper surface, hairy on the lower surface. **FLOWER CLUSTERS:** 6–16 inches long, erect, open and airy, with widely spreading or slightly drooping branches mostly in whorls; tiny, red or purple spikelets are widely spaced along the branches. **FLOWERING:** Jul–Oct. **RANGE:** FL west to TX, north to NE and NH. **NOTES:** A fresh flower cluster feels greasy or waxy if pulled through your fist. **SIMILAR TO** Johnson Grass (*Sorghum halepense*), a Eurasian plant that aggressively invades disturbed areas; it is a larger grass with wider leaves that have a thick, white midvein.

Sea Oats

Uniola paniculata Linnaeus
Grass Family (Poaceae)

PERENNIAL GRASS found on beaches and dunes and in swales between dunes on barrier islands. **STEMS:** Up to 8 feet tall, stout, spreading by extensive networks of underground stems. **LEAVES:** Blades up to 2 feet long and ½ inch wide, alternate, tough, tapering and curled at the tip, with rough margins. **FLOWER CLUSTERS:** Up to 2 feet long and 6 inches wide, nodding at the top of the stem, green to golden brown, with many drooping branches bearing flattened seed heads up to 1 inch long. **FLOWERING:** Jun–Nov. **RANGE:** FL west to TX, north to DE. **NOTES:** Sea Oats form large, deeply rooted colonies that promote dune formation and lessen erosion during storms; picking or destroying Sea Oats is illegal in most of the states where it occurs. **SIMILAR TO** River Oats (*Chasmanthium latifolium*), page 411.

Southern Wild Rice

Zizania aquatica Linnaeus
Grass Family (Poaceae)

ANNUAL GRASS found in tidally influenced freshwater marshes and in quiet water at the edges of streams in Georgia's coastal counties. **STEMS:** Up to 10 feet tall, stout. **LEAVES:** Blades up to 5 feet long and 2 inches wide, alternate, with finely and sharply toothed edges. **FLOWER CLUSTERS:** Up to 2 feet tall, held erect at the top of the stem, much-branched; the upper branches angle stiffly upward and bear the female, seed-producing flowers; lower branches are widely spreading with drooping branchlets and pollen-producing flowers. **FLOWERING:** May–Oct. **RANGE:** FL west to LA, north to WI and ME. **SIMILAR TO** Water Millet (*Zizaniopsis miliacea*), another large grass with saw-toothed leaves; it occurs in freshwater wetlands near the coast and in a few inland counties. All the branches in its flower clusters are identical, angled upward with outwardly curving tips.

Pickerelweed

Pontederia cordata
Pickerelweed Family (Pontederiaceae)

PERENNIAL HERB rooted in bottoms of ponds, lakes, and wet ditches, mostly in the Coastal Plain. **STEMS:** Up to 4 feet tall, smooth, hollow, bearing 1 stem leaf and a crowded flower cluster up to 6 inches high. **LEAVES:** Blades up to 8¾ inches long and 4¾ inches wide, with long leaf stalks; the blades vary from narrowly lance-shaped to heart-shaped to arrowhead-shaped. **FLOWERS:** About ½ inch wide, blue-violet, with 6 spreading lobes, the upper lobe with a yellow blotch; open for only 1 day. **FLOWERING:** May–Oct. **RANGE:** FL west to TX, north to MN and NS. **SIMILAR TO** no other wildflower in Georgia. Pickerelweed is related to the highly destructive exotic pest plant Water Hyacinth (*Eichhornia crassipes*); it is a smaller, floating plant with pale purple flowers, inflated leaf stalks, and round leaf blades.

Smooth Carrion Flower

Smilax herbacea Linnaeus
Smilax Family (Smilacaceae)

PERENNIAL HERBACEOUS VINE found in moist, deciduous forests in north Georgia and a few Coastal Plain counties. **STEMS:** Up to 10 feet long, smooth, without prickles, climbing by long tendrils over other plants. **LEAVES:** Up to 4¾ inches long and 3½ inches wide, alternate, oval or heart-shaped, entire, smooth, waxy white on the lower surface, deciduous. **FLOWERS:** In round clusters on long stalks, each flower less than ¼ inch wide, green, foul-smelling, with 6 spreading tepals. **FRUIT:** About ⅓ inch wide, round, dark blue with a waxy white coating when ripe. **FLOWERING:** May–Jun. **RANGE:** GA and AL, north to QC and ME. **NOTES:** Leaves may appear whorled near the tip of the stem. **SIMILAR TO** Downy Carrion Flower (*Smilax pulverulenta*), which occurs in similar habitats; its lower leaf surfaces are bright green and shiny, often hairy. Also see Wild Yam (*Dioscorea villosa*), page 379, which lacks tendrils.

Huger's Carrion Flower

Smilax hugeri (Small) J. B. S. Norton ex Pennell
SYNONYM: *Smilax ecirrata* (Engelmann ex Kunth) S. Watson var. *hugeri* (Small) H. E. Ahles
Smilax Family (Smilacaceae)

PERENNIAL HERB found in moist, upland, hardwood forests in north Georgia and a few Coastal Plain counties. **STEMS:** Up to 3 feet tall, erect, spineless, smooth. **LEAVES:** 2–3½ inches long and ½–3 inches wide, alternate but clustered near the top of the stem and appearing whorled from above, oblong with a rounded tip, smooth on the upper surface, pale green and hairy on the lower surface. **FLOWERS:** In round, long-stalked clusters, each flower ¼ inch wide, green, with 6 spreading tepals. **FLOWERING:** Mar–Apr. **RANGE:** GA, FL, AL, TN, NC, and SC. **SIMILAR TO** Biltmore Carrion Flower (*Smilax biltmoreana*), which occurs in dry to moist, upland forests in north Georgia; the lower leaf surfaces are smooth and waxy white.

Blaspheme Vine

Smilax laurifolia Linnaeus
Smilax Family (Smilacaceae)

PERENNIAL VINE found in swamps, floodplains, and other wetlands throughout Georgia. **STEMS:** More than 16 feet long, woody, smooth, branching and forming dense tangles, with curling tendrils and large, scattered prickles. **LEAVES:** Up to 5 inches long and 2½ inches wide, evergreen, alternate, oblong, entire, leathery, smooth, with a prominent midvein on the lower surface; stalks are twisted so leaf is held erect. **FLOWERS:** Held in rounded clusters in the leaf axils, each flower long-stalked and about ⅓ inch wide, yellowish-green, with 6 spreading tepals. **FRUIT:** About ¼ inch long, round, shiny black when ripe. **FLOWERING:** Jul–Aug. **RANGE:** FL west to TX, north to OK and NJ; Bahamas, Cuba. **SIMILAR TO** Common Greenbrier (*Smilax rotundifolia*), which occurs in forests throughout Georgia; its leaves are broadly oval to heart-shaped, with reddish stalks; prickles are only on the leaf nodes.

Sarsaparilla Vine

Smilax pumila Walter
Smilax Family (Smilacaceae)

PERENNIAL VINE found in upland, hardwood forests and maritime forests in the Coastal Plain and lower Piedmont. **STEMS:** Up to 20 inches long or tall, slender, usually trailing across the ground, sometimes climbing by tendrils, densely hairy, spineless. **LEAVES:** Up to 4 inches long and 3 inches wide, evergreen, alternate, oval to oblong with a heart-shaped base and rounded tip, entire, upper surface dark green with 3 prominent veins, lower surface densely white-hairy. **FLOWERS:** Held in clusters in the axils of the leaves, each flower about ¼ inch wide, yellowish-white, with 6 spreading tepals. **FRUIT:** About ¼ inch long, bright red when ripe, oval, pointed. **FLOWERING:** Oct–Nov. **RANGE:** FL west to TX, north to AR and SC. **SIMILAR TO** other low plants with heart-shaped leaves but distinguished by its hairy stems and hairy leaves.

Jackson-Briar

Smilax smallii Morong
Smilax Family (Smilacaceae)

PERENNIAL VINE found in bottomland forests throughout Georgia. **STEMS:** Climbing by long tendrils and forming clumps in the crowns of tall trees, with a few scattered prickles. **LEAVES:** Up to 2½ inches long and 1¼ inches wide, alternate, evergreen, oval to lance-shaped, dark green and shiny on the upper surface, pale green below. **FLOWERS:** Held in round clusters on long stalks, each flower ¼–½ inch wide, pale yellowish-green, with 6 narrow, spreading tepals. **FRUIT:** About ¼ inch wide, round, dark red when ripe. **FLOWERING:** Jun–Jul. **RANGE:** FL west to TX, north to OK and NJ. **NOTES:** All *Smilax* have female and male flowers on separate plants. **SIMILAR TO** Coral Greenbriar (*Smilax walterii*), which occurs mostly in swamps in the Coastal Plain; its leaves are deciduous and leathery, and its fruits are bright coral-red.

Southern Bog Asphodel

Triantha racemosa (Walter) Small
SYNONYM: *Tofieldia racemosa* (Walter) B. S. P.
False Asphodel (Tofieldiaceae) or Lily (Liliaceae) Family

PERENNIAL HERB found in wet savannas, seepage slopes, and bogs in the Coastal Plain. **STEMS:** 8–28 inches tall, covered with glandular hairs. **LEAVES:** Mostly basal, up to 14 inches long and ¼ inch wide, erect, grasslike. **FLOWERS:** In a cylindrical cluster at the top of the stem with 15–80 flowers, each flower about ⅓ inch wide, with 6 white, blunt, spreading tepals and 6 stamens; the flower stalks are glandular-hairy. **FLOWERING:** Jun–early Aug. **RANGE:** FL west to TX, north to NJ (skips SC and NC). **SIMILAR TO** Crow Poison (*Stenanthium densum*), page 393.

Trilliums (Trilliaceae Family)

Georgia has more species of trilliums than any other state—22 at last count, and at least 1 new species has been discovered and not yet named. Trilliums are found in most of Georgia—from high mountain slopes to deep Coastal Plain ravines, almost always in moist, shady forests.

All trilliums have several traits in common. Their flowering stems arise from an underground stem called a rhizome. At the top of the stem are 3 leaves in a whorl and a single flower. About half of our trilliums are toadshades, which have their flower resting right on the top of the stem where the leaves arise. The other half, the wakerobins, have flowers on stalks that are either held erect or are nodding below the leaves. Regardless of their position, the flowers always have 3 green sepals, 3 colorful petals, 6 stamens, and a 3-lobed ovary. Different trillium species have distinctive flower smells. Those with a spicy or sweet fragrance are probably pollinated by bees and beetles. Those that smell like rotting fruit or moldy cheese may be pollinated by flies and beetles.

Trilliums have an interesting life history. After flowering and fruiting, the aboveground plant disappears, persisting through the late summer, fall, and winter as an underground rhizome. Seeds shed in the summer germinate the following spring and, within a year or two, send up a single, spatula-shaped seed leaf for 1 year's growing season. The next year the plant produces a true leaf; then, for several more years, the plant bears 3 leaves. After 5–7 years, the plant produces a flowering stalk and a flower and, conditions permitting, continues to do so for the rest of its life. Their seeds are dispersed by ants (see "The Skinny on Fat Bodies—Elaiosomes," p. 44). Some trillium plants may live to be hundreds of years old, with the older parts of the rhizome dying back and the new, growing tip sending up a flowering stem each year.

Catesby's Trillium

Trillium catesbaei Elliott
Trillium Family (Trilliaceae)

PERENNIAL HERB found in acidic soils in cove forests and oak-pine forests in north Georgia and several Coastal Plain counties. **STEMS:** Up to 18 inches tall. **LEAVES:** Up to 6 inches long and 2 inches wide, elliptic, solid green. **FLOWERS:** Held on stalks curved below the leaves, with 3 pink (rarely white), strongly curved petals 1½–2 inches long, 3 green, pink-tinged sepals, 6 yellow, twisted anthers, and a white ovary. **FLOWERING:** Mar–Jun. **RANGE:** GA, AL, SC, and NC. **SIMILAR TO** Edna's Trillium (*Trillium persistens*), a federally listed species found in northeast Georgia; its leaves are less than 3½ inches long; the flowers are white (turning pink with age) and are held above the leaves on leaning stalks.

Purple Toadshade, Sweet Betsy

Trillium cuneatum Rafinesque
Trillium Family (Trilliaceae)

PERENNIAL HERB found in moist, hardwood forests, mostly in north Georgia. **STEMS:** 6–18 inches tall. **LEAVES:** 3–7 inches long and 3–5 inches wide, broadly oval with an abruptly pointed tip, sometimes wider than long and overlapping; mottled with 2 or 3 shades of green. **FLOWERS:** 1½–2¾ inches long, without stalks, with 3 maroon, bronze, or yellowish petals held erect and usually overlapping at the base, 3 green, purple-tinged sepals, 6 straight, maroon stamens, and a maroon ovary; with a spicy-fruity fragrance. **FLOWERING:** Mid-Mar–Apr. **RANGE:** GA west to MS, north to KY and NC. **SIMILAR TO** Yellow Trillium (*Trillium luteum*), page 424; and Spotted Trillium (*Trillium maculatum*), page 424.

Trailing Trillium, Decumbent Trillium

Trillium decumbens Harbison
Trillium Family (Trilliaceae)

PERENNIAL HERB found in moist forests in northwest Georgia and rarely in the upper Coastal Plain. **STEMS:** 2–8 inches long, S-shaped, resting on the ground, hidden by the leaves. **LEAVES:** Up to 5 inches long and 2¾ inches wide, broadly oval to nearly round, mottled green and bronze with large silvery green patches, resting on the ground. **FLOWERS:** 2–3 inches long, without stalks, with 3 maroon petals, 3 green- or maroon-streaked sepals, 6 straight, dark maroon stamens, and a maroon ovary; with a fetid odor. **FLOWERING:** Mid-Mar–Apr. **RANGE:** GA, AL, and TN. **SIMILAR TO** Relict Trillium (*Trillium reliquum*), a federally listed species with S-shaped stems that occurs in rich, moist forests in the lower Piedmont and upper Coastal Plain; its mottled leaves have a central silver stripe, and the putrid-smelling flowers are usually less than 2 inches long.

Red Trillium, Red Wake-Robin

Trillium erectum Linnaeus
Trillium Family (Trilliaceae)

PERENNIAL HERB found in moist, high-elevation forests in the Blue Ridge. **STEMS:** 6–24 inches tall. **LEAVES:** 2–8 inches long and wide, roughly diamond-shaped with rounded sides and a pointed tip, solid green. **FLOWERS:** 1½–3½ inches wide, flat in profile, held perpendicular to the leaves on an erect stalk, with 3 lance-shaped petals, 3 green-and-maroon-tinged sepals, 6 yellow or maroon anthers, and a maroon ovary; with a "wet dog" or fishy odor. **FLOWERING:** Mid-Apr–May. **RANGE:** GA north to MI and NB. **NOTES:** Petals may be maroon, white, yellow, greenish, or speckled maroon and cream. **SIMILAR TO** Sweet White Trillium (*Trillium simile*), a Georgia Special Concern species that occurs on moist slopes in the Blue Ridge; its flowers are white with a cup-shaped profile and a faint green apple smell; they bloom late Mar–Apr.

Large-Flowered White Trillium

Trillium grandiflorum (Michaux) Salisbury
Trillium Family (Trilliaceae)

PERENNIAL HERB found in rich, moist forests in the Blue Ridge. **STEMS:** 6–20 inches tall. **LEAVES:** 2½–8 inches long and 1½–4¾ inches wide, broadly oval or lance-shaped, widest below the middle with an abruptly pointed tip, solid green. **FLOWERS:** 1½–3 inches long, held above the leaves on a stalk up to 4 inches long, with 3 white petals that overlap at the base then flare outward to form a funnel shape, 3 green sepals, 6 straight, pale yellow anthers, and a white ovary. **FLOWERING:** Apr–May. **RANGE:** GA and AL, north to MN, ON, and QC. **NOTES:** Flowers turn pink with age; they have no fragrance. **SIMILAR TO** no other wildflower in Georgia.

Yellow Trillium

Trillium luteum (Muhlenberg) Harbison
SYNONYM: *Trillium cuneatum* Rafinesque var. *luteum* (Muhlenberg) Ahles
Trillium Family (Trilliaceae)

PERENNIAL HERB found in moist, upland, hardwood forests over mafic or limestone bedrock in the Blue Ridge mountains and foothills of north Georgia. **STEMS:** 6–16 inches tall. **LEAVES:** Up to 7 inches long and 4 inches wide, broadly oval with a pointed tip, weakly mottled with pale green. **FLOWERS:** 2–2½ inches long, without stalks, with 3 yellow, erect petals, 3 green, spreading sepals, 6 pale yellow anthers, and a pale green ovary; smells like Lemon Pledge. **FLOWERING:** Apr–May. **RANGE:** Southern Appalachian Mountains and foothills in GA, TN, KY, and NC. **SIMILAR TO** Pale Yellow Trillium (*Trillium discolor*), a Georgia Special Concern species that occurs in ravines near the Savannah River; its flowers are pale yellow or cream-colored with dark stamens; 1 petal has a distinctly pointed tip.

Spotted Trillium

Trillium maculatum Rafinesque
Trillium Family (Trilliaceae)

PERENNIAL HERB found in rich, hardwood forests in the Coastal Plain. **STEMS:** 5–16 inches tall. **LEAVES:** 3–6 inches long and 2–2½ inches wide, broadly oval to nearly round, mottled with 2 or 3 shades of green, without a central silvery green stripe. **FLOWERS:** 1½–2¾ inches tall, without stalks, with 3 erect, maroon or yellow petals that are narrowed at the base so the stamens are visible in side view, 3 spreading, green or maroon sepals, 6 dark maroon anthers, and a dark purple ovary; with a spicy-fruity fragrance. **FLOWERING:** Late Feb–Apr. **RANGE:** GA, FL, AL, and SC. **SIMILAR TO** Lance-leaf Trillium (*Trillium lancifolium*), a Georgia Special Concern species that occurs in rich forests over basic soils; its leaves are less than 1½ inches wide and drooping, and the petals are very narrow and often twisted.

Southern Nodding Trillium

Trillium rugelii Rendle
Trillium Family (Trilliaceae)

PERENNIAL HERB found in rich forests on moist lower slopes in the upper Piedmont and Ridge & Valley. **STEMS:** 6–16 inches tall. **LEAVES:** 2½–6 inches long and wide, wider than long, diamond-shaped with rounded angles, solid green. **FLOWERS:** 1–2 inches long, facing downward, held on stalks nodding above or curved below the leaves, with 3 white, oval, strongly curved petals, 3 green sepals that are much smaller than the petals, 6 stamens with maroon anthers and white filaments, and a white ovary often tinged with maroon. **FLOWERING:** Apr–early May. **RANGE:** GA, AL, TN, SC, and NC. **SIMILAR TO** Bent Trillium (*Trillium flexipes*), a Georgia Special Concern species that occurs on limestone-based slopes in the Cumberland Plateau; its flowers are on slightly nodding stalks and have creamy white anthers.

Underwood's Trillium

Trillium underwoodii Small
Trillium Family (Trilliaceae)

PERENNIAL HERB found in moist, deciduous forests in southwest Georgia. **STEMS:** 3–8 inches tall, shortest during flowering, then elongating and lifting the leaves well above the ground after flowering. **LEAVES:** 2½–4¾ inches long and 2–3 inches wide, elliptic, mottled with 4 or 5 shades of green with a silvery green stripe along the midvein, the tips nearly touching the ground during flowering. **FLOWERS:** 1–2 inches long, without stalks, with 3 maroon or yellowish-green petals, 3 green or maroon sepals, 6 maroon stamens, and a dark purple ovary; smells like moldy cheese. **FLOWERING:** Mid-Jan–mid-Apr. **RANGE:** GA, FL, and AL. **SIMILAR TO** Chattahoochee Trillium (*Trillium decipiens*), a Georgia Special Concern species that occurs in ravines in southwest Georgia and the Fall Line counties. Its stems are 7–17 inches tall and always hold the leaves above the ground; its flowers smell like overripe bananas.

Painted Trillium

Trillium undulatum Willdenow
Trillium Family (Trilliaceae)

PERENNIAL HERB found in acidic soils of rhododendron thickets, mountain bogs, and high-elevation oak forests in the Blue Ridge. **STEMS:** 4–16 inches tall. **LEAVES:** 4¾–7 inches long and 3–8 inches wide, broadly oval, solid blue-green, with short stalks. **FLOWERS:** Held above the leaves on an erect stalk, with 3 white, wavy-margined petals 1–2 inches long with red blazes at the base, 3 green or maroon sepals, 6 lavender anthers, and a white ovary; without a fragrance. **FRUIT:** A red, ½-inch-wide berry. **FLOWERING:** Late Apr–May. **RANGE:** GA north to MI, QC, and NS. **SIMILAR TO** no other wildflower in Georgia.

Vasey's Trillium

Trillium vaseyi Elliott
Trillium Family (Trilliaceae)

PERENNIAL HERB found in moist, rich forests in the Blue Ridge and, rarely, in the Piedmont and inner Coastal Plain. **STEMS:** 12–28 inches tall. **LEAVES:** Up to 8 inches long and 8 inches wide, broadly oval to diamond-shaped with rounded angles, often wider than long, solid green. **FLOWERS:** Held on stalks curved below the leaves, with 3 oval, curved, maroon petals up to 2½ inches long, 3 spreading, green sepals, 6 long, yellow anthers, and a purplish-black ovary; smells like old roses. **FLOWERING:** Late Apr–early Jun. **RANGE:** GA, AL, TN, SC, and NC. **SIMILAR TO** Southern Red Trillium (*Trillium sulcatum*), a Georgia Special Concern species that occurs in moist, hardwood forests on the Cumberland Plateau; its flowers are held on tall, erect stalks, and the sepals have tips tapered like the bow of a canoe.

American Bur-Reed

Sparganium americanum Nuttall
Cat-Tail Family (Typhaceae)

PERENNIAL HERB found in shallow water of streams, ponds, and marshes throughout Georgia. **STEMS:** Up to 3 feet tall, stout, branched, smooth, slightly zigzag, round in cross-section, usually in large patches. **LEAVES:** Up to 3 feet long and ¾ inch wide, alternate, erect, narrowly strap-shaped, entire, spongy, smooth. **FLOWER CLUSTERS:** Round, green, up to 1 inch wide; male-flowered heads at the top are smaller than the bristly, female-flowered heads below. **FLOWERS:** Tiny, with 5 translucent tepals; long, white pistils radiate from the female-flowered heads. **FRUIT:** In round, burlike heads up to 1 inch wide, the pointed beaks of the individual fruits giving the heads a spiny look. **FLOWERING:** May–Sep. **RANGE:** FL west to TX, north to MN and NL. **SIMILAR TO** no other wildflower in Georgia.

Common Cat-Tail

Typha latifolia Linnaeus
Cat-Tail Family (Typhaceae)

PERENNIAL HERB found in shallow freshwater ponds, marshes, and wet ditches throughout Georgia. **STEMS:** 5–10 feet tall, stout, smooth, forming dense stands in disturbed wetlands. **LEAVES:** Up to 10 feet long and ½–1 inch wide, alternate, entire, loosely twisted, more or less flat, smooth. **FEMALE FLOWER CLUSTERS:** Up to 7 inches long and 1¼ inches wide, cylindrical, pale green when immature, dark brown in flower, fluffy and white in fruit. **MALE FLOWER CLUSTERS:** Immediately above the female spike, up to 7 inches long, narrow, yellowish-brown. **FLOWERING:** May–Jul. **RANGE:** Throughout North America, and worldwide. **SIMILAR TO** Southern Cat-tail (*Typha domingensis*), which occurs in fresh and brackish marshes in Georgia's coastal counties. Its leaves are rounded on the back and less than ½ inch wide; the female spikes are separated from the male spike by a short length of stem.

Yellow-Eyed Grasses

Xyris species
Yellow-Eyed Grass Family (Xyridaceae)

PERENNIAL HERBS found in open, sunny wetlands and pine flatwoods throughout Georgia, mostly in the Coastal Plain. **STEMS:** Very short up to 4 feet tall, slender and flexible, often winged, rough, bumpy, or twisted. **LEAVES:** Up to 20 inches long (depending on species), narrow, linear, overlapping at the base and forming tufts. **FLOWER CLUSTERS:** A conelike spike, oval or round, ¼–1 inch long, held at the top of the stem, with many brown-and-green, overlapping bracts, each of which contains 1 flower. **FLOWERS:** Up to 1½ inches wide, with 3 showy, yellow (rarely white), rounded petals (3 tiny, papery sepals are hidden under the bract). Flowers usually open in the morning and wither by early afternoon, only 1 or 2 open each day. **FLOWERING:** Summer–fall. **RANGE:** *Xyris* species occur almost worldwide and are especially abundant in the New World tropics. **NOTES:** There are about 20 species of yellow-eyed grasses in Georgia, and identification is fairly technical.

COASTAL PLAIN YELLOW-EYED GRASS (*Xyris ambigua*), another common species, has slightly twisted stems up to 3 feet tall, leaves with pale tan or pinkish bases, and pale brown or tan spikes up to 1 inch long.

PINELAND YELLOW-EYED GRASS (*Xyris caroliniana*) is one of the most common species of *Xyris* in the Coastal Plain; it occurs in somewhat drier habitats. Its twisted stems are up to 3½ feet tall. The leaves are strongly twisted and are swollen and dark reddish-brown at the base. Its spike is about ¾ inch long.

RICHARD'S YELLOW-EYED GRASS (*Xyris jupicai*), often found in disturbed wetlands or roadside ditches, is the most common yellow-eyed grass in Georgia. Its stem is up to 3 feet tall, and its leaves are shiny yellow-green with tan to brown bases. The spikes are about ½ inch long, often with withered flowers still attached.

COLOR THUMBNAILS

WHITE

If you don't find your flower here, check yellow, pink, and blue.

Eastern Bishopweed p. 36	Indian Hemp p. 40	Wild Sarsaparilla p. 41	Devil's Walking Stick p. 41
Dune Pennywort p. 42	Milkweeds pp. 47–53	Yarrow p. 56	White Snakeroot p. 56
Plantain Pussy-toes p. 57	Pale Indian-plantain p. 58	Groundsel Tree p. 58	Southern Doll's-daisy p. 62
False Boneset p. 63	Sun-bonnets p. 66	Horseweed p. 69	Appalachian Flat-topped White Aster, p. 73
False Daisy p. 74	Fleabanes pp. 76–77	Robin's Plantain p. 76	Thoroughworts pp. 77–80

Wood Asters pp. 80–81	Woolly-white p. 92	Ox-eye Daisy p. 96	Elegant Blazing-star p. 97
Narrow-leaf Barbara's Buttons, p. 101	Salt-and-pepper p. 102	Climbing Hempweed p. 102	Whorled Wood Aster p. 103
Wild Quinine p. 105	Rabbit-tobacco p. 108	Blackroot p. 109	White-topped Asters pp. 112–113
Fall-flowering white asters, pp. 120–121	Frostweed p. 124	Umbrella Leaf p. 128	May-apple p. 128
Wild Comfrey p. 130	Spring Forget-me-not p. 132	Crucifers pp. 133–139	Lobelias pp. 145–146

Sticky Mouse-ear Chickweed, p. 149	Sandworts pp. 150–151	Dune Whitlow-wort p. 152	Starry Campion p. 153
Starry Chickweed p. 154	Pineland Scaly-pink p. 154	Gopher-apple p. 155	Sweet-pepperbush p. 157
Hedge Bindweed p. 158	Common Dodder p. 158	Wild Sweet Potato p. 160	Coastal Plain Dawnflower, p. 161
Diamorpha p. 162	Mountain Stonecrop p. 162	Bur-cucumber p. 163	Buckwheat Tree p. 164
Galax p. 164	Round-leaved Sundew, p. 166	Pipsissewa p. 167	Trailing Arbutus p. 167

Indian Pipes p. 170	Sweet Azalea p. 171	Blueberries pp. 174–176	Tread-softly p. 178
Silver Croton p. 178	Glade Rush-foil p. 179	Flowering Spurge p. 180	Spotted Spurge p. 181
Cumberland Spurge p. 181	White Wild Indigo p. 186	Summer Farewell p. 191	Elliott's Milk-pea p. 193
Bush clovers pp. 196–197	White Sweetclover p. 200	Clovers pp. 207–209	Carolina Vetch p. 209
Dutchman's Britches p. 211	Screwstem Bartonia p. 212	Striped Gentian p. 213	Rose-gentians pp. 216–217

Climbing Hydrangea p. 220	Wild Hydrangea p. 220	Oak-leaf Hydrangea p. 221	Mock-orange p. 221
Virginia-willow p. 228	Southern Horsebalm p. 231	Musky Mint p. 234	Virginia Bugle-weed p. 236
Basil Bee-balm p. 236	Savanna Mountain-mint, p. 240	Blue Sage p. 241	Caribbean Miterwort p. 251
Rustweed p. 252	Toothcup p. 254	Mallows pp. 255–256	Maryland Meadow-beauty, p. 259
Little Floating Hearts p. 261	White Waterlily p. 262	Enchanter's Night-shade, p. 263	Eastern Willow-herb p. 263

Slender Gaura p. 264	Bear Corn p. 272	Beech Drops p. 273	Cow-wheat p. 273
One-flowered Cancer-root, p. 274	Bloodroot p. 277	Grass-of-Parnassus p. 277	Ditch Stonecrop p. 279
American Lopseed p. 279	Pokeweed p. 280	Milkworts pp. 285–289	Dog-tongue p. 290
Climbing Buckwheat p. 291	Smartweeds pp. 291–293	Virginia Jumpseed p. 293	Southern Jointweed p. 294
Sandhill Jointweed p. 294	Eastern Shooting-star p. 299	Water Pimpernel p. 299	Doll's Eyes p. 300

Black Cohosh p. 301	Wood Anemone p. 302	Tall Thimbleweed p. 302	Virgin's Bower p. 304
Mountain Meadow-rue, p. 306	Skunk Meadow-rue p. 307	Rue-anemone p. 308	Tassel-rue p. 308
New Jersey Tea p. 309	Goat's Beard p. 311	Wild Strawberry p. 311	White Avens p. 312
Indian Physic p. 313	Chickasaw Plum p. 314	Blackberries pp. 315–316	Cleavers p. 317
Long-leaf Bluet p. 319	Innocence p. 319	Partridge-berry p. 321	Rough Mexican Clover, p. 321

Lizard's Tail p. 324	Brook Saxifrage p. 324	Hairy Alumroot p. 325	Cliff Saxifrage p. 325
Early Saxifrage p. 326	Foamflower p. 326	White Turtlehead p. 328	Virginia Hedge-hyssop, p. 328
Common Axil-flower p. 330	Many-flowered Beard-tongue, p. 332	Shaggy Hedge-hyssop p. 333	Moth Mullein p. 334
Culver's Root p. 335	Jimson Weed p. 336	American Black Nightshade, p. 338	Carolina Horse-nettle p. 338
Corn Salad p. 341	White Vervain p. 345	Violets pp. 347–354	Georgia Bear-grass p. 356

Yuccas pp. 357–358	Arrowheads, Southern Water Plantain, pp. 359–360	False Garlic p. 361	Hammock Spider-lily p. 362
Atamasco-lily p. 362	Canada May-flower p. 365	Solomon's Plume p. 366	Solomon's Seal p. 366
White-top Sedge p. 378	Flattened Pipewort p. 380	White Bog Buttons p. 380	Redroot p. 381
Blue-eyed grasses pp. 385–386	Speckled Wood-lily p. 389	Fly Poison p. 392	Devil's Bit p. 393
Crow Poison p. 393	Feather-bells p. 394	Northern White Colic Root, p. 395	Downy Rattlesnake-orchid, p. 400

Small Green Wood Orchid, p. 404

Snowy Orchid p. 404

Ladies' Tresses pp. 405–407

Southern Bog Asphodel, p. 420

Trilliums pp. 421–426

YELLOW

If you don't find your flower here, check white and green.

Meadow Parsnips pp. 37–38	Mountain Golden Alexander, p. 38	Climbing Dogbane p. 40	Honeycomb Heads p. 59
Green-eyes p. 60	Devil's Beggar-ticks p. 61	Smooth Beggar-ticks p. 61	Pineland Rayless Goldenrod, p. 62
Sea Ox-eye Daisy p. 63	Green-and-gold p. 66	Golden-asters p. 67	Bull Thistle p. 68
Tickseeds pp. 70–71	Plains Tickseed p. 72	Slender Scratch-daisy p. 72	Carolina Goldentop p. 82

Sandhill Blanket-flower, p. 83	Bitter-weed p. 84	Autumn Sneezeweed p. 85	Southern Sneezeweed p. 85
Sunflowers pp. 86–90	Common Camphor-weed, p. 90	Hawkweeds p. 91	Hairy Cat's-ear p. 92
Dwarf-dandelions p. 94	Wild Lettuce p. 95	Ragworts pp. 104–105	Sticky Golden-aster p. 106
Grass-leaved Golden-aster, p. 106	Tall Rattlesnake-root p. 108	False Dandelion p. 109	Gray-headed Coneflower, p. 110
Black-eyed Susans pp. 110–111	Cutleaf Coneflower p. 111	Rosin-weeds pp. 113–114	Bear's-foot p. 114

Goldenrods pp. 115–118	Sow Thistle p. 118	Squarehead p. 123	Wing-stem p. 124
Jewelweed p. 126	Blue Cohosh p. 127	Cross-vine p. 129	Coastal Plain Puccoon, p. 131
Virginia Marbleseed p. 132	Wild Radish p. 139	Cacti p. 141	Coastal Plain Nailwort, p. 152
Sun-roses p. 156	Creeping Cucumber p. 163	Queen's Delight p. 182	Bearded Milk-vetch p. 185
Wild indigos pp. 187–188	Partridge Pea p. 189	Rabbit Bells p. 190	Showy Rattlebox p. 190

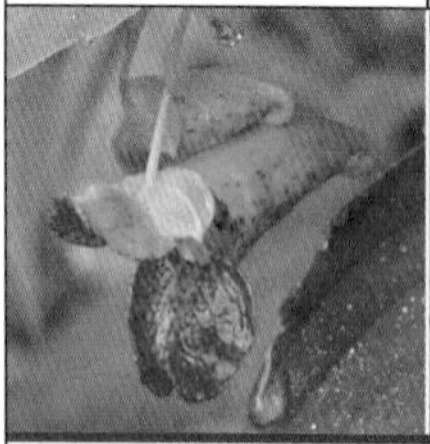

Black Medic p. 200	Dollar Weed p. 203	Erect Snoutbean p. 203	Wild Senna p. 204
Pencil Flower p. 205	Virginia Goat's Rue p. 206	Large Hop Clover p. 207	Yellow Fumewort p. 211
Painted Buckeye p. 219	St. Andrew's Cross p. 224	St. John's-worts pp. 224–227	Pineweed p. 225
Northern Horsebalm p. 231	Spotted Bee-balm p. 238	Yellow Butterwort p. 247	Bladderworts pp. 247–249
Yellow Flaxes p. 250	Yellow Jessamine p. 251	Arrow-leaf Sida p. 257	Yellow Meadow-beauty, p. 259

Coral Beads p. 261	Spatterdock p. 262	Seedboxes, Water-primroses, pp. 264–265	Water-willow p. 265
Sundrops, Evening-primroses, pp. 266–267	False Foxgloves p. 271	Bear Corn p. 272	Cow-wheat p. 273
Lousewort p. 274	Black Senna p. 275	Yellow Wood Sorrel p. 276	Yellow Passion-flower p. 278
Tall Pine Barren Milkwort, p. 286	Candy-root p. 288	Common Purslane p. 297	Loosestrifes p. 298
Hooked Buttercup p. 306	Southern Agrimony p. 310	Southern Barren Strawberry, p. 312	Cinquefoils pp. 313–314

Pitcherplants pp. 322–323	Woolly Mullein p. 334	Ground Cherries, pp. 336–337	Piriqueta p. 339
Violets pp. 349–353	Golden Club p. 364	Arrow-arum p. 365	Solomon's Seal p. 366
Yellow Sunny-bells p. 367	Yellow Mandarin p. 369	Perfoliate Bellwort p. 369	Wild Oats p. 370
Redroot p. 381	Star-grasses p. 382	Yellow Flag p. 384	Hedgehog Wood Rush, p. 387
Dimpled Trout Lily p. 389	Indian Cucumber- root, p. 391	Yellow Colic Root p. 395	Puttyroot p. 397

Crested Coral-root p. 401	Large Whorled Pogonia, p. 402	Toothache Grass p. 412	Yellow Indian Grass p. 415
Yellow Trillium p. 424	Yellow-eyed grasses p. 428	Southern Bush-honeysuckle, p. 147	Yellow Honeysuckle p. 148
Bristly Buttercup p. 305			

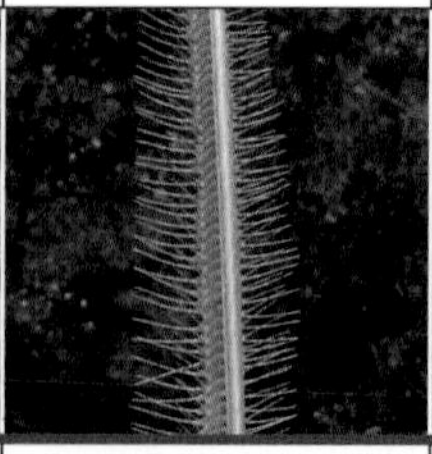

ORANGE

If you don't find your flower here, check red and yellow.

RED AND MAROON

If you don't find your flower here, check pink, purple, and orange.

Woody Glasswort p. 27	Small-flowered Pawpaw, p. 28	Meadow parsnips pp. 37–38	Wild Ginger p. 43
Large-flowered Heartleaf, p. 45	Few-flowered Milkweed, p. 49	Climbing Milkweed p. 53	Blanket-flower p. 84
Round-leaf Sunflower p. 89	Cross-vine p. 129	Water-shield p. 140	Sweet Shrub p. 142
Cardinal Flower p. 144	Coral Honeysuckle p. 148	Fire Pink p. 153	Strawberry-bush p. 155

Hairy Pinweed p. 157 	Red Morning-glory p. 159 	Pine-sap p. 168 	Flame Azalea p. 171 
Short-stalked Copper-leaf, p. 177 	Maroon Sandhills Spurge, p. 180 	Ground-nut p. 185 	Coral Bean p. 193 
Spiked Goat's Rue p. 206 	Crimson Clover p. 208 	Red Buckeye p. 219 	Trailing Ratany p. 228 
Scarlet Wild Basil p. 230 	Scarlet Bee-balm p. 237 	Indian Pink p. 252 	Toothcup p. 254 
Coral Beads p. 261 	Lousewort p. 274 	Ditch Stonecrop p. 279 	Standing Cypress p. 282 

Docks p. 295	Eastern Columbine p. 303	Northern Leather-flower, p. 304	Yellow-root p. 309
Hairy Bedstraw p. 318	Parrot Pitcherplant p. 323	Figwort p. 333	Spring Coral-root p. 399
Crane-fly Orchid p. 407	Purple Love Grass p. 412	Purple Top p. 416	Trilliums pp. 422–426
Trilliums pp. 422–426			

PINK, MAGENTA, AND LAVENDER

If you don't find your flower here, check red, purple, and white.

American Water-willow, p. 21	Large Sea-purslane p. 24	Clasping Milkweed p. 47	Carolina Milkweed p. 48
Pineywoods Milk-weed, p. 49	Michaux's Milkweed p. 50	Whorled Milkweed p. 52	Musk Thistle p. 64
Chaffheads, pp. 64–65	Thistles p. 68	Georgia Tickseed p. 71	Eastern Purple Cone-flower, p. 73
Elephant's Foot pp. 74–75	Fleabanes pp. 76–77	Joe-pye weeds pp. 82–83	Blazing-stars pp. 96–100

Rose-rush p. 101	Narrow-leaf Barbara's Buttons, p. 101	Camphorweed p. 107	Stinking Camphorweed, p. 107
Ironweeds pp. 125–126	Pineland Cress p. 140	Deptford Pink p. 150	Strawberry-bush p. 155
Hedge Bindweed p. 158	Railroad Vine p. 160	Pink Sundew p. 165	Tracy's Dew-threads p. 166
Trailing Arbutus p. 167	Pine-sap p. 168	Hairy Wicky p. 169	Mountain Laurel p. 169
Shining Fetterbush p. 170	Piedmont Azalea p. 172	Rhododendrons, pp. 172–173	Hog Peanut p. 184

Ground-nut p. 185	Spurred Butterfly Pea p. 188	Pigeon Wing p. 189	Beggar-ticks pp. 191–192
Eastern Milk-pea p. 194	Carolina Wild Indigo p. 195	Japanese Clover p. 195	Everlasting Pea p. 196
Lespedezas pp. 196–198	Lady Lupine p. 199	Sensitive Brier p. 201	Sampson's Snakeroot p. 201
Wild Bean p. 202	Sand Bean p. 205	Spiked Goat's Rue p. 206	Virginia Goat's Rue p. 206
Rabbit-foot Clover p. 207	Wild White Clover p. 208	Carolina Vetch p. 209	Pennywort p. 214

Rose-pinks pp. 215–217	Rose-gentians pp. 215–217	Carolina Crane's-bill p. 218	Wild Geranium p. 218
Maple-leaf Waterleaf p. 222	Marsh St. John's-wort p. 227	Trailing Ratany p. 228	Downy Wood Mint p. 229
Georgia Basil p. 230	Whorled Horsebalm p. 232	Stone Mint p. 232	Coastal Plain Balm p. 233
Gill-over-the-ground p. 233	American Pennyroyal p. 234	Henbit p. 235	Purple Dead-nettle p. 235
Wild Bergamot p. 237	Spotted Bee-balm p. 238	Beefsteak Plant p. 238	Obedient Plant p. 239

Mountain-mints pp. 240–241	Florida Betony p. 244	American Germander p. 245	Purple Bladderwort p. 249
Columbian Waxweed p. 253	Swamp Loosestrife p. 253	Southern Winged Loosestrife, p. 254	Smooth Rose Mallow p. 255
Comfort-root p. 256	Meadow-beauties pp. 258–260	Bog Meadow-beauty p. 260	Eastern Willow-herb p. 263
Slender Gaura p. 264	Showy Evening-primrose, p. 268	False Foxgloves pp. 269–270	Violet Wood Sorrel p. 276
American Lopseed p. 279	Sea Lavender p. 281	Phloxes pp. 282–284	Drumheads p. 285

Milkworts pp. 286–289	Pink Milkwort p. 287	Whorled Milkwort p. 289	Gay-wings p. 289
Smartweeds pp. 291–292	Arrow-leaf Tearthumb, p. 293	Jointweeds p. 294	Eastern Spring-beauty p. 296
Fame-flower p. 296	Kiss-me-quick p. 297	Round-lobed Hepatica, p. 301	Southern Leather-flower, p. 303
Northern Leather-flower, p. 304	Carolina Rose p. 315	Purple-flowering Raspberry, p. 316	Rough Buttonweed p. 317
Field Madder p. 318	Long-leaf Bluet p. 319	Beard-tongues pp. 331–332	American Beauty-berry, p. 342

Verbenas pp. 342–343	Frog-fruit p. 343	Vervains pp. 344–345	Wild Onion p. 361
Grass-leaf Roseling p. 372	Annual Blue-eyed Grass, p. 386	Bog Moss p. 392	Grass-pinks pp. 397–398
Large Spreading Pogonia, p. 398	Pink Lady's Slipper p. 399	Showy Orchis p. 400	Lily-leaved Tway-blade, p. 402
Rose Pogonia p. 405	Purple Love Grass p. 412	Muhly Grass p. 413	Catesby's Trillium p. 421

BLUE TO BLUISH-PURPLE

If you don't find your flower here, check purple and white.

Blue Twinflower p. 20	Carolina Scaly-stem p. 20	Carolina Wild Petunia, p. 21	Savanna Eryngo p. 33
Creeping Eryngo p. 33	Blue stars p. 39	Bachelor's Buttons p. 65	Mistflower p. 69
Creeping Aster p. 81	Woodland Lettuce p. 95	Purple Asters pp. 119–123	Wild Comfrey p. 130
Turnsole p. 130	American Bell-flower p. 143	Southern Harebell p. 143	Lobelias pp. 144–146

Clasping Venus's Looking-glass, p. 146	Wahlenbergia p. 147	Ivy-leaf Morning-glory, p. 159	Hairy Cluster-vine p. 161
Blue Wild Indigo p. 186	Sundial Lupine p. 199	Buckroot, Hoary Scurf-pea, p. 202	American Wisteria p. 210
Soapwort Gentian p. 213	Stiff Gentian p. 214	Skyflower p. 222	Phacelias p. 223
Blue Sage p. 241	Lyre-leaved Sage p. 242	Nettle-leaf Sage p. 242	Skullcaps pp. 243–244
Blue Curls p. 245	Blue Butterwort p. 246	Florida Bluehearts p. 272	Blue Phlox p. 283

Blue Monks-hood
p. 300

Blue Larkspur
p. 305

Bluets
pp. 319–320

Blue Water-hyssop
p. 327

Flatrock Pimpernel
p. 329

Japanese Mazus
p. 329

Winged Monkey-flower, p. 330

Toad-flax
p. 331

Corn Speedwell
p. 335

Violets
pp. 347–354

Nodding Nixie
p. 368

Northern Blue Thread, p. 368

Spiderworts
pp. 371–373

Crested Iris
p. 383

Blue Flag
p. 385

Blue-eyed Grasses
pp. 385–386

Pickerelweed
p. 417

PURPLE AND VIOLET

If you don't find your flower here, check blue, red, and pink.

Blue Twinflower p. 20	American Water-willow, p. 21	Carolina Wild Petunia, p. 21	Wild Ginger p. 43
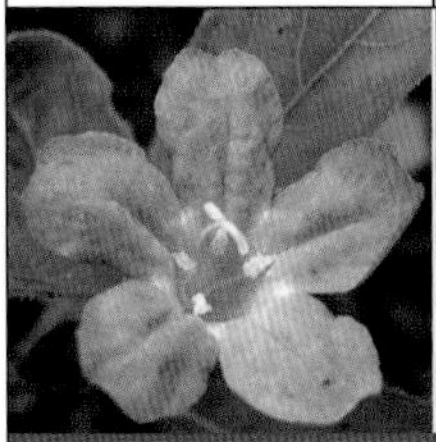	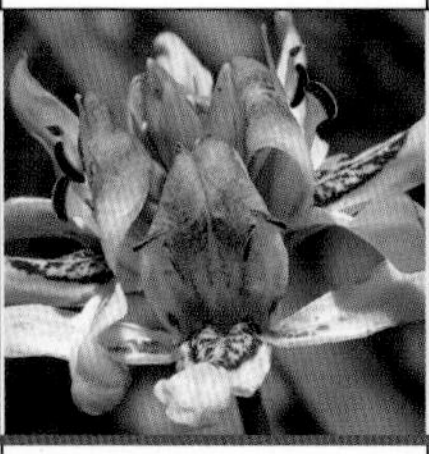	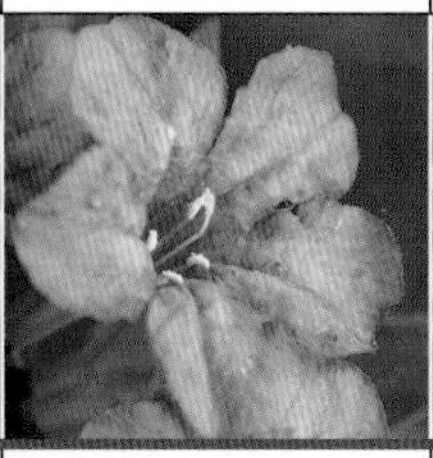	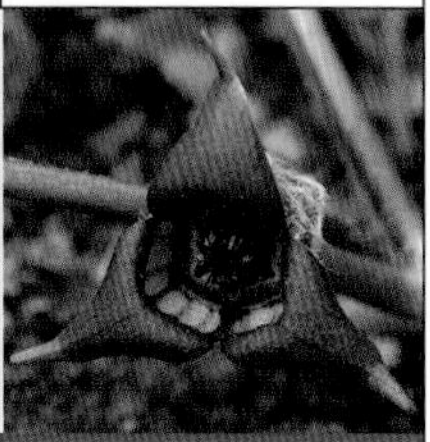
Heartleaf p. 43	Large-flowered Heartleaf, p. 45	Dutchman's Pipe p. 45	Clasping Milkweed p. 47
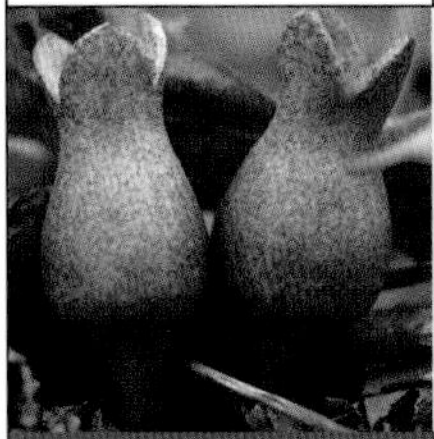	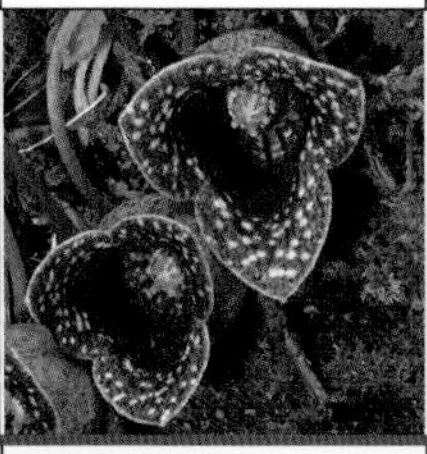	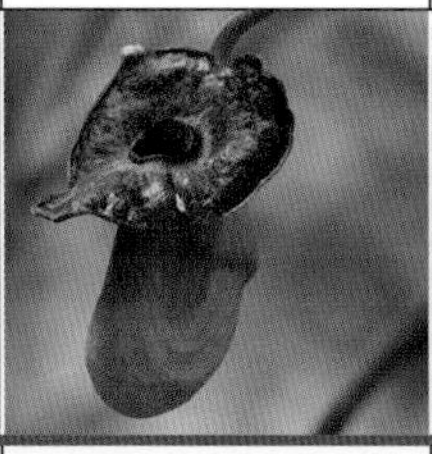	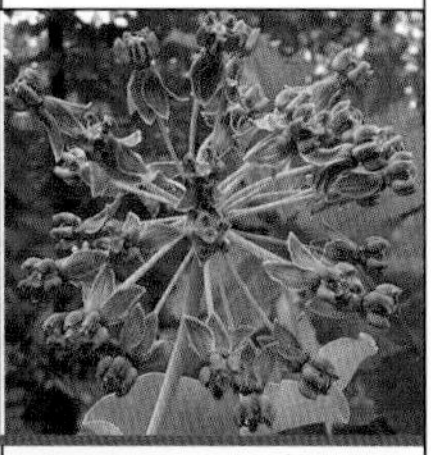
Carolina Milkweed p. 47	Pineywoods Milkweed, p. 48	Climbing Milkweed p. 53	Musk Thistle p. 64
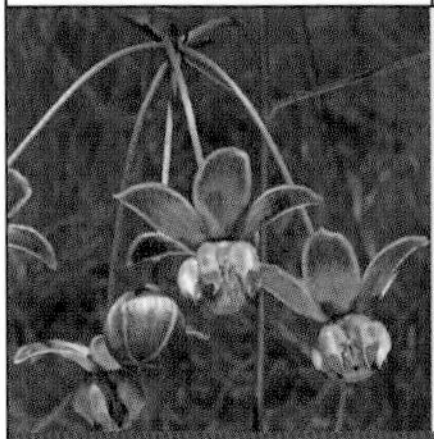	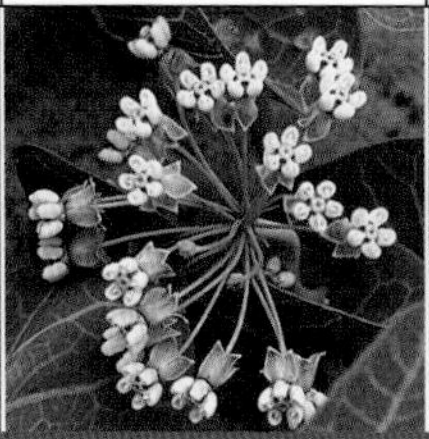	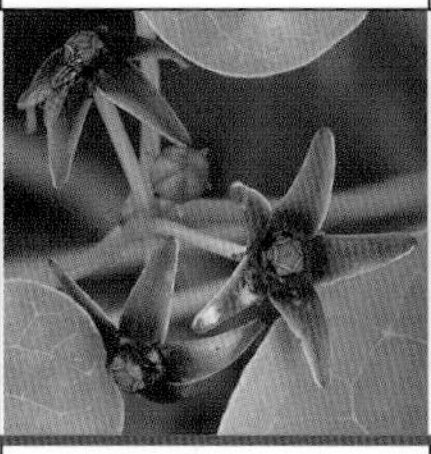	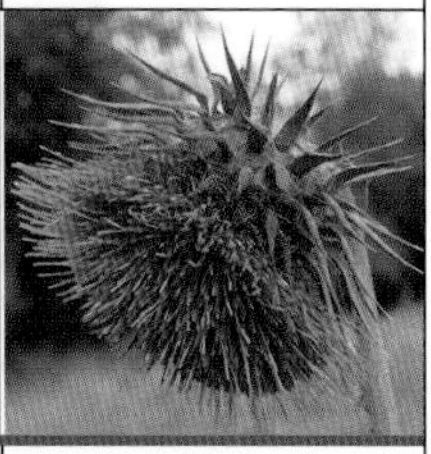
Vanilla-leaf p. 64	Bull Thistle p. 68	Mistflower p. 69	Purple Coneflower p. 73
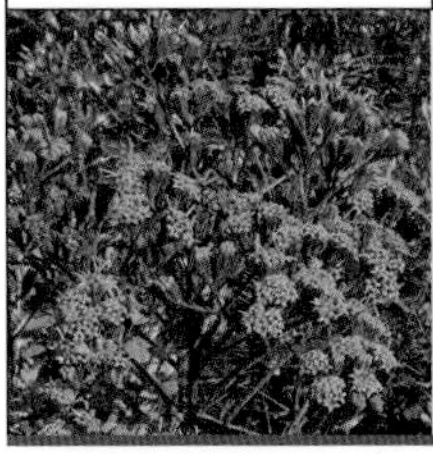	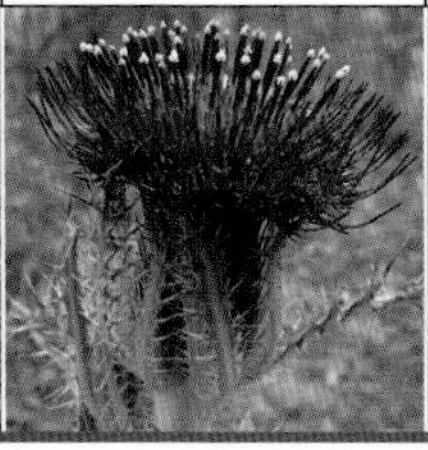	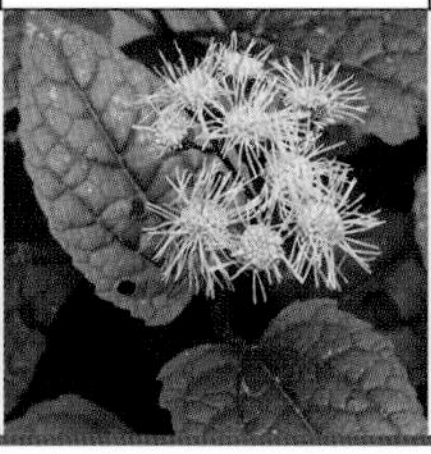	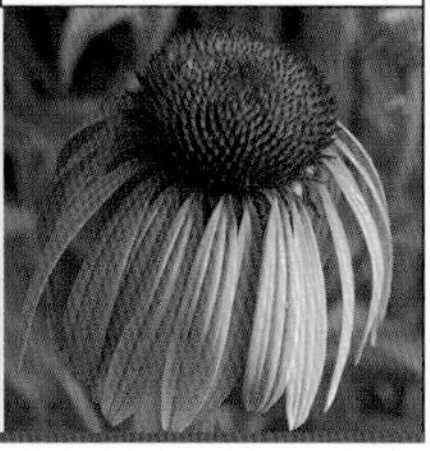

Creeping Aster p. 81	Joe-pye Weeds pp. 82–83	Stiff-leaved Aster p. 93	Woodland Lettuce p. 95
Blazing-stars pp. 96–100	Purple Asters pp. 119–123	Ironweeds pp. 125–126	Pineland Cress p. 140
Lobelias pp. 144–146	Venus's Looking-glass p. 146	Tall Indigo-bush p. 184	Sampson's Snakeroot p. 201
Winter Vetch p. 210	American Wisteria p. 210	Soapwort Gentian p. 213	Stiff Gentian p. 214
Pennywort p. 214	Maple-leaf Waterleaf p. 222	Phacelias p. 223	Gill-over-the-ground p. 233

Purple Dead-nettle p. 235	Wild Bergamot p. 237	Beefsteak Plant p. 238	Self-heal p. 239
Lyre-leaved Sage p. 242	Nettle-leaf Sage p. 242	Skullcaps pp. 243–244	Blue Curls p. 245
Purple Bladderwort p. 249	Florida Bluehearts p. 272	May-pop p. 278	Sea Lavender p. 281
Blue Phlox p. 283	Blue Monks-hood p. 300	Southern Leather- flower, p. 303	Yellow-root p. 309
Purple-flowering Raspberry, p. 316	Japanese Mazus p. 329	Toad-flax p. 331	Southern Beard- tongue, p. 331

Jimson Weed
p. 336
Verbenas
pp. 342–345
Brazilian Vervain
p. 344
Violets
pp. 347–354
Jack-in-the-pulpit
p. 364
Nodding Nixie
p. 368
Northern Blue
Thread, p. 368
Spiderworts
pp. 371–373
Irises
pp. 383–385
Blue-eyed grasses
pp. 385–386
Lily-leaved Tway-
blade, p. 402
Crane-fly Orchid
p. 407
Purple Love Grass
p. 412
Purple Top
p. 416
Pickerelweed
p. 417

GREEN AND BROWN

If you don't find your flower here, check white, yellow, and maroon.

Spiny Amaranth p. 25	Lamb's-quarters p. 25	Russian-thistle p. 26	Woody Glasswort p. 27
Southern Sea-blite p. 27	Coin-leaf p. 30	Rattlesnake Master p. 34	Canada Sanicle p. 37
Dune Pennywort p. 42	Ginseng p. 42	Heartleaf p. 43	Dutchman's Pipe p. 45
Large-flowered Green Milkweed, p. 48	Whorled Milkweed p. 52	Green-flowered Milkweed, p. 53	Climbing Milkweed p. 53

Coastal Swallow-wort p. 54	Dune Marsh-elder p. 93	Saltwort p. 127	Blue Cohosh p. 127
Dune Whitlow-wort p. 152	Coastal Plain Nailwort, p. 152	Strawberry-bush p. 155	Hairy Pinweed p. 157
Bur-cucumber p. 163	Cumberland Spurge p. 181	Queen's Delight p. 182	Nettle-leaf Noseburn p. 182
Screwstem Bartonia p. 212	Columbo p. 212	Marsh Seedbox p. 266	Yellow Passion-flower p. 278
Long-bracted Plantain, p. 280	Virginia Plantain p. 281	Climbing Buckwheat p. 291	Curly Dock p. 295

Meadow-rues pp. 306–307	Figwort p. 333	False Nettle p. 339	Wood Nettle p. 340
Clearweed p. 341	Green Violet p. 346	Rattlesnake Master p. 356	Jack-in-the-pulpit p. 364
Arrow-arum p. 365	Spanish Moss p. 367	Sedges pp. 374–379	Flat-sedges pp. 376–377
Three-way Sedge p. 378	Woolly Bulrush p. 379	Wild Yam p. 379	Soft Rush p. 387
Hedgehog Wood Rush, p. 387	Fly Poison p. 392	Appalachian Bunch-flower, p. 394	Puttyroot p. 397

Water-spider Orchid
p. 401
Large Whorled
Pogonia, p. 402
Green Adder's-mouth
Orchid, p. 403
Small Green Wood
Orchid, p. 404
Crane-fly Orchid
p. 407
Grasses
pp. 408–417
Greenbriars
pp. 418–420
American Bur-reed
p. 427
Common Cat-tail
p. 427

GLOSSARY

Alternate. A leaf arrangement in which the leaves are attached singly to the stem, not in pairs or whorls. Compare with opposite leaves.
Annual. A plant that completes its life cycle in one season.
Anther. The pollen-bearing portion of a stamen.
Awn. A needle-like bristle at the tip of a leaf or other structure, especially in grass flowers.
Axil. The angle between a stem and the base of a leaf stalk or leaf. There is usually a tiny leaf bud or flower bud at the base of the angle.

Banner petal. The large upper petal of a pea flower.
Barbed. Having a short, sharp, hooked tip.
Basal rosette. A cluster of leaves at ground level that radiate from the base of the stem.
Beak. A narrow, elongated tip.
Berry. A juicy, fleshy fruit with several to many seeds.
Biennial. A plant that completes its life cycle in two seasons, usually forming a basal leaf rosette the first year and producing a stem, flowers, and fruits the second year.
Blade. The flat, expanded portion of a leaf.
Bract. A small, leaflike structure at the base of a flower or flower cluster.
Bulb. An underground storage organ composed of fleshy, overlapping layers, like an onion.

Calyx. The collective term for the sepals. Usually green but sometimes colorful, the calyx surrounds the base of the flower and often protects the other flower parts when in bud. The number of sepals in a calyx usually equals the number of petals.
Capsule. A dry fruit that splits open when mature.
Clasping. Nearly surrounding, as when a leaf base nearly surrounds the stem.
Compound leaf. A leaf made up of two to many smaller leaflets.
Corm. An upright, solid underground storage organ.
Cross-pollination. The transfer of pollen from the flowers of one plant to the flowers of another, which results in genetic mixing. Self-pollination occurs when a flower pollinates itself or when pollen is moved from flower to flower on the same plant, which does not result in genetic mixing.

Deciduous. Having leaves that drop in the fall and winter in temperate climates.
Disk. The central portion of most Aster Family flower heads; it bears very small disk flowers.
Dissected. Finely divided into slender segments.

Elliptic. A leaf shape that is broadest in the middle and tapers equally at both ends.
Entire. A leaf margin that is not toothed, lobed, or otherwise interrupted.
Evergreen. Having leaves that overwinter and persist for more than one growing season.

Falls. The colorful, showy sepals of an iris flower.
Fringed. Very finely cut along the edge or edged with hairs or bristles.

Gland. A plant structure that secretes sticky or oily substances.
Gland dots. Tiny, flat or pitted areas on plant surfaces that may be colorful or translucent. Although the function of glandular dots is not completely understood, they are usually associated with strong smells and tastes that deter plant-eating animals.
Gland-tipped hairs. Hairs with tiny, ball-shaped glands on the tips.
Glandular-hairy. Covered with hairs bearing tiny, ball-shaped glands on the tips.

Hairy. Covered to a greater or lesser extent with hairs. There are hundreds of technical terms that describe plant hairs.
Heart-shaped. A leaf shape that is broadly oval with a pointed tip and a notch at the base. The technical term is "cordate."
Herb. A soft-tissued plant without any woody stems, usually dying back in the fall and winter.
Herbaceous. Having nonwoody, soft tissues.

Inflorescence. A flower cluster.
Irregular flower. Flower with bilateral symmetry, meaning that the petals come in different sizes and shapes and form a flower that can be divided into equal halves (mirror images) along only one line. The technical term for this flower pattern is "zygomorphic." Examples are orchids, violets, peas, mints, and lobelias.

Lance-shaped. A leaf shape that tapers from the base to a pointed tip, with the widest point nearer the base.
Lateral. To the side of.
Latex. White, milky, often sticky juice found in about 10% of all plants. Distinct from sap, latex deters plant-eating animals by its sticky texture and toxic content.
Leaflet. A distinct and separate segment of a compound leaf.
Lenticel. A small, raised, spongy or corky place on a stem or other plant part through which gases move back and forth between the atmosphere and the interior of the plant.
Linear. A leaf shape that is narrow with parallel sides.
Lip. The lower, usually larger petal on irregularly shaped flowers. It is usually colored or patterned to attract pollinators that use the lip as a "landing platform."
Lobe. A rounded or pointed division or segment of a flower, leaf, or other plant part.

Mafic. A type of bedrock that is rich in magnesium and iron. Georgia's mafic rocks are often amphibolite, which weathers to nutrient-rich soils.
Margin. The edge of a leaf or other flat plant part.
Mealy. Covered with a white, grainy powder.
Midvein. The central vein of a leaf.
Mucilage. A sticky, slimy, or gelatinous substance secreted by plants.

Oblong. A leaf shape that is 2–4 times longer than wide, with nearly parallel sides and a blunt or rounded tip.
Ocrea. A thin sheath that wraps around the leaf node and the adjacent portion of stem in the Smartweed Family (Polygonaceae). Plural: ocreae.
Opposite. A leaf arrangement in which the leaves are attached to the stem in pairs. Compare with alternate leaves.

Oval. A leaf that is egg-shaped in outline, broader at one end than the other. In this guide, "oval" includes the technical terms "ovate" (broader at the base) and "obovate" (broader at the tip).
Ovary. The enlarged base of the pistil that contains the eggs (ovules), the future seeds.

Palate. A raised portion on the lower lip of butterwort and bladderwort flowers.
Pea-flower shape. A flower shape typical of many members of the Pea Family (Fabaceae), consisting of a large banner petal, two wing petals (often spreading), and a canoe-shaped keel petal.
Perennial. A plant that lives three or more years, either going dormant during the winter or persisting aboveground as an evergreen.
Petals. The usually brightly colored parts of a flower that attracts pollinators.
Petiole. A leaf stalk.
Pistil. The part of the flower that develops into the fruit. Usually located at the center of the flower, it consists of an ovary (which contains ovules or eggs), a stigma (the sticky tip that captures the pollen), and a style that connects the stigma to the ovary.
Pith. The spongy tissue in the center of a plant's stem.
Pod. In this guide, a beanlike fruit that splits lengthwise to expose numerous seeds in a row. Technical terms include "legume" for the Bean Family and "siliques" and "silicles" for the Mustard Family.

Ray flowers. The strap-shaped, petal-like flowers that form a whorl around the perimeter of an Aster Family flower head. They usually surround a central disk of tiny, tubular flowers but occasionally comprise the entire head.
Regular flower. Flower with radial symmetry, meaning that the petals are all more or less alike and radiate symmetrically from the center of the flower. The flower can be divided into equal halves (mirror images) along several lines. The technical term for this flower pattern is "actinomorphic." Examples include Bloodroot, buttercups, cacti, evening-primroses, lilies, and roses.
Rhizome. Underground stem.
Rosette. See basal rosette.

Sepals. Flower parts that make up the calyx, which surrounds the base of the flower and often protects the other flower parts when in bud; usually green but sometimes colorful. The number of sepals in a calyx usually equals the number of petals.
Sheath. A more or less tubular portion of the leaf surrounding the adjacent stem.
Shrub. A woody plant with multiple stems that is less than 20 feet tall at maturity.
Smooth. Not hairy. The technical term is "glabrous."
Spadix. A type of flower cluster that contains many flowers held on a fleshy column. The spadix is usually surrounded by a colorful, leaflike "spathe." The "Jack" in Jack-in-the-pulpit is a spadix.
Spathe. A large, leaflike, sometimes colorful bract that partially encloses a flower cluster. The "pulpit" in Jack-in-the-pulpit is a spathe.
Spur. A slender, hollow, projecting structure, usually at the base or back of a flower, that produces and holds nectar. While probing the hollow spurs for nectar, pollinators come in contact with the flower's pollen-bearing and pollen-receiving structures.
Stamen. The part of the flower that produces the pollen, consisting of a filament (stalk) and an anther.
Stigma. The top of the pistil (the fruit-producing organ of a flower) that captures pollen from insects, birds, or the wind. It is usually sticky.
Stipules. A pair of small, leaflike structures at the base of the leaf stalk in some plants, notably those in the Rose and Bean Families.

Style. The central part of the pistil that conveys pollen from the stigma to the ovary.
Succulent. Juicy and fleshy.

Tendril. A curling structure, modified from a leaf, that helps a plant climb other plants or other types of support.
Tepals. A term used for petals and sepals that look more or less alike and cannot be easily distinguished, as in the lilies.
Toothed. A leaf margin that has points or scallops.
Trailing. Lying flat across the ground but not rooting at the nodes.
Tuber. A thick underground stem that stores carbohydrates.

Umbel. A domed or flat-topped flower cluster with all the flower stalks arising from the same point, like the ribs of an umbrella. Often the flower stalks are topped with smaller umbels called umbelets.

Waxy white. Covered with a white or gray, waxy substance that gives the plant a blue-green or gray-green look. The technical term is "glaucous."
Whorl. Three or more leaves, petals, or bracts that rise in a circle.

REFERENCES

Bidartondo, M. I. 2005. The Evolutionary Ecology of Mycoheterotrophy. New Phytologist 167: 335–352.
Blankespoor, J. 2012. Spring Ephemerals and Elaiosomes. http://blog.chestnutherbs.com/spring-ephemerals-and-elaiosomes.
Brown, P. M., and S. N. Folsom. 2004. Wild Orchids of the Southeastern United States, North of Peninsular Florida. University Press of Florida, Gainesville.
Buchmann, S. L., and G. P. Nabhan. 1997. The Forgotten Pollinators. Island Press, Washington, DC.
Carter, J. R. 2005. An Introduction to the Sedges of Georgia. Tipularia—the Journal of the Georgia Botanical Society 20: 15–44.
Chafin, L. G. 2000. Field Guide to the Rare Plants of Florida. Florida Natural Areas Inventory, Tallahassee.
Chafin, L. G. 2007. Field Guide to the Rare Plants of Georgia. State Botanical Garden of Georgia and University of Georgia Press, Athens.
Clark, J. 2004. Wildflowers of Pigeon Mountain, Lookout Mountain, Cloudland Canyon State Park, and Chickamauga National Military Park in Northwest Georgia. Waldenhouse Publishers, Walden, TN.
Cronquist, A. 1980. Vascular Flora of the Southeastern United States. Vol. 1, Asteraceae. University of North Carolina Press, Chapel Hill.
De Luca, P. A., and M. Vallejo-Marín. 2013. What's the "Buzz" About? The Ecology and Evolutionary Significance of Buzz Pollination. Current Opinion in Plant Biology 16(4): 429–435.
Drake, J. 2011. Gentians of the Eastern United States. Breath O'Spring Publishing, Suwannee, GA.
Drake, J. 2011. *Platanthera*, Orchids on the Fringe. Tipularia: Journal of the Georgia Botanical Society 26: 2–16.
Drake, J. 2014. Lilies in the Wild and in the Garden. Breath O'Spring Publishing, Suwannee, GA.
Duncan, W. H., and M. B. Duncan. 1987. Seaside Plants of the Gulf and Atlantic Coasts. Smithsonian Institution Press, Washington, DC.
Duncan, W. H., and M. B. Duncan. 1999. Wildflowers of the Eastern United States. University of Georgia Press, Athens.
Edwards, L., J. Ambrose, and L. K. Kirkman. 2013. Natural Communities of Georgia. University of Georgia Press, Athens.
Elpel, T. J. 2013. Botany in a Day: The Patterns Method of Plant Identification. HOPS Press, Pony, MT.
FNA. 1997. Flora of North America. Vol. 3, Magnoliophyta: Magnoliidae and Hamamelidae. Oxford University Press, New York.
FNA. 2000. Flora of North America. Vol. 22, Magnoliophyta: Alismatidae, Arecidae, Commelinidae (in Part), and Zingiberidae. Oxford University Press, New York.
FNA. 2003. Flora of North America. Vol. 23, Magnoliophyta: Commelinidae (in Part), Cyperaceae, Part 1. Oxford University Press, New York.

FNA. 2003. Flora of North America. Vol. 25, Magnoliophyta: Commelinidae (in Part), Poaceae. Oxford University Press, New York.
FNA. 2003. Flora of North America. Vol. 26, Magnoliophyta: Liliidae: Liliales and Orchidales. Oxford University Press, New York.
FNA. 2005. Flora of North America. Vol. 5, Magnoliophyta: Caryophyllidae, Part 2: Caryophyllaceae, Plumbaginaceae, and Polygonaceae. Oxford University Press, New York.
FNA. 2006. Flora of North America. Vol. 19, Magnoliophyta: Asteridae, Part 6: Asteraceae, Part 1. Oxford University Press, New York.
FNA. 2006. Flora of North America. Vol. 20, Magnoliophyta: Asteridae, Part 7: Asteraceae, Part 2. Oxford University Press, New York.
FNA. 2006. Flora of North America. Vol. 22, Magnoliophyta: Asteridae, Part 8: Asteraceae, Part 3. Oxford University Press, New York.
FNA. 2006. Flora of North America. Vol. 24, Magnoliophyta: Commelinidae (in Part), Poaceae. Oxford University Press, New York. http://herbarium.usu.edu/webmanual/default.htm.
Foote, L. E., and S. B. Jones Jr. 1989. Native Shrubs and Woody Vines of the Southeast. Timber Press, Portland, WA.
Giannasi, D. E., and W. B. Zomlefer. 2010 and Ongoing. Vascular Plant Atlas of Georgia. University of Georgia Herbarium, Plant Biology Department, University of Georgia, Athens.
Godfrey, R. K. 1988. Trees, Shrubs, and Woody Vines of Northern Florida and Adjacent Georgia and Alabama. University of Georgia Press, Athens.
Godfrey, R. K., and J. W. Wooten. 1979. Aquatic and Wetland Plants of Southeastern United States. Vol. 1, Monocotyledons. University of Georgia Press, Athens.
Godfrey, R. K., and J. W. Wooten. 1981. Aquatic and Wetland Plants of Southeastern United States. Vol. 2, Dicotyledons. University of Georgia Press, Athens.
Gracie, C. 2012. Spring Wildflowers of the Northeast: A Natural History. Princeton University Press, Princeton, NJ.
Grant, V. 1950. Pollination of *Calycanthus occidentalis*. American Journal of Botany 37(4): 294–297.
Harris, J. G., and M. W. Harris. 2000. Plant Identification Terminology: An Illustrated Glossary. Spring Lake Publishing, Payson, UT.
Horak, D. 2004. Orchids and Their Pollinators. Brooklyn Botanic Garden, NY. http://www.bbg.org/gardening/article/orchids_and_their_pollinators.
Horn, D., T. Cathcart, T. E. Hemmerly, and D. Duhl. 2005. Wildflowers of Tennessee, the Ohio Valley, and the Southern Appalachians. Lone Pine Publishing, Auburn, WA.
Isely, D. 1990. Vascular Flora of the Southeastern United States. Vol. 3, Part 2, Leguminosae (Fabaceae). University of North Carolina Press, Chapel Hill.
Klooster, M. R., and T. D. Culley. 2010. Population Genetic Structure of the Mycoheterotroph *Monotropa hypopitys* L. (Ericaceae) and Differentiation between Red and Yellow Color Forms. International Journal of Plant Sciences 17: 167–174.
Mauritz, S. G. 2010. Fearless Latin: A Gardener's Introduction to Botanical Nomenclature. CreateSpace Independent Publishing Platform, Charleston.
McCormack, J. H. 1975. Beetle Pollination of *Calycanthus floridus*: Pollinator Behavior as a Function of the Volatile Flower Oils. Ph.D. dissertation. University of Michigan, Ann Arbor.
Moisset, B. Pollinator of the Month: Yucca Moths. http://www.fs.fed.us/wildflowers/pollinators/pollinator-of-the-month/yucca_moths.shtml.
Murdy, W. H., and E. B. Carter. 2000. Guide to the Plants of Granite Outcrops. University of Georgia Press, Athens.
Nelson, G. 2005. East Gulf Coastal Plain Wildflowers. Globe Pequot Press, Guilford, CT.
Nelson, G. 2006. Atlantic Coastal Plain Wildflowers. Globe Pequot Press, Guilford, CT.

Nourse, H. 2009. Polygalaceae: The Milkwort Family in Georgia. Tipularia: Journal of the Georgia Botanical Society 24: 17–23.
Nourse, H., and C. Nourse. 2007. Favorite Wildflower Walks in Georgia. University of Georgia Press, Athens.
Olmstead, R. G. 2002. Whatever Happened to the Scrophulariaceae? Fremontia 30(2): 13–22.
Patrick, T. 2007. Trilliums of Georgia. Tipularia: Journal of the Georgia Botanical Society 22: 3–22.
Porcher, R. D., and D. A. Rayner. 2001. Guide to the Wildflowers of South Carolina. University of South Carolina Press, Columbia.
Radford, A. E., H. E. Ahles, and C. R. Bell. 1968. Manual of the Vascular Flora of the Carolinas. University of North Carolina Press, Chapel Hill.
Schnell, D. E. 2002. Carnivorous Plants of the United States and Canada. Second edition. Timber Press, Portland, OR.
Sorrie, B. A. 2011. Field Guide to the Wildflowers of the Sandhills Region: North Carolina, South Carolina, Georgia. University of North Carolina Press, Chapel Hill.
Spira, T. P. 2011. Wildflowers and Plant Communities of the Southern Appalachian Mountains and Piedmont. University of North Carolina Press, Chapel Hill.
Stearn, W. T. 2004. Botanical Latin. Timber Press, Portland, OR.
Stearn, W. T. 1992. Stearn's Dictionary of Plant Names for Gardeners. Cassell Publishers, London.
Stevens, P. F. 2001 and ongoing. Angiosperm Phylogeny Website. Version 12, July 2012 [continuously updated]. http://www.mobot.org/MOBOT/research/APweb/.
Szabolcs, L. B., A. D. Govec, A. M. Latimerd, J. D. Majerc, and R. R. Dunn. 2010. Convergent Evolution of Seed Dispersal by Ants, and Phylogeny and Biogeography in Flowering Plants: A Global Survey. Perspectives in Plant Ecology, Evolution, and Systematics 12(1): 43–55.
Taylor, W. K. 1992. Guide to Florida Wildflowers. Taylor Publishing Company, Dallas, TX.
Taylor, W. K. 1998. Florida Wildflowers in Their Natural Communities. University Press of Florida, Gainesville.
Tenaglia, D. 2006. Missouri Plants: Photographs and Descriptions of Flowering and Non-flowering Plants of Missouri. htttp://www.missouriplants.com.
Weakley, A. S. 2004. The Curious Case of the Disappearing Asters. North Carolina Botanical Garden Newsletter 2: 9.
Weakley, A. S. 2005. Why Are Plant Names Changing So Much? Native Plants Journal 6(1): 52–58.
Weakley, A. S. 2012, 2014. Flora of the Southern and Mid-Atlantic States: Working Draft. University of North Carolina Herbarium, Chapel Hill.
Wunderlin, R. P., and B. F. Hansen. 2003. Guide to the Vascular Plants of Florida. Second edition. University Press of Florida, Gainesville.
Zomlefer, W. B. 1995. Guide to Flowering Plant Families. University of North Carolina Press, Chapel Hill.

IMAGE CREDITS

All photographs are by Hugh and Carol Nourse unless acknowledged here. The author thanks all the photographers and illustrators for their contributions.

T = top; M = middle; B = bottom; L = left; R = right; I = inset; CC = Creative Commons

Adamantios (CC lic. 3.0), 16
Mac Alford, 379T
Michael Apel (CC lic. 2.0), 427B
Anna Armitage, Texas A&M University of Galveston, 127T
Steven J. Baskauf, bioimages.vanderbilt.edu, 176B
BJO (CC lic. 2.0), 88T
Keith Bradley, 20B, 364B
Suzanne Cadwell, 39B, 68T, 113B, 293B
Cdc25A (CC lic. 3.0), 400BI
Dendroica Cerulea (CC lic. 2.0), 117B, 155T
Linda Chafin, 168
Carmen Champagne, 227B
Roy Cohutta, 377T
Cotinis (Flickr Commons; CC lic. 2.0), 44L
Jan Coyne, 95T, 261TI
Alan Cressler, 21T, 23T, 24B, 28T, 30T, 35B, 41T, 42B, 45T, 47B, 50B, 53T, 61T, 64B, 67T, 69BI, 70B, 72B, 77T, 85T, 85B, 89B, 92T, 95B, 97T, 101T, 102T, 103T, 105T, 119B, 121TL, 122T, 122B, 126T, 131T, 132T, 134T, 140BI, 141B, 151T, 151B, 156T, 158B, 160B, 161B, 166T, 166TI, 169T, 170T, 176T, 178B, 181B, 183L, 189B, 191T, 193B, 197B, 208T, 214B, 215B, 216T, 216B, 217T, 218T, 219T, 220T, 221B, 223T, 224T, 224B, 228B, 230T, 234T, 238T, 244T, 247T, 249T, 249B, 250T, 253B, 254T, 255B, 256B, 260B, 262T, 263T, 275T, 275M, 275B, 277B, 279BI, 287B, 288B, 289B, 293BI, 296B, 298T, 299B, 303B, 304T, 307B, 307BI, 312B, 315T, 317B, 319B, 333T, 339T, 346, 347T, 348T, 356T, 358L, 358R, 359, 359I, 361B, 365T, 367B, 368T, 368B, 372T, 378T, 378B, 379M, 379B, 392T, 393B, 394T, 399T, 401T, 402B, 404B, 405T, 406B, 412T, 416B, 417T, 418B, 427T, 428
Dalgial (CC lic. 3.0), 295T
Dendrofil (CC lic. 3.0), 57T, 57TI
Eleanor Dietrich, 49B, 59B, 65T, 71B, 78B, 83B, 161T, 165, 165I, 166B, 166BI, 217B, 251B, 269T, 303T, 323B, 332B, 349B, 373T, 380T
Jacqueline Donnelly, 177, 177I
Jim Drake, 213B, 214T
K. Draper (CC. lic.), 345T
Maja Dumat (CC lic. 2.0), 357BI
José María Escolano (CC lic. 2.0), 135T
Dan Evans, Bugwood.org, 196B
Gary Fleming, 95M, 157T, 202T, 380B
Fornax (CC lic. 3.0), 149B
Jim Fowler, 148T
Sam Fraser-Smith (CC lic.), 63T
GFDL (GNU free doc. lic. 1.2), 331T
John Gwaltney, 250B, 252TI, 285T, 386T
Jean C. Putnam Hancock, 29, 47T, 55, 258T, 396, 408
Bobby Hattaway, 195T, 226T, 252T, 356B
Christa F. Hayes, 54T
Bernd Haynold, 64T
Sonnia Hill, 50T, 411B
Hans Hillewaert, 280TR
James Holland, 180B, 204B, 240T
Jason Hollinger (CC lic. 2.0), 402T
Ellen Honeycutt, 164B, 221T
S. Honeytart (CC lic. 3.0), 324B
Eric Hunt, Arkansas Native Plant Society, 289T

Margie Hunter, 108T
Marco Iocchi (CC lic. 3.0), 207B
IROZ (CC lic. 3.0), 159B
Jomegat (CC lic. 3.0), 391B
Douglas W. Jones, 44R
Mary Keim, 103B, 123T, 410B, 416T
Gary Knight, 59T
Linda Lee, USC Herbarium, 152T, 152TI, 152B
Loret, 333B
Joseph A. Marcus, Lady Bird Johnson Wildflower Center, 62B
Janie Marlowe, 356BI
Hal Massie, 34B, 82T, 87B, 100T, 150B
Joshua Mayer (CC lic. 2.0), 209B
Frank Mayfield, 279TI
George F. Mayfield (CC lic. 2.0), 361T
Dan Mullen, 305B
Daniel Nickrent, Southern Illinois University Phyto-Images, 290M, 292B
Allan Norcross, NHGardensolutions, 81T
Jerry Oldenettel (CC lic. 2.0), 118T, 310I, 318T, 318TI
Martin Olsson (CC lic. 2.5; GNU free doc. lic.), 290T
John Oyston (CC lic. 1.0), 309B
Cary Paynter, N.C. Native Plant Society, 107B
Pieter Pelser, phytoimages.siu.edu, 412B
Bob Peterson (CC lic. 2.0), 156B, 163T, 163TI, 234BI, 382T
Jeffrey S. Pippen, www.jeffpippen.com, 191B, 191BI
Tom Potterfield (CC lic. 2.0), 341T
Corey Raimond, 212T
Karan A. Rawlins, UGA, Bugwood.org, 48T, 205TI
Julio Reis (CC lic. 3.0), 26B
Fritz Flohr Reynolds, 32T, 73T, 202B, 306T, 315B, 324T, 382B
Jason Sharp, 27B
Bob Smith, 291B
David Smith, Delawarewildflowers.org, 281B, 293T
Elise Smith, 121BL
Bruce A. Sorrie, 117T, 247B, 341, 344T
SPHL (GNU free doc. lic.), 75BI
Forest and Kim Starr (CC lic. 3.0), 30B, 297B, 343B, 410T
Dan Tenaglia, 23B, 24T, 25B, 60B, 66T, 69B, 79B, 83T, 84T, 84B, 91T, 92B, 94T, 108B, 112T, 116B, 125T, 131B, 133, 137B, 138T, 139B, 147T, 154B, 164T, 180T, 181T, 182B, 185B, 190T, 192T, 194T, 195B, 197T, 197TI, 201B, 207T, 208B, 209T, 227T, 232B, 235T, 236T, 242T, 244B, 253T, 257T, 257B, 263B, 266T, 266TI, 270B, 279T, 280B, 284B, 287T, 294T, 294B, 295B, 297T, 310, 318B, 321B, 328B, 330T, 335T, 336T, 338T, 344B, 345B, 345BI, 360B, 373B, 374, 375B, 376T, 376B, 377B, 384B, 387T, 403T, 407T, 419B, 419BI
Tigerente (GNU free doc. lic. 1.2), 200T
Thayne Tuason (CC lic. 2.0), 210T
James Van Kley, 33T
Rebekah D. Wallace, UGA, Bugwood.org, 93B, 139T, 281T
Richard and Teresa Ware, 25T, 26T, 33B, 40B, 41B, 47T, 49T, 52B, 61B, 62T, 63B, 67B, 68B, 72T, 74T, 76T, 77B, 81B, 86T, 86B, 87T, 93T, 96B, 100B, 101B, 109T, 110B, 114T, 114TI, 114B, 116T, 118B, 125B, 130B, 138B, 140T, 144B, 150T, 154BI, 183R, 184T, 189T, 192B, 192BI, 203B, 222T, 229, 235B, 238B, 241T, 242B, 243B, 254B, 258B, 259T, 260T, 264T, 264B, 265B, 266B, 268, 270T, 271T, 271B, 272T, 276B, 278B, 280TL, 284T, 290B, 291T, 292T, 300T, 307B, 308B, 308BI, 311B, 314T, 327, 329B, 330B, 335B, 343T, 348B, 350T, 350B, 359B, 359BI, 360T, 375, 381T, 385B, 394B, 395T, 405B, 409T, 409B, 415T, 417B, 420B
George Williams, 200B
Irvine Wilson, Va. Dept. of Conservation and Recreation, 179B
Ilona L. Woolcarderbee (CC lic. 2.0), 418T

INDEX

NOTE: Species featured with photographs are in bold.